THE MATHEMATICS
OF DIFFUSION

發行所：狀元出版社
登記證字號：行政院新聞局局版
　　　　　　台業字第0785號
發行人：廖燦景
印刷者：東南印製廠有限公司
　地址：台北市西園路2段294巷15弄17號
總經銷：歐亞書局有限公司
　地址：台北市新生南路3段102號
　電話：3213233・3213141
　郵政劃撥：0006307-6
中　華　民　國　73 年　8 月

THE MATHEMATICS
OF DIFFUSION

BY

J. CRANK

BRUNEL UNIVERSITY
UXBRIDGE

SECOND EDITION

Oxford University Press, Ely House, London W.1

LONDON   OXFORD   GLASGOW   NEW YORK   TORONTO   MELBOURNE
WELLINGTON   CAPE TOWN   IBADAN   NAIROBI   DAR ES SALAAM
LUSAKA   ADDIS ABABA·   DELHI   BOMBAY   CALCUTTA   MADRAS
KARACHI   KUALA   LUMPUR   SINGAPORE   JAKARTA   HONG KONG
TOKYO

ISBN 0 19 853344 6

FIRST EDITION 1956
SECOND EDITION 1975
Reprinted 1976

PREFACE TO SECOND EDITION

IN preparing this second edition I have tried to incorporate as much new material as possible but to preserve the character of the original volume. The book contains a collection of mathematical solutions of the differential equations of diffusion and methods of obtaining them. They are discussed against a background of some of the experimental and practical situations to which they are relevant. Little mention is made of molecular mechanisms,. and I have made only fleeting excursions into the realms of irreversible thermodynamics. These I hope are self-explanatory. A number of general accounts of the subject are already available, but very few mathematical solutions of the equations of non-equilibrium thermodynamics have been obtained for practical systems.

During the last 15–20 years the widespread occurrence of concentration-dependent diffusion has stimulated the development of new analytical and numerical solutions. The time-lag method of measuring diffusion coefficients has also been intensively investigated and extended. Similarly, a lot of attention has been devoted to moving-boundary problems since the first edition was published. These and other matters have now been included by extensive revision of several chapters. Also, the chapter dealing with the numerical solution of the diffusion equations has been completely rewritten and brought up to date. It seems unbelievable now that most of the calcula- tions in the first edition were carried out on desk calculating machines.

Two entirely new chapters have been added. In one are assembled some of the mathematical models of non-Fickian, or anomalous diffusion occurring mainly in solvent–polymer systems in the glassy state. The other attempts a systematic review of diffusion in heterogeneous media, both laminates and particulates. A succession of improved solutions are described to the problem of diffusion in a medium in which are embedded discrete particles with different diffusion properties.

I have resisted the temptation to lengthen appreciably the earlier chapters. The enlarged edition of Carslaw and Jaeger's book *Conduction of heat in solids* contains a wealth of solutions of the heat-flow equations for constant heat parameters. Many of them are directly applicable to diffusion problems, though it seems that some non-mathematicians have difficulty in making the necessary conversions. For them I have included a brief 'translator's guide'. A few new solutions have been added, however, some of them in the context in which they arose, that is the measurement of diffusion coefficients.

I should like to express my appreciation to the Vice Chancellor and Council of Brunel University for so readily agreeing to my application for extended leave without which I could not have undertaken the preparation of a second edition. I am deeply grateful to my academic colleagues who shared my administrative responsibilities and particularly to Professor Peter Macdonald who so willingly and effectively assumed the role of Acting Head of the School of Mathematical Studies.

I am most grateful to Mrs. Joyce Smith for all the help she gave me, not least by typing the manuscript and checking the proofs. Mr. Alan Moyse kept me well supplied with the seemingly innumerable books, journals, and photostat copies which I requested.

I owe a great deal to friendly readers who have pointed out mistakes in the first edition and made helpful suggestions for the second. In particular I have benefited from discussions with my friend and former colleague, Dr. Geoffrey Park. I had an invaluable introduction to the literature on which Chapter 12 is based from Mr. W. M. Woodcock, who came to me for help but, in fact, gave far more than he received. Finally, I have appreciated the understanding help and guidance afforded me by members of staff of the Clarendon Press.

Uxbridge J. C.
October 1973

ACKNOWLEDGEMENTS

I wish to acknowledge the permission of the authors or publishers of the following journals or books to reproduce the figures and tables specified: Fig. 4.9, Barrer, R. M. *Trans. Faraday Soc.*; Fig. 10.8, Frensdorff, H. K. *J. Poly. Sci.*; Fig. 11.1, Rogers, C. E. 'Physics and Chemistry of the Organic Solid State'. J. Wiley & Sons Inc.; Fig. 11.2, *J. Poly. Sci.*; Fig. 12.1, Jefferson, T. B. *Ind. & Eng. Chem.*; Fig. 12.2, Tsao, G. *Ind. & Eng. Chem.*; Fig. 12.3, Cheng, S. C. & Vachon, R. I. *J. Heat & Mass Transfer*; Fig. 13.11, *J. Inst. Maths. Applics.*; Table 7.1, Lee, C. F. *J. Inst. Maths. Applics.*; Table 7.6, Wilkins, J. E. *J. Soc. Ind. Appl. Maths.*; Table 7.7, Philip, J. R. *Aust. J. Phys.*; Table 9.1, Wilkins, J. E. *J. Soc. Ind. Appl. Maths.*; and Table 10.2, Hansen, C. M. *I. & E. C. Fundamentals*.

PREFACE

A MORE precise title for this book would be 'Mathematical solutions of the diffusion equation', for it is with this aspect of the mathematics of diffusion that the book is mainly concerned. It deals with the description of diffusion processes in terms of solutions of the differential equation for diffusion. Little mention is made of the alternative, but less well developed, description in terms of what is commonly called 'the random walk', nor are theories of the mechanism of diffusion in particular systems included.

The mathematical theory of diffusion is founded on that of heat conduction and correspondingly the early part of this book has developed from 'Conduction of heat in solids' by Carslaw and Jaeger. These authors present many solutions of the equation of heat conduction and some of them can be applied to diffusion problems for which the diffusion coefficient is constant. I have selected some of the solutions which seem most likely to be of interest in diffusion and they have been evaluated numerically and presented in graphical form so as to be readily usable. Several problems in which diffusion is complicated by the effects of an immobilizing reaction of some sort are also included. Convenient ways of deriving the mathematical solutions are described.

When we come to systems in which the diffusion coefficient is not constant but variable, and for the most part this means concentration dependent, we find that strictly formal mathematical solutions no longer exist. I have tried to indicate the various methods by which numerical and graphical solutions have been obtained, mostly within the last ten years, and to present, again in graphical form, some solutions for various concentration-dependent diffusion coefficients. As well as being useful in themselves these solutions illustrate the characteristic features of a concentration-dependent system. Consideration is also given to the closely allied problem of determining the diffusion coefficient and its dependence on concentration from experimental measurements. The diffusion coefficients measured by different types of experiment are shown to be simply related. The final chapter deals with the temperature changes which sometimes accompany diffusion.

In several instances I have thought it better to refer to an easily accessible book or paper rather than to the first published account, which the reader might find difficult to obtain. Ease of reference usually seemed of primary importance, particularly with regard to mathematical solutions.

I should like to express my thanks to my friend and colleague, Mr. A. C. Newns, who read the typescript and made many valuable comments and suggestions, and also to Mrs. D. D. Whitmore, who did most of the calculations and helped to correct the proofs and compile the index. I am grateful to Miss D. Eldridge who, by patient and skilful typing, transformed an almost illegible manuscript into a very clear typescript for the printer. I should also like to thank the following who readily gave permission to use material from various publications: Professor R. M. Barrer, Mr. M. B. Coyle, Dr. P. V. Danckwerts, Dr. L. D. Hall, Dr. P. S. H. Henry, Professor J. C. Jaeger, Dr. G. S. Park, Dr. R. H. Stokes, Dr. C. Wagner, and the publishers of the following journals, *British Journal of Applied Physics*, *Journal of Chemical Physics*, *Journal of Metals*, *Journal of Scientific Instruments*, *Philosophical Magazine*, *Proceedings of the Physical Society*, *Transactions of the Faraday Society*. Finally, it is a pleasure to thank those members of the staff of the Clarendon Press who have been concerned with the production of this book for the kindness and consideration they have shown to me.

Maidenhead   J. C.
December 1955

CONTENTS

Contents

1

THE DIFFUSION EQUATIONS

1.1. The diffusion process

DIFFUSION is the process by which matter is transported from one part of a system to another as a result of random molecular motions. It is usually illustrated by the classical experiment in which a tall cylindrical vessel has its lower part filled with iodine solution, for example, and a column of clear water is poured on top, carefully and slowly, so that no convection currents are set up. At first the coloured part is separated from the clear by a sharp, well-defined boundary. Later it is found that the upper part becomes coloured, the colour getting fainter towards the top, while the lower part becomes correspondingly less intensely coloured. After sufficient time the whole solution appears uniformly coloured. There is evidently therefore a transfer of iodine molecules from the lower to the upper part of the vessel taking place in the absence of convection currents. The iodine is said to have diffused into the water.

If it were possible to watch individual molecules of iodine, and this can be done effectively by replacing them by particles small enough to share the molecular motions but just large enough to be visible under the microscope, it would be found that the motion of each molecule is a random one. In a dilute solution each molecule of iodine behaves independently of the others, which it seldom meets, and each is constantly undergoing collision with solvent molecules, as a result of which collisions it moves sometimes towards a region of higher, sometimes of lower, concentration, having no preferred direction of motion towards one or the other. The motion of a single molecule can be described in terms of the familiar 'random walk' picture, and whilst it is possible to calculate the mean-square distance travelled in a given interval of time it is not possible to say in what direction a given molecule will move in that time.

This picture of random molecular motions, in which no molecule has a preferred direction of motion, has to be reconciled with the fact that a transfer of iodine molecules from the region of higher to that of lower concentration is nevertheless observed. Consider any horizontal section in the solution and two thin, equal, elements of volume one just below and one just above the section. Though it is not possible to say which way any particular iodine molecule will move in a given interval of time, it can be said that on the average a definite fraction of the molecules in the lower element of volume will cross the section from below, and the same fraction of molecules in the upper element will cross the section from above, in a given time. Thus, simply because there are more iodine molecules in the lower element than in the upper one, there is a net transfer from the lower to the upper side of the section as a result of random molecular motions.

1.2. Basic hypothesis of mathematical theory

Transfer of heat by conduction is also due to random molecular motions, and there is an obvious analogy between the two processes. This was recognized by Fick (1855), who first put diffusion on a quantitative basis by adopting the mathematical equation of heat conduction derived some years earlier by Fourier (1822). The mathematical theory of diffusion in isotropic substances is therefore based on the hypothesis that the rate of transfer of diffusing substance through unit area of a section is proportional to the concentration gradient measured normal to the section, i.e.

$$F = - D \, \partial C / \partial x, \tag{1.1}$$

where F is the rate of transfer per unit area of section, C the concentration of diffusing substance, x the space coordinate measured normal to the section, and D is called the diffusion coefficient. In some cases, e.g. diffusion in dilute solutions, D can reasonably be taken as constant, while in others, e.g. diffusion in high polymers, it depends very markedly on concentration. If F, the amount of material diffusing, and C, the concentration, are both expressed in terms of the same unit of quantity, e.g. gram or gram molecules, then it is clear from (1.1) that D is independent of this unit and has dimensions (length)^2 (time)$^{-1}$, e.g. cm^2 s^{-1}. The negative sign in eqn (1.1) arises because diffusion occurs in the direction opposite to that of increasing concentration.

It must be emphasized that the statement expressed mathematically by (1.1) is in general consistent only for an isotropic medium, whose structure and diffusion properties in the neighbourhood of any point are the same relative to all directions. Because of this symmetry, the flow of diffusing substance at any point is along the normal to the surface of constant concentration through the point. As will be seen later in § 1.4 (p. 5), this need not be true in an anisotropic medium for which the diffusion properties depend on the direction in which they are measured.

1.3. Differential equation of diffusion

The fundamental differential equation of diffusion in an isotropic medium is derived from eqn (1.1) as follows.

Consider an element of volume in the form of a rectangular parallelepiped whose sides are parallel to the axes of coordinates and are of lengths 2 dx, 2 dy, 2 dz. Let the centre of the element be at P(x, y, z), where the concentration of diffusing substance is C. Let ABCD and A'B'C'D' be the faces perpendicular to the axis of x as in Fig. 1.1. Then the rate at which diffusing substance enters the element through the face ABCD in the plane $x - dx$ is given by

$$4\,dy\,dz\left(F_x - \frac{\partial F_x}{\partial x}dx\right),$$

where F_x is the rate of transfer through unit area of the corresponding plane through P. Similarly the rate of loss of diffusing substance through the face A'B'C'D' is given by

$$4\,dy\,dz\left(F_x + \frac{\partial F_x}{\partial x}dx\right).$$

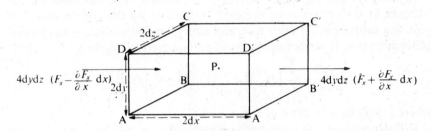

FIG. 1.1. Element of volume.

The contribution to the rate of increase of diffusing substance in the element from these two faces is thus equal to

$$-8\,dx\,dy\,dz\frac{\partial F_x}{\partial x}.$$

Similarly from the other faces we obtain

$$-8\,dx\,dy\,dz\frac{\partial F_y}{\partial y} \quad \text{and} \quad -8\,dx\,dy\,dz\frac{\partial F_z}{\partial z}.$$

But the rate at which the amount of diffusing substance in the element increases is also given by

$$8\,dx\,dy\,dz\frac{\partial C}{\partial t},$$

and hence we have immediately

$$\frac{\partial C}{\partial t} + \frac{\partial F_x}{\partial x} + \frac{\partial F_y}{\partial y} + \frac{\partial F_z}{\partial z} = 0. \tag{1.2}$$

If the diffusion coefficient is constant, F_x, F_y, F_z are given by (1.1), and (1.2) becomes

$$\frac{\partial C}{\partial t} = D\left(\frac{\partial^2 C}{\partial x^2} + \frac{\partial^2 C}{\partial y^2} + \frac{\partial^2 C}{\partial z^2}\right),\tag{1.3}$$

reducing simply to

$$\frac{\partial C}{\partial t} = D\frac{\partial^2 C}{\partial x^2}\tag{1.4}$$

if diffusion is one-dimensional i.e. if there is a gradient of concentration only along the x-axis. Expressions (1.1) and (1.4) are usually referred to as Fick's first and second laws of diffusion, since they were first formulated by Fick (1855) by direct analogy with the equations of heat conduction.

In many systems, e.g. the interdiffusion of metals or the diffusion of organic vapours in high-polymer substances, D depends on the concentration of diffusing substance C. In this case, and also when the medium is not homogeneous so that D varies from point to point, eqn (1.2) becomes

$$\frac{\partial C}{\partial t} = \frac{\partial}{\partial x}\left(D\frac{\partial C}{\partial x}\right) + \frac{\partial}{\partial y}\left(D\frac{\partial C}{\partial y}\right) + \frac{\partial}{\partial z}\left(D\frac{\partial C}{\partial z}\right),\tag{1.5}$$

where D may be a function of x, y, z, and C.

If D depends on the time during which diffusion has been taking place but not on any of the other variables, i.e.

$$D = f(t),$$

then on introducing a new time-scale T such that

$$dT = f(t)\,dt,$$

the diffusion equation becomes

$$\frac{\partial C}{\partial T} = \frac{\partial^2 C}{\partial x^2} + \frac{\partial^2 C}{\partial y^2} + \frac{\partial^2 C}{\partial z^2},\tag{1.6}$$

which is the same as (1.3) for a constant diffusion coefficient equal to unity.

1.3.1. Diffusion in a cylinder and sphere

Other forms of the above equations follow by transformation of coordinates, or by considering elements of volume of different shape. Thus by putting

$$x = r\cos\theta,$$

$$y = r\sin\theta,$$

or by considering an element of volume of a cylinder of sides dr, $r\,d\theta$, dz, we obtain the equation for diffusion in a cylinder,

$$\frac{\partial C}{\partial t} = \frac{1}{r}\left\{\frac{\partial}{\partial r}\left(rD\frac{\partial C}{\partial r}\right) + \frac{\partial}{\partial\theta}\left(\frac{D}{r}\frac{\partial C}{\partial\theta}\right) + \frac{\partial}{\partial z}\left(rD\frac{\partial C}{\partial z}\right)\right\}, \tag{1.7}$$

in terms of the cylindrical coordinates r, θ, z. The corresponding equation for a sphere in terms of spherical polar coordinates r, θ, ϕ is obtained by writing

$$x = r \sin\theta \cos\phi,$$
$$y = r \sin\theta \sin\phi,$$
$$z = r \cos\theta,$$

or by considering an element of volume of a sphere of sides dr, $r\,d\theta$, $r\sin\theta\,d\phi$. It is

$$\frac{\partial C}{\partial t} = \frac{1}{r^2}\left\{\frac{\partial}{\partial r}\left(Dr^2\frac{\partial C}{\partial r}\right) + \frac{1}{\sin\theta}\frac{\partial}{\partial\theta}\left(D\sin\theta\frac{\partial C}{\partial\theta}\right) + \frac{D}{\sin^2\theta}\frac{\partial^2 C}{\partial\phi^2}\right\}. \tag{1.8}$$

The simplified forms of (1.7) and (1.8) for purely radial diffusion, e.g. in a long cylinder where end effects are negligible or in a spherically symmetrical system, are given in Chapters 5 and 6, where some solutions of the differential equations are to be found. All these diffusion equations can be expressed in terms of the nomenclature of vector analysis as

$$\frac{\partial C}{\partial t} = \text{div}\,(D\,\text{grad}\,C).$$

1.4. Anisotropic media

Anisotropic media have different diffusion properties in different directions. Some common examples are crystals, textile fibres, and polymer films in which the molecules have a preferential direction of orientation. For such media it is not always true, as was stated in § 1.2 (p. 2) for isotropic media, that the direction of flow of diffusing substance at any point is normal to the surface of constant concentration through the point. This means that (1.1) must be replaced in general by the assumptions

$$\left.\begin{aligned}
-F_x &= D_{11}\frac{\partial C}{\partial x} + D_{12}\frac{\partial C}{\partial y} + D_{13}\frac{\partial C}{\partial z}\\[4pt]
-F_y &= D_{21}\frac{\partial C}{\partial x} + D_{22}\frac{\partial C}{\partial y} + D_{23}\frac{\partial C}{\partial z}\\[4pt]
-F_z &= D_{31}\frac{\partial C}{\partial x} + D_{32}\frac{\partial C}{\partial y} + D_{33}\frac{\partial C}{\partial z}
\end{aligned}\right\} \tag{1.9}$$

so that F_x, for example, depends not only on $\partial C/\partial x$ but also on $\partial C/\partial y$ and $\partial C/\partial z$. The Ds have the significance that $D_{13}\,\partial C/\partial z$, for example, is the contribution to the rate of transfer in the x-direction due to the component of concentration gradient in the z-direction. Substituting from (1.9) for the Fs in (1.2) we obtain

$$\frac{\partial C}{\partial t} = D_{11}\frac{\partial^2 C}{\partial x^2} + D_{22}\frac{\partial^2 C}{\partial y^2} + D_{33}\frac{\partial^2 C}{\partial z^2} + (D_{23} + D_{32})\frac{\partial^2 C}{\partial y\,\partial z}$$

$$+ (D_{31} + D_{13})\frac{\partial^2 C}{\partial z\,\partial x} + (D_{12} + D_{21})\frac{\partial^2 C}{\partial x\,\partial y}, \tag{1.10}$$

if the Ds are taken as constant. The extension to non-constant Ds is obvious from (1.5). A transformation to rectangular coordinates ξ, η, ζ can be found which reduces (1.10) to

$$\frac{\partial C}{\partial t} = D_1\frac{\partial^2 C}{\partial \xi^2} + D_2\frac{\partial^2 C}{\partial \eta^2} + D_3\frac{\partial^2 C}{\partial \zeta^2}. \tag{1.11}$$

This is the same transformation as that by which the ellipsoid

$$D_{11}x^2 + D_{22}y^2 + D_{33}z^2 + (D_{23} + D_{32})yz + (D_{31} + D_{13})zx$$

$$+ (D_{12} + D_{21})xy = \text{constant} \tag{1.12}$$

is reduced to

$$D_1\xi^2 + D_2\eta^2 + D_3\zeta^2 = \text{constant}. \tag{1.13}$$

The new axes may be called the principal axes of diffusion and D_1, D_2, D_3 the principal diffusion coefficients. If we make the further transformation

$$\xi_1 = \xi\sqrt{(D/D_1)}, \qquad \eta_1 = \eta\sqrt{(D/D_2)}, \qquad \zeta_1 = \zeta\sqrt{(D/D_3)}, \tag{1.14}$$

where D may be chosen arbitrarily, (1.11) becomes

$$\frac{\partial C}{\partial t} = D\left(\frac{\partial^2 C}{\partial \xi_1^2} + \frac{\partial^2 C}{\partial \eta_1^2} + \frac{\partial^2 C}{\partial \zeta_1^2}\right). \tag{1.15}$$

This has the same form as eqn (1.3) for isotropic media, and hence certain problems in anisotropic media can be reduced to corresponding problems in isotropic media. Whether or not this can be done in a given case depends on the boundary conditions. Thus it is possible when the medium is infinite, or when it is bounded by planes perpendicular to the principal axes of diffusion so that the boundary conditions are of the familiar form $C = \text{constant}$, $\xi = 0$, $\xi = l$, $t > 0$, for example, and similarly for η and ζ. The problem of diffusion into an anisotropic cylinder which has its axis along ξ and is bounded by planes perpendicular to ξ reduces to the corresponding problem in an isotropic cylinder provided $D_2 = D_3$.

Certain properties deduced by Carslaw and Jaeger (1959, p. 46) indicate the physical significance of the ellipsoid and also of the principal axes of diffusion. Thus it can be shown that the square of the radius vector of the ellipsoid in any direction is inversely proportional to the diffusion coefficient normal to the surfaces of constant concentration at points where their normals are in that direction. Hence the diffusion coefficient D_n at right angles to surfaces whose normals have direction cosines l, m, n relative to the principal axes of diffusion is given by

$$D_n = l^2 D_1 + m^2 D_2 + n^2 D_3. \tag{1.16}$$

Carslaw and Jaeger further show that if there is symmetry about the planes $\xi = 0$ and $\eta = 0$, then the general relationships (1.9) for the Fs reduce to

$$-F_\xi = D_1 \, \partial C/\partial \xi, \qquad -F_\eta = D_2 \, \partial C/\partial \eta, \qquad -F_\zeta = D_3 \, \partial C/\partial \zeta. \tag{1.17}$$

This simplification also occurs for other types of crystallographic symmetry. It means that the flow through a surface perpendicular to a principal axis of diffusion is proportional simply to the concentration gradient normal to the surface as is the case for isotropic media.

1.4.1. *Significance of measurements in anisotropic media*

Since in the majority of experiments designed to measure a diffusion coefficient the flow is arranged to be one-dimensional, it is worth while to see how such measurements are affected by anisotropy. If the diffusion is one-dimensional in the sense that a concentration gradient exists only along the direction of x, it is clear from (1.10), since both C and $\partial C/\partial x$ are everywhere independent of y and z, that the diffusion is governed by the simple equation

$$\frac{\partial C}{\partial t} = D_{11} \frac{\partial^2 C}{\partial x^2}, \tag{1.18}$$

and D_{11} is the diffusion coefficient measured. If the direction of diffusion is osen to be that of a principal axis, then D_{11} is equal to one or other of the principal diffusion coefficients D_1, D_2, or D_3. Otherwise the coefficient $D_{11} = D_n$, related to D_1, D_2, D_3, by (1.16) is measured. This would be measured, for example, by an observation of the rate of flow through a plane sheet of a crystal cut so that its normal has direction cosines (l, m, n) relative to the principal axes of diffusion of the crystal. Similar remarks apply to a high polymer sheet in which there is both uniplanar and undirectional orientation, i.e., the molecules are arranged with their long axes lying mainly parallel to the plane of the sheet and all parallel to one direction in that plane. The principal axes of diffusion of such a sheet will be normal to the plane sheet, and along and perpendicular to the preferred direction of orientation in that plane. Even if a concentration gradient exists in one direction only, it is clear from (1.9) and (1.15) that the diffusion flow is not along this direction unless it coincides with a principal axis of diffusion.

1.4.2. *Conversion of heat flow to diffusion solutions*

Carslaw and Jaeger (1959) and other books contain a wealth of solutions of the heat-conduction equation. There is general awareness among scientists and engineers that the phenomena of heat flow and diffusion are basically the same. Nevertheless, many non-mathematicians experience difficulty in making the changes of notation needed to transcribe from one set of solutions to the other. In this section we examine in detail the correspondence between the physical parameters, the variables, and the equations and boundary conditions which occur in heat-flow and diffusion problems. We take the one-dimensional case with constant properties as an illustration.

(i) *The equations*. Diffusion theory is based on Fick's two equations (1.1) and (1.4), where C is the concentration of diffusant expressed, say, in mass per unit volume, and D is the diffusion coefficient. The two corresponding equations in heat flow are

$$F = -K\, \partial\theta/\partial x, \tag{1.19}$$

$$\frac{\partial\theta}{\partial t} = \left(\frac{K}{c\rho}\right)\frac{\partial^2\theta}{\partial x^2}, \tag{1.20}$$

where θ is temperature, K is the heat conductivity, ρ is density, and c specific heat, so that ρc is the heat capacity per unit volume. The space coordinate is x and t is time. In (1.19), F is the amount of heat flowing in the direction of x increasing per unit time through unit area of a section which is normal to the direction of x.

(ii) *Variables and parameters*. In order that the two sets of equations should correspond we may identify concentration C with temperature θ, and take $D = K$ in (1.1) and (1.19) and $D = K/(c\rho)$ in (1.4) and (1.20). The two together mean $c\rho = 1$. This is a consequence of our having identified C with θ. The diffusing substance' in heat flow is heat not temperature. The factor $c\rho$ is needed to convert temperature to the amount of heat per unit volume; but concentration is, by definition, the amount of diffusing substance per unit volume and so no conversion factor is needed, i.e. $c\rho = 1$. It is usual to write $K/(c\rho) = k$, the heat diffusivity. All three of K, k, $c\rho$ appear in heat-flow equations, and to convert to diffusion terms we take

$$D = K = k \quad \text{and} \quad c\rho = 1. \tag{1.21}$$

(iii) *Boundary conditions*. (a) Prescribed surface temperature corresponds to prescribed concentration just within the surface, say of a plane sheet. If the surface is in contact with a vapour at pressure p, there is some relation between p and C such as $C = Sp$, where S is the solubility. There is no ana-logue to S in heat flow.

(b) Prescribed heat flux corresponds to flux of diffusant and we have

$$-K\,\partial\theta/\partial x = F(t), \quad \text{in heat,} \tag{1.22a}$$

$$-D\,\partial C/\partial x = F(t), \quad \text{in diffusion,} \tag{1.22b}$$

where F is in general a known function of time but may be constant.

(c) A heat-insulated surface corresponds to an impermeable surface and is the special case of (1.22a, b) with $F = 0$, i.e.

$$\partial C/\partial x = \partial\theta/\partial x = 0.$$

(d) What is referred to as a 'radiation boundary condition' in heat flow usually means that the heat flux across unit area of the surface is proportional to the difference between the surface temperature θ_s and the temperature θ_0 of the outside medium, i.e. is given by $H(\theta_s - \theta_0)$. But the rate of heat loss from unit area of a surface is $-K\,\partial\theta/\partial n$ in the direction of the normal $\mathbf{n}$, measured away from the surface, so that the boundary condition is

$$K\,\partial\theta/\partial n + H(\theta_s - \theta_0) = 0,$$

i.e.

$$\partial\theta/\partial n + h(\theta_s - \theta_0) = 0,$$

where $h = H/K$. Sometimes this is referred to as Newton's law of cooling. It corresponds to surface evaporation in diffusion, and we have

$$\partial C/\partial n + \alpha(C_s - C_0) = 0$$

where $\alpha = h = H/D$.

If the surface is perpendicular to the x-direction, as an example, $\partial C/\partial n = \partial C/\partial x$ if n is along the direction of x *increasing* but $\partial C/\partial n = -\partial C/\partial x$ if along x *decreasing*. Thus for a slab between $x = 0$ and $x = 1$ we have

$$\partial c/\partial x + \alpha(C_s - C_0) = 0, \qquad x = 1,$$

but

$$-\partial c/\partial x + \alpha(C_s - C_0) = 0, \qquad x = 0.$$

(e) A perfect conductor of heat is always at a uniform temperature and so is equivalent in this respect to a well-stirred fluid. Thus, a boundary condition describing thermal contact with a perfect conductor also describes diffusion of solute from a well-stirred solution or vapour. Conservation of heat from a well-stirred fluid of fixed volume V and uniform temperature θ_s, a function of time, gives as a boundary condition on the surface $x = 0$ of a medium, $x > 0$, in contact with the fluid

$$c\rho V\,\partial\theta_s/\partial t = K\,\partial\theta/\partial x, \qquad x = 0,$$

where c, ρ have their usual meaning for the fluid. Correspondingly for diffusion we have

$$V \, \partial C_s / \partial t = D \, \partial C / \partial x, \qquad x = 0.$$

In addition, θ_s will be the temperature just within the surface but the two concentrations may be related by some isotherm equation. Care is needed with algebraic signs as in the previous section.

(f) The conservation principle applied at the interface between two media of different properties leads immediately to boundary conditions

$$\theta_1 = \theta_2, \qquad K_1 \, \partial \theta_1 / \partial x = K_2 \, \partial \theta_2 / \partial x$$

in heat flow and

$$C_1 = PC_2 + Q, \qquad D_1 \, \partial C_1 / \partial x = D_2 \, \partial C_2 / \partial x,$$

in diffusion, where the suffices 1 and 2 denote the two media and P and Q are constant.

(g) If heat is produced in a medium, e.g. as a result of an exothermic reaction, at a rate A per unit volume, this must be added to the right side of eqn. (1.20), which can be written

$$c\rho \frac{\partial \theta}{\partial t} = K \frac{\partial^2 \theta}{\partial x^2} + A.$$

The diffusion equation (1.4) must be similarly modified if the diffusing substance is created or removed as diffusion proceeds. We identify A as the rate of creation per unit volume and put $c\rho = 1$, $K = D$ as usual.

(h) Problems in heat flow may involve moving boundaries on which phase changes occur, accompanied by the absorption or liberation of latent heat. In the first case, the relevant feature is that latent heat is removed instantaneously from the heat-conduction process, in which it takes no further part. The diffusion counterpart is the immobilizing of diffusing molecules on fixed sites or in holes. The velocity of the transformation boundary at $X(t)$ is related to the difference between the rate of heat arriving and leaving it by conduction by a condition

$$-K_1 \, \partial \theta_1 / \partial x + K_2 \, \partial \theta_2 / \partial x = L\rho \, dX/dt,$$

where L is the latent heat per unit mass and ρ the density, assuming no volume changes accompany the transformation. The condition in a diffusion problem follows by writing $\theta = C, D = K, \rho = 1$ as usual, and L becomes the capacity of the immobilizing sites in unit volume for trapped diffusing molecules.

Useful collections of mathematical solutions of the diffusion equations are to be found in books by Barrer (1951), Jost (1952), and Jacobs (1967). Jacob's solutions are of particular interest to biologists and biophysicists.

2

METHODS OF SOLUTION WHEN THE
DIFFUSION COEFFICIENT IS CONSTANT

2.1. Types of solution

GENERAL solutions of the diffusion equation can be obtained for a variety of initial and boundary conditions provided the diffusion coefficient is constant. Such a solution usually has one of two standard forms. Either it is comprised of a series of error functions or related integrals, in which case it is most suitable for numerical evaluation at small times, i.e. in the early stages of diffusion, or it is in the form of a trigonometrical series which converges most satisfactorily for large values of time. When diffusion occurs in a cylinder the trigonometrical series is replaced by a series of Bessel functions. Of the three methods of solution described in this chapter, the first two illustrate the physical significance of the two standard types of solution. The third, employing the Laplace transform, is essentially an operator method by which both types of solution may be obtained. It is the most powerful of the three, particularly for more complicated problems. The methods are presented here as simply as possible. The fuller treatments necessary to make the discussion mathematically rigorous are to be found in works on heat conduction, e.g. Carslaw and Jaeger (1959).

2.2. Method of reflection and superposition

2.2.1. *Plane source*

It is easy to see by differentiation that

$$C = \frac{A}{t^{\frac{1}{2}}} \exp\left(-x^2/4Dt\right), \tag{2.1}$$

where A is an arbitrary constant, is a solution of

$$\frac{\partial C}{\partial t} = D\frac{\partial^2 C}{\partial x^2}, \tag{2.2}$$

which is the equation for diffusion in one dimension when D is constant. The expression (2.1) is symmetrical with respect to $x = 0$, tends to zero as x approaches infinity positively or negatively for $t > 0$, and for $t = 0$ it vanishes everywhere except at $x = 0$, where it becomes infinite. The total amount of substance M diffusing in a cylinder of infinite length and unit cross-section is given by

$$M = \int_{-\infty}^{\infty} C\,dx, \tag{2.3}$$

and if the concentration distribution is that of expression (2.1) we see, on writing

$$x^2/4Dt = \xi^2, \qquad dx = 2(Dt)^{\frac{1}{2}}\,d\xi, \tag{2.4}$$

that

$$M = 2AD^{\frac{1}{2}} \int_{-\infty}^{\infty} \exp(-\xi^2)\,d\xi = 2A(\pi D)^{\frac{1}{2}}. \tag{2.5}$$

Expression (2.5) shows that the amount of substance diffusing remains constant and equal to the amount originally deposited in the plane $x = 0$. Thus, on substituting for A from (2.5) in eqn (2.1), we obtain

$$C = \frac{M}{2(\pi Dt)^{\frac{1}{2}}} \exp(-x^2/4Dt), \tag{2.6}$$

and this is therefore the solution which describes the spreading by diffusion of an amount of substance M deposited at time $t = 0$ in the plane $x = 0$. Fig. 2.1 shows typical distributions at three successive times.

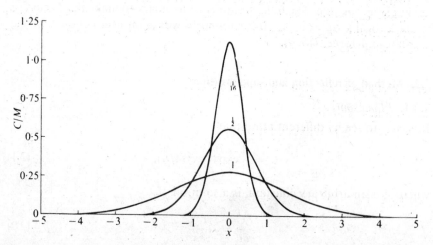

FIG. 2.1. Concentration–distance curves for an instantaneous plane source. Numbers on curves are values of Dt.

2.2.2. Reflection at a boundary

Expression (2.6) can be used to build up solutions of other problems in linear flow by introducing the concept of reflection at a boundary. Thus, in the problem just considered, half the diffusing substance moves in the direction of positive x and the other half along negative x. If, however, we have a semi-infinite cylinder extending over the region $x > 0$ and with an impermeable boundary at $x = 0$, all the diffusion occurs in the direction of positive x. We can consider the solution for negative x to be reflected in the plane $x = 0$ and superposed on the original distribution in the region $x > 0$. Since the original solution was symmetrical about $x = 0$ the concentration distribution for the semi-infinite cylinder is given by

$$C = \frac{M}{(\pi Dt)^{\frac{1}{2}}} \exp(-x^2/4Dt). \tag{2.7}$$

This procedure of reflection and superposition is mathematically sound, for reflection at $x = 0$ means the adding of two solutions of the diffusion equation. Since this equation is linear the sum of the two solutions is itself a solution, and we see that (2.7) satisfies the condition that the total amount of diffusing substance remains constant at M. Furthermore, the condition to be satisfied at the impermeable boundary is

$$\partial C/\partial x = 0, \qquad x = 0, \tag{2.8}$$

since this is the mathematical condition for zero flow across a boundary. As $\partial C/\partial x$ is zero at $x = 0$ in the original solution (2.6), it is clearly still zero after reflection and superposition.

2.2.3. *Extended initial distributions*

So far we have considered only cases in which all the diffusing substance is concentrated initially in a plane. More frequently in practice, however, the initial distribution occupies a finite region and we have an initial state such as that defined by

$$C = C_0, \qquad x < 0, \qquad C = 0, \qquad x > 0, \qquad t = 0. \tag{2.9}$$

This is the initial distribution, for example, when a long column of clear water rests on a long column of solution, or when two long metal bars are placed in contact end to end. The solution to such a problem is readily deduced by considering the extended distribution to be composed of an infinite number of line sources and by superposing the corresponding infinite number of elementary solutions. With reference to Fig. 2.2, consider the diffusing substance in an element of width $\delta\xi$ to be a line source of strength $C_0\,\delta\xi$. Then, from (2.6) the concentration at point P, distance ξ from the element, at time t is

$$\frac{C_0\,\delta\xi}{2(\pi Dt)^{\frac{1}{2}}} \exp(-\xi^2/4Dt),$$

and the complete solution due to the initial distribution (2.9) is given by summing over successive elements $\delta\xi$, i.e. by

$$C(x, t) = \frac{C_0}{2(\pi Dt)^{\frac{1}{2}}} \int_x^\infty \exp(-\xi^2/4Dt)\,d\xi = \frac{C_0}{\pi^{\frac{1}{2}}} \int_{x/2\sqrt{(Dt)}}^\infty \exp(-\eta^2)\,d\eta, \quad (2.10)$$

where $\eta = \xi/2\sqrt{(Dt)}$.

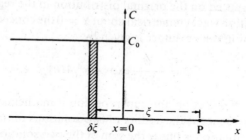

FIG. 2.2. Extended initial distribution.

A standard mathematical function, of which extensive tables are available, is the error function, usually written as erf z, where

$$\text{erf } z = \frac{2}{\pi^{\frac{1}{2}}} \int_0^z \exp(-\eta^2)\,d\eta. \quad (2.11)$$

This function has the properties

$$\text{erf}(-z) = -\text{erf } z, \qquad \text{erf}(0) = 0, \qquad \text{erf}(\infty) = 1, \quad (2.12)$$

and hence, since

$$\frac{2}{\pi^{\frac{1}{2}}} \int_z^\infty \exp(-\eta^2)\,d\eta = \frac{2}{\pi^{\frac{1}{2}}} \int_0^\infty \exp(-\eta^2)\,d\eta - \frac{2}{\pi^{\frac{1}{2}}} \int_0^z \exp(-\eta^2)\,d\eta$$

$$= 1 - \text{erf } z = \text{erfc } z, \quad (2.13)$$

where erfc is referred to as the error-function complement, the solution (2.10) of the diffusion problem is usually written in the form

$$C(x, t) = \tfrac{1}{2}C_0 \,\text{erfc}\frac{x}{2\sqrt{(Dt)}}. \quad (2.14)$$

Convenient tables of the error function are those of the Works Project Association (1941) and shorter tables are to be found, for example, in Milne-Thomson and Comrie (1944). Table 2.1, taken from Carslaw and Jaeger (1959), is sufficient for many practical purposes. The form of the concentration distribution is shown in Fig. 2.3. It is clear from (2.14) that $C = \tfrac{1}{2}C_0$ at $x = 0$ for all $t > 0$.

The error function therefore enters into the solution of a diffusion problem as a consequence of summing the effect of a series of line sources, each yielding an exponential type of distribution.

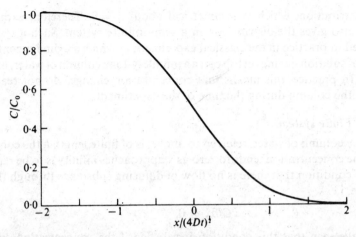

FIG. 2.3. Concentration–distance curve for an extended source of infinite extent.

In the same way, we can study the diffusion of a substance initially confined in the region $-h < x < +h$ as in Fig. 2.4. Here the integration is from $x-h$ to $x+h$ instead of from x to ∞ as in (2.10), leading immediately to the result

$$C = \tfrac{1}{2}C_0\left\{\operatorname{erf}\frac{h-x}{2\sqrt{(Dt)}} + \operatorname{erf}\frac{h+x}{2\sqrt{(Dt)}}\right\}. \tag{2.15}$$

The concentration distribution at successive times is shown in Fig. 2.4. It is clear that the system can be cut in half by a plane at $x = 0$ without affecting

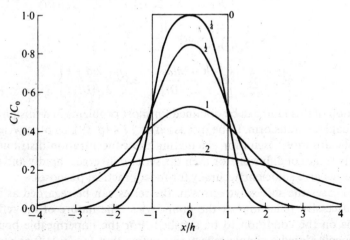

FIG. 2.4. Concentration–distance curves for an extended source of limited extent. Numbers on curves are values of $(Dt/h^2)^{\frac{1}{2}}$.

the distribution, which is symmetrical about $x = 0$. Therefore expression (2.15) also gives the distribution in a semi-infinite system. Such a system is realized in practice in the classical experiment in which a cylinder contains a layer of solution having on top of it an infinitely-long column of water, initially clear. In practice, this means that concentration changes do not reach the top of the column during the time of the experiment.

2.2.4. *Finite systems*

If the column of water, referred to above, is of finite length l, the condition that the concentration tends to zero as x approaches infinity is to be replaced by the condition that there is no flow of diffusing substance through the top surface, i.e.

$$\partial C/\partial x = 0, \qquad x = l. \tag{2.16}$$

We have seen that this condition is satisfied if the concentration curve is considered to be reflected at the boundary and the reflected curve superposed on the original one. In the finite system we are considering now, the curve reflected at $x = l$ is reflected again at $x = 0$, and then at $x = l$, and so on, the result of each successive reflection being superposed on the original curve (2.15). Since the original solution is the sum of two error functions, the complete expression for the concentration in the finite system is an infinite series of error functions or error-function complements so that

$$
\begin{aligned}
C = \tfrac{1}{2}C_0 \Bigg\{ & \operatorname{erfc} \frac{x-h}{2\sqrt{(Dt)}} - \operatorname{erfc} \frac{x+h}{2\sqrt{(Dt)}} + \operatorname{erfc} \frac{2l-h-x}{2\sqrt{(Dt)}} \\
& - \operatorname{erfc} \frac{2l+h-x}{2\sqrt{(Dt)}} + \operatorname{erfc} \frac{2l-h+x}{2\sqrt{(Dt)}} - \operatorname{erfc} \frac{2l+h+x}{2\sqrt{(Dt)}} \\
& + \operatorname{erfc} \frac{4l-h-x}{2\sqrt{(Dt)}} - \operatorname{erfc} \frac{4l-x+h}{2\sqrt{(Dt)}} + \ldots \Bigg\}
\end{aligned}
$$

$$
= \tfrac{1}{2}C_0 \sum_{n=-\infty}^{\infty} \left\{ \operatorname{erf} \frac{h+2nl-x}{2\sqrt{(Dt)}} + \operatorname{erf} \frac{h-2nl+x}{2\sqrt{(Dt)}} \right\}. \tag{2.17}
$$

A solution of this kind can be obtained for most problems in diffusion by use of the Laplace transform, to be discussed in § 2.4 (p. 19), or otherwise. Such solutions are most useful for calculating the concentration distribution in the early stages of diffusion, for then the series converges rapidly and two or three terms give sufficient accuracy for most practical purposes.

In all cases the successive terms in the series can be regarded as arising from successive reflections at the boundaries. The nature of the reflection depends on the condition to be satisfied. For the impermeable boundary already considered a simple reflection ensures that $\partial C/\partial x = 0$ as required. Another boundary condition which occurs frequently is of the type $C = 0$, in which case it is necessary to change the sign of the concentration when it is reflected at the boundary. A further example of the use of this method is given by Jost (1952). For more complicated problems, however, the reflection and superposition method soon becomes unwieldy and results are more readily obtained by other methods.

2.3. Method of separation of variables

A standard method of obtaining a solution of a partial differential equation is to assume that the variables are separable. Thus we may attempt to find a solution of (2.2) by putting

$$C = X(x)T(t), \tag{2.18}$$

where X and T are functions of x and t respectively. Substitution in (2.2) yields

$$X\frac{dT}{dt} = DT\frac{d^2X}{dx^2},$$

which may be rewritten

$$\frac{1}{T}\frac{dT}{dt} = \frac{D}{X}\frac{d^2X}{dx^2}, \tag{2.19}$$

so that we have on the left-hand side an expression depending on t only, while the right-hand side depends on x only. Both sides therefore must be equal to the same constant which, for the sake of the subsequent algebra, is conveniently taken as $-\lambda^2 D$. We have, therefore, two ordinary differential equations

$$\frac{1}{T}\frac{dT}{dt} = -\lambda^2 D, \tag{2.20}$$

and

$$\frac{1}{X}\frac{d^2X}{dx^2} = -\lambda^2, \tag{2.21}$$

of which solutions are

$$T = e^{-\lambda^2 Dt}, \tag{2.22}$$

and

$$X = A\sin\lambda x + B\cos\lambda x, \tag{2.23}$$

leading to a solution of (2.2) of the form

$$C = (A\sin\lambda x + B\cos\lambda x)\exp(-\lambda^2 Dt), \tag{2.24}$$

where A and B are constants of integration. Since (2.2) is a linear equation, the most general solution is obtained by summing solutions of type (2.24), so that we have

$$C = \sum_{m=1}^{\infty} (A_m \sin \lambda_m x + B_m \cos \lambda_m x) \exp(-\lambda_m^2 Dt), \qquad (2.25)$$

where A_m, B_m, and λ_m are determined by the initial and boundary conditions for any particular problem. Thus if we are interested in diffusion out of a plane sheet of thickness l, through which the diffusing substance is initially uniformly distributed and the surfaces of which are kept at zero concentration, the conditions are

$$C = C_0, \quad 0 < x < l, \qquad t = 0 \qquad (2.26)$$

$$C = 0, \qquad x = 0, \qquad x = l, \qquad t > 0. \qquad (2.27)$$

The boundary conditions (2.27) demand that

$$B_m = 0, \qquad \lambda_m = m\pi/l, \qquad (2.28)$$

and hence the initial condition (2.26) becomes

$$C_0 = \sum_{1}^{\infty} A_m \sin(m\pi x/l), \quad 0 < x < l. \qquad (2.29)$$

By multiplying both sides of (2.29) by $\sin(p\pi x/l)$ and integrating from 0 to l using the relationships

$$\int_0^l \sin\frac{p\pi x}{l} \sin\frac{m\pi x}{l} dx = \begin{cases} 0, & m \neq p, \\ \tfrac{1}{2}l, & m = p, \end{cases} \qquad (2.30)$$

we find that terms for which m is even vanish, and

$$A_m = 4C_0/m\pi, \quad m = 1, 3, 5, \dots.$$

The final solution is therefore

$$C = \frac{4C_0}{\pi} \sum_{n=0}^{\infty} \frac{1}{2n+1} \exp\{-D(2n+1)^2\pi^2 t/l^2\} \sin\frac{(2n+1)\pi x}{l}, \qquad (2.31)$$

where $2n+1$ has been substituted for m for convenience so that n takes values $0, 1, 2, \dots$. This trigonometrical-series type of solution converges satisfactorily for moderate and large times, and it is then used for numerical evaluation in preference to the error-function type of solution discussed earlier in § 2.2.4 (p. 16).

In (2.29) the initial distribution is expressed as a sum of sine functions. This reveals the physical significance of the trigonometrical series in (2.31), each term of which corresponds to a term in the Fourier series (2.29) by which the initial distribution can be represented.

2.4. Method of the Laplace transform

The Laplace transformation is a mathematical device which is useful for the solution of various problems in mathematical physics. Application of the Laplace transform to the diffusion equation removes the time variable, leaving an ordinary differential equation the solution of which yields the transform of the concentration as a function of the space variables x, y, z. This is then interpreted, according to certain rules, to give an expression for the concentration in terms of x, y, z and time, satisfying the initial and boundary condition. Historically the method may be regarded as derived from the operational methods introduced by Heaviside. Full accounts of the Laplace transform and its application have been given by Carslaw and Jaeger (1941), Churchill (1944), and others. Shorter accounts by Jaeger (1949) and Tranter (1951) are also available. Here we shall deal only with its application to the diffusion equation, the aim being to describe rather than to justify the procedure.

The solution of many problems in diffusion by this method calls for no mathematics beyond ordinary calculus. No attempt is made here to explain its application to the more difficult problems for which the theory of functions of a complex variable must be used, though solutions to problems of this kind are quoted in later chapters. The fuller accounts should be consulted for the derivation of such solutions.

2.4.1. Definition of the Laplace transform

Suppose $f(t)$ to be a known function of t for positive values of t. Then the Laplace transform $\bar{f}(p)$ of $f(t)$ is defined as

$$\bar{f}(p) = \int_0^\infty e^{-pt} f(t)\, dt, \tag{2.32}$$

where p is a number sufficiently large to make the integral (2.32) converge. It may be a complex number whose real part is sufficiently large, but in the present discussion it suffices to think of it simply as a real positive number. For example, if $f(t) = e^{2t}$, p must exceed 2. Unless it is necessary to emphasize that $\bar{f}$ is a function of p, just as f is a function of t, we shall usually denote the Laplace transform of f by $\bar{f}$.

Laplace transforms of common functions are readily constructed by carrying out the integration in (2.32) as in the following examples:

$$f(t) = 1, \qquad \bar{f}(p) = \int_0^\infty e^{-pt}\, dt = 1/p, \tag{2.33}$$

$$f(t) = e^{at}, \qquad \bar{f}(p) = \int_0^\infty e^{-pt} e^{at}\, dt = \int_0^\infty e^{-(p-a)t}\, dt = \frac{1}{p-a}, \tag{2.34}$$

$$f(t) = \sin \omega t, \qquad \bar{f}(p) = \int_0^\infty e^{-pt} \sin \omega t\, dt = \frac{\omega}{p^2 + \omega^2}. \tag{2.35}$$

Extensive tables or dictionaries of Laplace transforms are available, some in the works referred to above. A short table of transforms occurring frequently in diffusion problems is reproduced from Carslaw and Jaeger's book (1959) in Table 2.2.

2.4.2. *Semi-infinite medium*

As an example of the application of the Laplace transform, consider the problem of diffusion in a semi-infinite medium, $x > 0$, when the boundary is kept at a constant concentration C_0, the initial concentration being zero throughout the medium. We need a solution of

$$\frac{\partial C}{\partial t} = D\frac{\partial^2 C}{\partial x^2}, \qquad (2.36)$$

satisfying the boundary condition

$$C = C_0, \qquad x = 0, \qquad t > 0, \qquad (2.37)$$

and the initial condition

$$C = 0, \qquad x > 0, \qquad t = 0. \qquad (2.38)$$

On multiplying both sides of (2.36) by e^{-pt} and integrating with respect to t from 0 to ∞ we obtain

$$\int_0^\infty e^{-pt}\frac{\partial^2 C}{\partial x^2}\,dt - \frac{1}{D}\int_0^\infty e^{-pt}\frac{\partial C}{\partial t}\,dt = 0. \qquad (2.39)$$

If we assume that the orders of differentiation and integration can be interchanged, and this can be justified for the functions in which we are interested, then

$$\int_0^\infty e^{-pt}\frac{\partial^2 C}{\partial x^2}\,dt = \frac{\partial^2}{\partial x^2}\int_0^\infty C e^{-pt}\,dt = \frac{\partial^2 \bar{C}}{\partial x^2}. \qquad (2.40)$$

Also, integrating by parts, we have

$$\int_0^\infty e^{-pt}\frac{\partial C}{\partial t}\,dt = [C e^{-pt}]_0^\infty + p\int_0^\infty C e^{-pt}\,dt = p\bar{C}, \qquad (2.41)$$

since the term in the square bracket vanishes at $t = 0$ by virtue of the initial condition (2.38) and at $t = \infty$ through the exponential factor. Thus (2.36) reduces to

$$D\frac{\partial^2 \bar{C}}{\partial x^2} = p\bar{C}. \qquad (2.42)$$

By treating the boundary condition (2.37) in the same way we obtain

$$\bar{C} = \int_0^\infty C_0 e^{-pt}\,dt = \frac{C_0}{p}, \qquad x = 0. \qquad (2.43)$$

Thus the Laplace transform reduces the partial differential equation (2.36) to the ordinary differential equation (2.42). The solution of (2.42) satisfying (2.43), and for which $\bar{C}$ remains finite as x approaches infinity is

$$\bar{C} = \frac{C_0}{p} e^{-qx}, \tag{2.44}$$

where $q^2 = p/D$. Reference to Table 2.2, item 8, shows that the function whose transform is given by (2.44) is

$$C = C_0 \operatorname{erfc} \frac{x}{2\sqrt{(Dt)}}, \tag{2.45}$$

where, as before,

$$\operatorname{erfc} z = 1 - \operatorname{erf} z. \tag{2.46}$$

It is easy to verify that (2.45) satisfies (2.36), (2.37), and (2.38) and that it is therefore the required solution of the diffusion problem.

2.4.3. Plane sheet

In the problem just considered the transform solution could be interpreted immediately by reference to the table of transforms. Consider now, as an example of a slightly more difficult problem in which this is not so, a plane sheet of thickness $2l$, whose surface are maintained at constant concentration C_0, and with zero concentration of diffusing substance throughout the sheet initially. Let the sheet occupy the region $-l \leqslant x \leqslant l$, so that there is symmetry about $x = 0$ and the boundary conditions may be written

$$C = C_0, \qquad x = l, \qquad t \geqslant 0, \tag{2.47}$$

$$\partial C/\partial x = 0, \qquad x = 0, \qquad t \geqslant 0. \tag{2.48}$$

Eqn (2.48) expresses the condition that there is no diffusion across the central plane of the sheet. It is often more convenient to use this condition and to consider only half the sheet, $0 \leqslant x \leqslant l$, instead of using the condition $C = C_0, x = -l$.

The equations for the Laplace transform $\bar{C}$ are

$$\frac{d^2\bar{C}}{dx^2} - q^2\bar{C} = 0, \qquad 0 < x < \tag{2.49}$$

with

$$d\bar{C}/dx = 0, \qquad x = 0, \tag{2.50}$$

and

$$\bar{C} = C_0/p, \qquad x = l, \tag{2.51}$$

where $q^2 = p/D$ as before. The solution of these is

$$\bar{C} = \frac{C_0 \cosh qx}{p \cosh ql}. \tag{2.52}$$

There are two methods of dealing with this transform solution, leading to the two standard types of solution we have already met. We shall first obtain a solution useful for small values of the time.

(i) *Expansion in negative exponentials.* We express the hyperbolic functions in (2.52) in terms of negative exponentials and expand in a series by the binomial theorem. Thus we obtain from (2.52),

$$\bar{C} = \frac{C_0(e^{qx} + e^{-qx})}{p \, e^{ql}(1 + e^{-2ql})} = \frac{C_0}{p} \{ e^{-q(l-x)} + e^{-q(l+x)} \} \sum_{n=0}^{\infty} (-1)^n e^{-2nql}$$

$$= \frac{C_0}{p} \sum_{n=0}^{\infty} (-1)^n e^{-q((2n+1)l-x)} + \frac{C_0}{p} \sum_{n=0}^{\infty} (-1)^n e^{-q((2n+1)l+x)}. \tag{2.53}$$

Thus, using item 8 of the table of transforms (Table 2.2), we obtain

$$C = C_0 \sum_{n=0}^{\infty} (-1)^n \operatorname{erfc} \frac{(2n+1)l - x}{2\sqrt{(Dt)}} + C_0 \sum_{n=0}^{\infty} (-1)^n \operatorname{erfc} \frac{(2n+1)l + x}{2\sqrt{(Dt)}}. \tag{2.54}$$

This is a series of error functions such as we obtained by the method of reflection and superposition. Successive terms are in fact the concentrations at depths $l-x$, $l+x$, $3l-x$, $3l+x$, ... in the semi-infinite medium. The series converges quite rapidly for all except large values of Dt/l^2. For example, we have for the concentration at the centre of the sheet ($x = 0$) when $Dt/l^2 = 1$

$$C/C_0 = 0.9590 - 0.0678 + 0.0008 = 0.8920, \tag{2.55}$$

and when $Dt/l^2 = 0.25$

$$C/C_0 = 0.3146 - 0.0001 = 0.3145. \tag{2.56}$$

(ii) *Expression in partial fractions.* It can be shown that if a transform $\bar{y}$ has the form

$$\bar{y} = \frac{f(p)}{g(p)}, \tag{2.57}$$

where $f(p)$ and $g(p)$ are polynomials in p which have no common factor, the degree of $f(p)$ being lower than that of $g(p)$, and if

$$g(p) = (p - a_1)(p - a_2) \dots (p - a_n), \tag{2.58}$$

where $a_1, a_2, \dots, a_n$ are constants which may be real or complex but must all be different, then the function $y(t)$ whose transform is $\bar{y}(p)$ is given by

$$y(t) = \sum_{r=1}^{n} \frac{f(a_r)}{g'(a_r)} e^{a_r t}. \tag{2.59}$$

Here $g'(a_r)$ denotes the value of $dg(p)/dp$ when $p = a_r$. A proof of this by Jaeger (1949) is reproduced in the Appendix to this chapter. It is derived by expressing (2.57) in partial fractions. Since the hyperbolic functions cosh z and sinh z can be represented by the following infinite products (see, e.g. Carslaw (1909, p. 275))

$$\cosh z = \left(1+\frac{4z^2}{\pi^2}\right)\left(1+\frac{4z^2}{3^2\pi^2}\right)\left(1+\frac{4z^2}{5^2\pi^2}\right)\cdots, \tag{2.60}$$

$$\sinh z = z\left(1+\frac{z^2}{\pi^2}\right)\left(1+\frac{z^2}{2^2\pi^2}\right)\left(1+\frac{z^2}{3^2\pi^2}\right)\cdots, \tag{2.61}$$

a quotient of these functions such as in (2.52) may still be regarded as being of the type (2.57) except that now $f(p)$ and $g(p)$ have an infinite number of factors. The $a_1, a_2, \ldots$ are the zeros of $g(p)$, i.e. solutions of the equation, $g(p) = 0$, and if these are all different it is plausible to assume that (2.59) still holds with $n = \infty$. The justification of this assumption involves the theory of functions of a complex variable in order to carry out a contour integration and is to be found in the fuller accounts of the subject. There is, in fact, a rigorous mathematical argument by which the use of (2.59) with $n = \infty$, can be justified in diffusion problems in a finite region only. It must not be applied to (2.44), for example, for the semi-infinite region. The above refers to a_1, $a_2, \ldots$ all different. The extension of (2.59) to cases in which $g(p)$ has repeated zeros, e.g. one of its factors is square, is given in the Appendix. Its application to an infinite number of factors is still justifiable.

We may now consider the application of (2.59) to (2.52). First the zeros of the denominator must be found. Clearly, $p = 0$ is a zero, and the other zeros are given by the values of q for which cosh $ql = 0$, i.e.

$$q = \pm\frac{(2n+1)\pi i}{2l}, \quad n = 0, 1, 2, \ldots \tag{2.62}$$

and hence

$$p = -\frac{D(2n+1)^2\pi^2}{4l^2}, \quad n = 0, 1, 2, \ldots. \tag{2.63}$$

To apply (2.59) to (2.52) we need

$$g'(p) = \frac{d}{dp}(p \cosh ql) = \cosh ql + \tfrac{1}{2}ql \sinh ql. \tag{2.64}$$

For the zero $p = 0$, $g'(p) = 1$. For the other zeros, given by (2.62) and (2.63), $\cosh ql = 0$, and

$$\sinh ql = \sinh \frac{(2n+1)\pi i}{2} = i \sin \frac{(2n+1)\pi}{2} = i(-1)^n, \qquad 2.65)$$

so that for these zeros, by substituting in (2.64) we obtain

$$g'(p) = \frac{(2n+1)\pi(-1)^{n+1}}{4}. \qquad (2.66)$$

Hence finally by inserting the zeros into (2.59) we obtain

$$C = C_0 - \frac{4C_0}{\pi} \sum_{n=0}^{\infty} \frac{(-1)^n}{2n+1} \exp\{-D(2n+1)^2\pi^2 t/4l^2\} \cos \frac{(2n+1)\pi x}{2l}. \qquad (2.67)$$

This is the trigonometrical-series type of solution obtained previously by the method of separation of the variables. The series converges rapidly for large values of t. Thus for the concentration at the centre of the sheet ($x = 0$) when $Dt/l^2 = 1$,

$$C/C_0 = 1 - 0.1080 = 0.8920, \qquad (2.68)$$

and when $Dt/l^2 = 0.25$

$$C/C_0 = 1 - 0.6872 + 0.0017 = 0.3145. \qquad (2.69)$$

2.5. Solutions in two and three dimensions

2.5.1. *Solutions expressed as the product of the solutions of simpler problems*

Consider the equation of diffusion

$$\frac{\partial^2 c}{\partial x_1^2} + \frac{\partial^2 c}{\partial x_2^2} + \frac{\partial^2 c}{\partial x_3^2} = \frac{1}{D} \frac{\partial c}{\partial t}, \qquad (2.70)$$

in the rectangular parallelepiped

$$a_1 < x_1 < b_1, \quad a_2 < x_2 < b_2, \quad a_3 < x_3 < b_3. \qquad (2.71)$$

For certain types of initial and boundary conditions, the solution of (2.70) is the product of the solutions of the three one-variable problems, and thus can be written down immediately if these are known. The following proof is given by Carslaw and Jaeger (1959, p. 33).

Suppose $c_r(x_r, t), r = 1, 2, 3$, is the solution of

$$\frac{\partial^2 c_r}{\partial x_r^2} = \frac{1}{D} \frac{\partial c_r}{\partial t}, \quad a_r < x_r < b_r, \qquad (2.72)$$

with boundary conditions

$$\alpha_r \frac{\partial c_r}{\partial x_r} - \beta_r c_r = 0, \qquad x_r = a_r, \qquad t > 0, \tag{2.73}$$

$$\alpha'_r \frac{\partial c_r}{\partial x_r} + \beta'_r c_r = 0, \qquad x_r = b_r, \qquad t > 0, \tag{2.74}$$

where the α_r and β_r are constants, either of which may be zero (so that the cases of zero surface concentration and no flow of heat at the surface are included) and with initial conditions

$$c_r(x_r, t) = C_r(x_r), \qquad t = 0, \qquad a_r < x_r < b_r. \tag{2.75}$$

Then the solution of (2.70) in the region defined by (2.71) with

$$c = C_1(x_1)C_2(x_2)C_3(x_3), \qquad t = 0, \tag{2.76}$$

and with boundary conditions

$$\alpha_r \frac{\partial c}{\partial x_r} - \beta_r c = 0, \qquad x_r = a_r, \qquad t > 0, \quad r = 1, 2, 3, \tag{2.77}$$

$$\alpha'_r \frac{\partial c}{\partial x_r} - \beta'_r c = 0, \qquad x_r = b_r, \qquad t > 0, \quad r = 1, 2, 3, \tag{2.78}$$

is

$$c = c_1(x_1, t)c_2(x_2, t)c_3(x_3, t). \tag{2.79}$$

For substituting (2.79) in (2.70) gives

$$c_2 c_3 \frac{\partial^2 c_1}{\partial x_1^2} + c_3 c_1 \frac{\partial^2 c_2}{\partial x_2^2} + c_1 c_2 \frac{\partial^2 c_3}{\partial x_3^2} - \frac{1}{D}\left(c_2 c_3 \frac{\partial c_1}{\partial t} + c_3 c_1 \frac{\partial c_2}{\partial t} + c_1 c_2 \frac{\partial c_3}{\partial t}\right) = 0, \tag{2.80}$$

using (2.72). Clearly the initial and boundary conditions (2.76), (2.77), and (2.78) are satisfied.

An essential condition is (2.76), namely that the initial condition must be expressible as a product of the initial conditions for the one-variable problems taken separately. Carslaw and Jaeger (1959) give solutions for a rectangular corner, rectangles, parallelepipeds, cylinders and some examples of isotherms are shown graphically.

2.5.2. A general relationship

Goldenberg (1963) derived a much more general relationship between the transient solutions of two-dimensional problems for an infinite cylinder of arbitrary cross-section, and the transient solutions of the corresponding three-dimensional problems in finite cylinders. The boundary conditions on the end faces of the cylinder may be of the type describing constant concentration, constant flux, or evaporation, and heat may be generated at a rate independent of time and the axial coordinate.

An example of Goldenberg's relationships is afforded by the homogeneous slab, bounded by the planes $z = 0$ and $z = l$, with heat produced in the cylindrical region R of arbitrary cross-section for time $t > 0$, at the constant rate A per unit time per unit volume.

The slab is initially at zero temperature with its faces maintained at zero temperature for $t > 0$. The corresponding two-dimensional problem is obtained when the thickness l of the slab is infinite and the solution is independent of z. Goldenberg (1963) shows that the solution $V(x, y, z, t)$ for the finite slab is related to $f(x, y, t)$ for the infinite slab by the expression

$$V = \frac{4}{\pi} \sum_{n=0}^{\infty} \frac{\sin\{(2n+1)\pi z/l\}}{2n+1} \left\{ e^{-\delta t}f + \delta \int_0^t e^{-\delta t}f\, dt \right\}, \qquad (2.81)$$

where $\delta = D\pi^2(2n+1)^2/l^2$ and D is the diffusion coefficient.

The same relationship is valid for a hollow cylinder of arbitrary cross-section and for the region external to a cylinder of arbitrary cross-section. Similar relationships hold in other situations discussed by Goldenberg.

2.6. Other solutions

Langford (1967) obtained new solutions of the one-dimensional heat equation for temperature and heat flux both prescribed at the same fixed boundary. They take the form of series of polynomial and quasi-polynomial solutions for plane sheets, cylinders, and spheres. They include as special cases some of the old or classical solutions. They also have applications to phase change problems with boundaries moving at a constant velocity.

APPENDIX TO CHAPTER 2

To deduce the function $y(t)$ whose Laplace transform $\bar{y}(p)$ is given by

$$\bar{y}(p) = \frac{f(p)}{g(p)}, \qquad (1)$$

we first put $\bar{y}(p)$ into partial fractions in the usual way by assuming

$$\frac{f(p)}{g(p)} = \sum_{r=1}^{n} \frac{A_r}{p-a_r} = \frac{A_1}{p-a_1} + \frac{A_2}{p-a_2} + \frac{A_3}{p-a_3} + \dots + \frac{A_n}{p-a_n}. \qquad (2)$$

Then

$$f(p) \equiv \sum_{r=1}^{n} A_r(p-a_1) \dots (p-a_{r-1})(p-a_{r+1}) \dots (p-a_n), \qquad (3)$$

and putting $p = a_r$ in this gives

$$f(a_r) = A_r(a_r - a_1) \dots (a_r - a_{r-1})(a_r - a_{r+1}) \dots (a_r - a_n) \quad (r = 1, 2, \dots, n). \quad (4)$$

Substituting for A_r from (4) in (2) gives

$$\bar{y}(p) = \sum_{r=1}^{n} \frac{1}{p - a_r} \frac{f(a_r)}{(a_r - a_1) \dots (a_r - a_{r-1})(a_r - a_{r+1}) \dots (a_r - a_n)}. \quad (5)$$

Now since

$$g(p) = (p - a_1)(p - a_2) \dots (p - a_n), \quad (6)$$

we have, on differentiating by the ordinary rule for differentiation of a product,

$$g'(p) = \sum_{r=1}^{n} (p - a_1) \dots (p - a_{r-1})(p - a_{r+1}) \dots (p - a_n). \quad (7)$$

Putting $p = a_r$ in this, gives

$$g'(a_r) = (a_r - a_1) \dots (a_r - a_{r-1})(a_r - a_{r+1}) \dots (a_r - a_n), \quad (8)$$

and using (8) in (5) gives a further form for $\bar{y}(p)$ namely

$$\bar{y}(p) = \sum_{r=1}^{n} \frac{f(a_r)}{(p - a_r)g'(a_r)}. \quad (9)$$

On applying item 3 of Table 2.2 to successive terms of (9) we obtain immediately

$$y(t) = \sum_{r=1}^{n} \frac{f(a_r)}{g'(a_r)} e^{a_r t}. \quad (10)$$

This result applies only to the case in which $g(p)$ has no repeated zeros, but it can readily be generalized for the case of repeated factors. Thus (10) implies that to each linear factor $p - a_r$ of the denominator of $\bar{y}(p)$ there corresponds a term

$$\frac{f(a_r)}{g'(a_r)} e^{a_r t} \quad (11)$$

in the solution. The generalization is that, to each squared factor $(p - b)^2$ of the denominator of $\bar{y}(p)$ there corresponds a term

$$\left[\frac{(p - b)^2 f(p)}{g(p)} \right]_{p = b} t\, e^{bt} + \left[\frac{d}{dp} \left\{ \frac{(p - b)^2 f(p)}{g(p)} \right\} \right]_{p = b} e^{bt} \quad (12)$$

in the solution. To each multiple factor $(p - c)^m$ of the denominator of $\bar{y}(p)$ there corresponds a term

$$\sum_{s=0}^{m-1} \left[\frac{d^s}{dp^s} \left\{ \frac{(p - c)^m f(p)}{g(p)} \right\} \right]_{p = c} \frac{t^{m-s-1}}{s!(m-s-1)!} e^{ct} \quad (13)$$

in the solution.

3

INFINITE AND SEMI-INFINITE MEDIA

3.1. Introduction

IN this and the following three chapters solutions of the diffusion equation are presented for different initial and boundary conditions. In nearly all cases the diffusion coefficient is taken as constant. In many cases the solutions are readily evaluated numerically with the help of tables of standard mathematical functions. Where this is not so, and where numerical evaluation is tedious, as many graphical and tabulated solutions as space permits are given.

3.2. Instantaneous sources

Under this heading are included all problems in which an amount of diffusing substance is deposited within a certain restricted region at time $t = 0$ and left to diffuse throughout the surrounding medium. For example, it may be located initially at a point, or in a plane, or within a sphere, when we have an instantaneous point, plane, or spherical source as the case may be.

The solution for an instantaneous plane source in an infinite medium has already been given in Chapter 2, eqn (2.6). The corresponding solution for an instantaneous point source on an infinite plane surface is obtained in the same way by recognizing that

$$C = \frac{A}{t} \exp\{-(x^2 + y^2)/4Dt\} \qquad (3.1)$$

is a solution of

$$\frac{\partial^2 C}{\partial x^2} + \frac{\partial^2 C}{\partial y^2} = \frac{1}{D} \frac{\partial C}{\partial t}, \qquad (3.2)$$

which is the equation for diffusion in two dimensions when the diffusion coefficient is constant. The arbitrary constant A is expressed in terms of M, the total amount of substance diffusing, by performing the integration

$$M = \int_{-\infty}^{\infty} \int_{-\infty}^{\infty} C \, dx \, dy = 4\pi DA, \qquad (3.3)$$

the concentration C being expressed in this problem as the amount of diffusing substance per unit area of surface. The concentration at a distance

r from a point source on an infinite plane surface is thus given by

$$C = \frac{M}{4\pi Dt} \exp\left(-r^2/4Dt\right). \tag{3.4}$$

The corresponding expression for a point source in an infinite volume is

$$C = \frac{M}{8(\pi Dt)^{\frac{3}{2}}} \exp\left(-r^2/4Dt\right). \tag{3.5}$$

By integrating the appropriate solution for a point source with respect to the relevant space variables, solutions may be obtained for line, surface, and volume sources. Thus for surface diffusion in the x, y plane due to a line source along the y-axis we have

$$C = \int_{-\infty}^{\infty} \frac{M}{4\pi Dt} \exp\left\{-(x^2+y^2)/4Dt\right\} dy = \frac{M}{2(\pi Dt)^{\frac{1}{2}}} \exp\left(-x^2/4Dt\right), \tag{3.6}$$

where now M is the amount of diffusing substance deposited initially per unit length of the line source. This is the same as expression (2.6) of Chapter 2 for a plane source of strength M per unit area in an infinite volume. The corresponding result for a line source of strength M per unit length in an infinite volume, obtained by integrating (3.5), is

$$C = \frac{M}{4\pi Dt} \exp\left(-r^2/4Dt\right), \tag{3.7}$$

which is the same expression as (3.4) for a point source on an infinite plane surface, though M has a different significance in the two cases. Results for a variety of sources are derived by Carslaw and Jaeger (1959, p. 255). The spherical and cylindrical sources are likely to be of practical interest. If the diffusing substance is initially distributed uniformly through a sphere of radius a, the concentration C at radius r, and time t is given by

$$C = \tfrac{1}{2}C_0\left\{\mathrm{erf}\,\frac{a+r}{2\sqrt{(Dt)}} + \mathrm{erf}\,\frac{a+r}{2\sqrt{(Dt)}}\right\}$$

$$-\frac{C_0}{r}\sqrt{\left(\frac{Dt}{\pi}\right)}\left[\exp\left\{-(a-r)^2/4Dt\right\} - \exp\left\{-(a+r)^2/4Dt\right\}\right], \tag{3.8}$$

where C_0 is the uniform concentration in the sphere initially. Expression (3.8) may easily be written in terms of the total amount of diffusing substance M, since

$$M = \tfrac{4}{3}\pi a^3 C_0. \tag{3.9}$$

The corresponding result for a cylinder of radius a may be written in the form

$$C = \frac{C_0}{2Dt} \exp\left(-r^2/4Dt\right) \int_0^a \exp\left(-r'^2/4Dt\right) I_0\left(\frac{rr'}{2Dt}\right) r'\, dr', \tag{3.10}$$

where I_0 is the modified Bessel function of the first kind of order zero. Tables of I_0 are available. The integral in (3.10) has to be evaluated numerically except on the axis $r = 0$, where (3.10) becomes

$$C = C_0\{1 - \exp(-a^2/4Dt)\}. \tag{3.11}$$

These expressions may be applied, for example, to the diffusion of a sphere or cylinder of solute into a large volume of solvent. Curves showing the concentration distribution at successive times are given in Figs. 3.1 and 3.2.

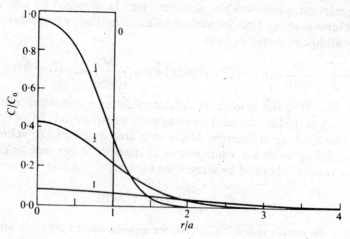

FIG. 3.1. Concentration distributions for a spherical source. Numbers on curves are values of $(Dt/a^2)^{\frac{1}{2}}$.

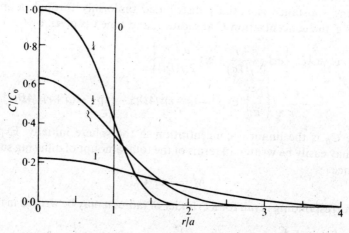

FIG. 3.2. Concentration distributions for a cylindrical source. Numbers on curves are values of $(Dt/a^2)^{\frac{1}{2}}$.

The expression (3.10) and the curves of Fig. 3.2 also apply to a circular disc source, of radius a, on an infinite plane surface, if C_0 denotes the uniform concentration of the diffusing substance over the region $0 < r < a$ initially. An alternative solution given by Rideal and Tadayon (1954) is

$$C = aC_0 \int_0^\infty J_1(ua)J_0(ur) \exp(-Dtu^2)\, du, \tag{3.10a}$$

where J_0 and J_1 are Bessel functions of the first kind and of order zero and one respectively. Tables of J_0 and J_1 are available. Rideal and Tadayon also give an expression for the total amount of diffusing substance Q remaining on the disc after time t, which is

$$Q = 2\pi a^2 C_0 \int_0^\infty \frac{J_1^2(ua)}{u} \exp(-Dtu^2)\, du. \tag{3.10b}$$

For small values of t, (3.10b) becomes

$$Q = \pi a^2 C_0 \left\{ 1 - \frac{2}{a}\left|\frac{Dt}{2\pi}\right|^{\frac{1}{2}} \right\}, \tag{3.10c}$$

and when t is large we have

$$Q = \pi a^4 C_0/(4Dt). \tag{3.10d}$$

Cases of an extended source in an infinite medium, where the diffusing substance initially occupies the semi-infinite region $x < 0$ or is confined to the region $-h < x < h$, have been considered (see eqns (2.14) and (2.15) in Chapter 2). The solution to the corresponding problem in which the region $-h < x < h$ is at zero concentration and $|x| > h$ at a uniform concentration C_0 initially is

$$C = \tfrac{1}{2}C_0 \left\{ \mathrm{erfc}\, \frac{h-x}{2\sqrt{(Dt)}} + \mathrm{erfc}\, \frac{h+x}{2\sqrt{(Dt)}} \right\}. \tag{3.12}$$

Knight and Philip (1973) have obtained an exact explicit solution for $D = a(1 - b^{-1}C)^{-2}$ with a and b positive constants and for an instantaneous distributed source at concentration b and the initial concentration uniform but arbitrary away from the source. The analysis holds for infinite and semi-infinite media, and physical applications are discussed.

3.2.1. *Continuous sources*

A solution for a continuous source, from which diffusing substance is liberated continuously at a certain rate, is deduced from the solution for the corresponding instantaneous source by integrating with respect to time t. Thus if diffusing substance is liberated continuously from a point in an infinite volume at the rate ϕ per second, the concentration at a point distant r

from the source at time t is obtained by integrating (3.5) and is

$$C = \frac{1}{8(\pi D)^{\frac{3}{2}}} \int_0^t \phi(t') \exp\left\{-r^2/4D(t-t')\right\} \frac{dt'}{(t-t')^{\frac{3}{2}}}. \tag{3.5a}$$

If ϕ is constant and equal to q, then

$$C = \frac{q}{4\pi Dr} \operatorname{erfc} \frac{r}{2\sqrt{(Dt)}}. \tag{3.5b}$$

Solutions for other continuous sources are obtained similarly and examples are given by Carslaw and Jaeger (1959, p. 261).

3.3. Semi-infinite media

The solution for a plane source deposited initially at the surface, $x = 0$, of a semi-infinite medium was given in Chapter 2, eqn (2.7), and that for the initial distribution $C = C_0, 0 < x < h, C = 0, x > h$, was seen to be given by eqn (2.15) for x positive.

The problem of the semi-infinite medium whose surface is maintained at a constant concentration C_0, and throughout which the concentration is initially zero, was handled by the method of the Laplace transform in Chapter 2, p. 21 (see eqn (2.45)). Other results of practical importance which may be obtained in the same way are given below.

(i) The concentration is C_0 throughout, initially, and the surface is maintained at a constant concentration C_1.

$$\frac{C - C_1}{C_0 - C_1} = \operatorname{erf} \frac{x}{2\sqrt{(Dt)}}. \tag{3.13}$$

The special case of zero surface concentration is immediately obvious. The rate of loss of diffusing substance from the semi-infinite medium when the surface concentration is zero, is given by

$$\left(D \frac{\partial C}{\partial x}\right)_{x=0} = \frac{DC_0}{\sqrt{(\pi Dt)}}, \tag{3.14}$$

so that the total amount M_t of diffusing substance which has left the medium at time t is given by integrating (3.14) with respect to t and is

$$M_t = 2C_0 \left(\frac{Dt}{\pi}\right)^{\frac{1}{2}}. \tag{3.15}$$

The same expression with C_0 replaced by C_1 gives the total amount taken up by the medium in time t if the initial concentration C_0 is zero. If the initial concentration is zero throughout the semi-infinite medium, and the surface concentration varies with time, solutions are still obtainable by the Laplace transform. Cases of practical interest are given below. Here M_t is used throughout to denote the total amount of diffusing substance which has entered the medium at time t.

(ii) $C_{x=0} = kt$, where k is a constant.

$$C = kt\left[\left(1 + \frac{x^2}{2Dt}\right)\operatorname{erfc}\frac{x}{2\sqrt{(Dt)}} - \frac{x}{\sqrt{(\pi Dt)}}\exp\left\{-x^2/4Dt\right\}\right]$$

$$= 4kt\,\mathrm{i}^2\operatorname{erfc}\frac{x}{2\sqrt{(Dt)}}. \tag{3.16}$$

$$M_t = \tfrac{4}{3}kt\left(\frac{Dt}{\pi}\right)^{\tfrac{1}{2}}. \tag{3.17}$$

The function $\mathrm{i}^2\operatorname{erfc}$ is defined and tabulated in Table 2.1, so that values of C may be written down immediately. The effect of an increasing surface concentration is shown in Fig. 3.3.

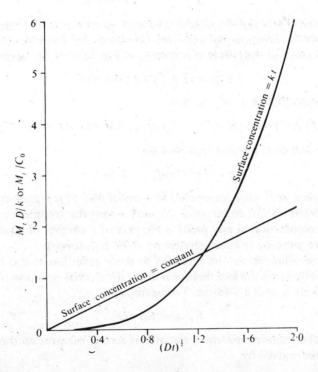

FIG. 3.3. Sorption curves for variable and constant surface concentrations in a semi-infinite medium.

(iii) $C_{x=0} = kt^{\frac{1}{2}}$, where k is a constant.

$$C = kt^{\frac{1}{2}} \left\{ \exp(-x^2/4Dt) - \frac{x\pi^{\frac{1}{2}}}{2\sqrt{(Dt)}} \operatorname{erfc} \frac{x}{2\sqrt{(Dt)}} \right\}$$

$$= k(\pi t)^{\frac{1}{2}} \operatorname{ierfc} \frac{x}{2\sqrt{(Dt)}}. \tag{3.18}$$

$$M_t = \tfrac{1}{2}kt(\pi D)^{\frac{1}{2}}. \tag{3.19}$$

The function ierfc is defined and tabulated in Table 2.1. In this case M is directly proportional to t and so the rate of uptake of diffusing substance is constant.

(iv) $C_{x=0} = kt^{\frac{1}{2}n}$, where k is a constant and n is any positive integer, even or odd.

$$C = k\Gamma(\tfrac{1}{2}n+1)(4t)^{\frac{1}{2}n} \operatorname{i}^n\!\operatorname{erfc} \frac{x}{2\sqrt{(Dt)}}. \tag{3.20}$$

$$M^t = \frac{k}{2^n}\sqrt{(Dt)}(4t)^{\frac{1}{2}n} \frac{\Gamma(\tfrac{1}{2}n+1)}{\Gamma(\tfrac{1}{2}n+\tfrac{3}{2})}. \tag{3.21}$$

The function $\operatorname{i}^n\!\operatorname{erfc}$ is defined and tabulated up to $n = 6$ in Table 2.1. $\Gamma(n)$ is the gamma function defined and tabulated, for example, by Milne–Thomson and Comrie (1944). If n is even, so that $\tfrac{1}{2}n = N$, an integer, then

$$\Gamma(\tfrac{1}{2}n+1) = \Gamma(N+1) = N!. \tag{3.22}$$

If n is odd, so that $\tfrac{1}{2}n = M - \tfrac{1}{2}$, then

$$\Gamma(\tfrac{1}{2}n+1) = \Gamma(M+\tfrac{1}{2}) = 1.3.5. \dots (2M-3)(2M-1)\pi^{\frac{1}{2}}/2^M. \tag{3.23}$$

Other properties of gamma functions are

$$\Gamma(n+1) = n\Gamma(n); \qquad \Gamma(\tfrac{1}{2}) = \pi^{\frac{1}{2}}. \tag{3.24}$$

A polynomial in $t^{\frac{1}{2}}$ may sometimes be a useful way of representing a given surface concentration empirically. In such a case the complete expression for the concentration at any point is the sum of a number of terms of type (3.20) corresponding to successive terms in the polynomial.

(v) These solutions can be extended to cover other initial and boundary conditions by using the fact that for a linear differential equation the sum of two solutions is itself a solution. For example, if

$$C_{x=0} = C_0 + kt \tag{3.25}$$

and the semi-infinite medium is initially at zero concentration throughout, the solution is given by

$$C = C_0 \operatorname{erfc} \frac{x}{2\sqrt{(Dt)}} + 4kt \operatorname{i}^2\!\operatorname{erfc} \frac{x}{2\sqrt{(Dt)}}, \tag{3.26}$$

since the first term on the right-hand side of (3.26) is the solution satisfying the conditions

$$
\left.
\begin{array}{ll}
C = C_0, & x = 0 \\
C = 0, & x > 0
\end{array}
\right\}, \tag{3.27}
$$

and the second term satisfies

$$
\left.
\begin{array}{ll}
C = kt, & x = 0 \\
C = 0, & x > 0
\end{array}
\right\}. \tag{3.28}
$$

In general the solution to the problem of the semi-infinite medium in which the surface concentration is given by $F(t)$ and in which the initial distribution is $f(x)$, is given by

$$
C = c_1 + c_2, \tag{3.29}
$$

where c_1 is a solution of the diffusion equation which satisfies

$$
\left.
\begin{array}{ll}
c_1 = 0, & t = 0 \\
c_1 = F(t), & x = 0
\end{array}
\right\}, \tag{3.30}
$$

and c_2 is another solution satisfying

$$
\left.
\begin{array}{ll}
c_2 = f(x), & t = 0 \\
c_2 = 0, & x = 0
\end{array}
\right\}. \tag{3.31}
$$

Clearly, with c_1 and c_2 so defined, the diffusion equation and the initial and boundary conditions are satisfied. Consider, as an example, the problem of desorption from a semi-infinite medium having a uniform initial concentration C_0, and a surface concentration decreasing according to (3.25), with k negative. The solution is

$$
C = C_0 + 4kt\, i^2 \mathrm{erfc}\, \frac{x}{2\sqrt{(Dt)}}. \tag{3.32}
$$

which is obtained by adding to (3.26) the solution satisfying

$$
\left.
\begin{array}{lll}
C = C_0, & x > 0, & t = 0 \\
C = 0, & x = 0, & t > 0
\end{array}
\right\}, \tag{3.33}
$$

i.e. by adding $C_0 \, \mathrm{erf}\, \{x/2\sqrt{(Dt)}\}$.

3.3.1. *Surface evaporation condition*

In some cases the boundary condition relates to the rate of transfer of diffusing substance across the surface of the medium. Thus, if a stream of dry air passes over the surface of a solid containing moisture, loss of moisture occurs by surface evaporation. Similarly if the solid is initially dry and the air contains water vapour, the solid takes up moisture. In each case the rate of exchange of moisture at any instant depends on the relative humidity of the air and the moisture concentration in the surface of the solid. The simplest reasonable assumption is that the rate of exchange is directly proportional to the difference between the actual concentration C_s in the surface at any time and the concentration C_0 which would be in equilibrium with the vapour pressure in the atmosphere remote from the surface. Mathematically this means that the boundary condition at the surface is

$$-D\frac{\partial C}{\partial x} = \alpha(C_0 - C_s), \quad x = 0, \tag{3.34}$$

where α is a constant of proportionality.

If the concentration in a semi-infinite medium is initially C_2 throughout, and the surface exchange is determined by (3.34), the solution is

$$\frac{C - C_2}{C_0 - C_2} = \operatorname{erfc}\frac{x}{2\sqrt{(Dt)}} - \exp(hx + h^2Dt)\operatorname{erfc}\left\{\frac{x}{2\sqrt{(Dt)}} + h\sqrt{(Dt)}\right\}, \tag{3.35}$$

where $h = \alpha/D$. The special cases of zero concentration in the medium initially ($C_2 = 0$), and evaporation into an atmosphere of zero relative humidity ($C_0 = 0$), are immediately obvious from (3.35). The rate at which the total amount M_t of diffusing substance in the semi-infinite medium per unit cross-sectional area changes is given by

$$\frac{dM_t}{dt} = -\left(D\frac{\partial C}{\partial x}\right)_{x=0} = \alpha(C_0 - C_s), \tag{3.36}$$

and, on substituting for C_s the value obtained from (3.35) by putting $x = 0$, after integration with respect to t we obtain for the total quantity of diffusing substance having crossed unit area of the surface,

$$M_t = \left(\frac{C_0 - C_2}{h}\right)\left\{\exp(h^2Dt)\operatorname{erfc}h\sqrt{(Dt)} - 1 + \frac{2}{\pi^{\frac{1}{2}}}h\sqrt{(Dt)}\right\} \tag{3.37}$$

If C_0 is greater than C_2 this amount is taken up by the medium; if C_0 is less than C_2 this amount is lost by evaporation from the surface. The expression (3.35) can be written in terms of any two of the dimensionless parameters

$$\frac{x}{2\sqrt{(Dt)}}, \quad h\sqrt{(Dt)}, \quad \text{or} \quad hx. \tag{3.38}$$

In Fig. 3.4 the ratio $(C - C_2)/(C_0 - C_2)$ is plotted as a function of $x/2\sqrt{(Dt)}$ for various values of $h\sqrt{(Dt)}$. In order to evaluate $hM_t/(C_0 - C_2)$ from (3.37), only one dimensionless parameter $h\sqrt{(Dt)}$ is needed. The relationship is readily evaluated from standard functions and is shown graphically in

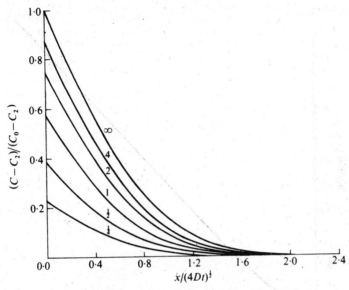

FIG. 3.4. Concentration distribution for a surface evaporation condition in a semi-infinite medium. Numbers on curves are values of $h(Dt)^{\frac{1}{2}}$.

Fig. 3.5. The evaluation for large $h\sqrt{(Dt)}$ is made easier by using the asymptotic formula

$$\exp{(z^2)}\,\mathrm{erfc}\,z = \frac{1}{\pi^{\frac{1}{2}}}\left(\frac{1}{z} - \frac{1}{2z^3} + \frac{1\,.\,3}{2^2\,.\,z^5}\cdots\right). \tag{3.39}$$

3.3.2. Square-root relationship

Expression (2.45) shows that the solution of the problem of diffusion into a semi-infinite medium having zero initial concentration and the surface of which is maintained constant, involves only the single dimensionless parameter

$$\frac{x}{2\sqrt{(Dt)}}. \tag{3.40}$$

It follows from this that

(i) the distance of penetration of any given concentration is proportional to the square root of time;

(ii) the time required for any point to reach a given concentration is proportional to the square of its distance from the surface and varies inversely as the diffusion coefficient;

(iii) the amount of diffusing substance entering the medium through unit area of its surface varies as the square root of time.

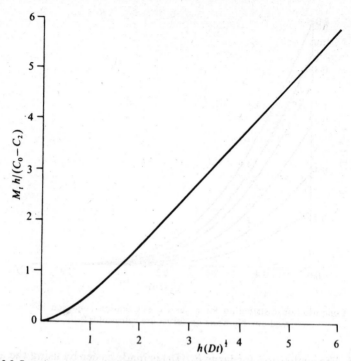

FIG. 3.5. Sorption curve for a surface evaporation condition in a semi-infinite medium.

These fundamental properties hold in general in semi-infinite media, provided the initial concentration is uniform and the surface concentration remains constant. They also hold for point and line sources on infinite surfaces or in infinite media, and also for the case of diffusion in an infinite medium where the diffusing substance is initially confined to the region $x < 0$. Clearly they do not apply to cases where parameters other than $x/2\sqrt{(Dt)}$ are involved, such as the width of an extended source or the rate of change of surface concentration, etc.

3.4. The infinite composite medium

Here we consider diffusion in systems in which two media are present. Suppose the region $x > 0$ is of one substance in which the diffusion coefficient is D_1, and in the region $x < 0$ the diffusion coefficient is D_2. In the simplest case, the initial conditions are that the region $x > 0$ is at a uniform concentration C_0, and in $x < 0$ the concentration is zero initially. If we write c_1 for the concentration in $x > 0$ and c_2 in $x < 0$ the boundary conditions at the interface $x = 0$ may be written

$$c_2/c_1 = k, \qquad x = 0, \tag{3.41}$$

$$D_1\,\partial c_1/\partial x = D_2\,\partial c_2/\partial x, \qquad x = 0, \tag{3.42}$$

where k is the ratio of the uniform concentration in the region $x < 0$ to that in $x > 0$ when final equilibrium is attained. The condition (3.42) expresses the fact that there is no accumulation of diffusing substance at the boundary. A solution to this problem is easily obtained by combining solutions for the semi-infinite medium so as to satisfy the initial and boundary conditions. We seek solutions of the type

$$c_1 = A_1 + B_1 \, \text{erf} \frac{x}{2\sqrt{(D_1 t)}}, \qquad x > 0, \tag{3.43}$$

$$c_2 = A_2 + B_2 \, \text{erf} \frac{|x|}{2\sqrt{(D_2 t)}}, \qquad x < 0, \tag{3.44}$$

which are known to satisfy the diffusion equations in the two regions. By choosing the constants A_1, B_1, A_2, B_2 to satisfy the initial conditions and (3.41), (3.42) we obtain

$$c_1 = \frac{C_0}{1 + k(D_2/D_1)^{\frac{1}{2}}} \left\{ 1 + k(D_2/D_1)^{\frac{1}{2}} \, \text{erf} \frac{x}{2\sqrt{(D_1 t)}} \right\}, \tag{3.45}$$

$$c_2 = \frac{kC_0}{1 + k(D_2/D_1)^{\frac{1}{2}}} \, \text{erfc} \frac{|x|}{2\sqrt{(D_2 t)}}. \tag{3.46}$$

Fig. 3.6 shows a typical concentration distribution for the case where $D_2 = 4D_1$ and $k = \frac{1}{2}$. Graphs for other cases are shown by Jost (1952) and by Barrer (1951). We may note that, as diffusion proceeds, the concentrations at the interface, $x = 0$, remain constant at the values

$$c_1 = \frac{C_0}{1 + k(D_2/D_1)^{\frac{1}{2}}}, \qquad c_2 = \frac{kC_0}{1 + k(D_2/D_1)^{\frac{1}{2}}}. \tag{3.47}$$

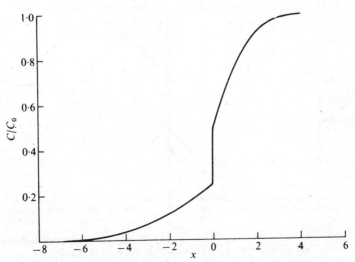

FIG. 3.6. Concentration distribution in a composite medium. $D_1 t = 1$, $D_2 = 4D_1$, $k = \frac{1}{2}$.

3.4.1. *Interface resistance*

If we have the same problem as on p. 38 except that there is a contact resistance at $x = 0$, then (3.41) is to be replaced by

$$D_1 \, \partial c_1 / \partial x + h(c_2 - c_1) = 0, \qquad x = 0, \tag{3.48}$$

while (3.42) still holds. The expressions for the concentrations in this case are

$$c_1 = \frac{C_0}{1 + (D_2/D_1)^{\frac{1}{2}}} \left[1 + \left(\frac{D_2}{D_1}\right)^{\frac{1}{2}} \left\{ \mathrm{erf} \frac{x}{2\sqrt{(D_1 t)}} \right. \right.$$

$$\left. \left. + \exp(h_1 x + h_1^2 D_1 t) \, \mathrm{erfc} \left(\frac{x}{2\sqrt{(D_1 t)}} + h_1 \sqrt{(D_1 t)} \right) \right\} \right], \tag{3.49}$$

$$c_2 = \frac{C_0}{1 + (D_2/D_1)^{\frac{1}{2}}} \left\{ \mathrm{erfc} \frac{|x|}{2\sqrt{(D_2 t)}} \right.$$

$$\left. - \exp(h_2 x + h_2^2 D_2 t) \, \mathrm{erfc} \left(\frac{|x|}{2\sqrt{(D_2 t)}} + h_2 \sqrt{(D_2 t)} \right) \right\}, \tag{3.50}$$

where

$$h_1 = \frac{h}{D_1} \{ 1 + (D_1/D_2)^{\frac{1}{2}} \}, \qquad h_2 = \frac{h}{D_2} \{ 1 + (D_2/D_1)^{\frac{1}{2}} \}. \tag{3.51}$$

The concentrations on either side of the interface are no longer constant but each approaches the equilibrium value $\frac{1}{2}C_0$ relatively slowly. This, and the general distribution at successive times, is illustrated in Fig. 3.7 for the case in which $D_1 = D_2$ and $h = \frac{1}{10}D_1$.

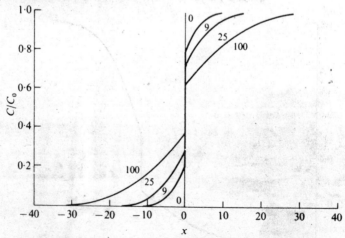

FIG. 3.7. Concentration distribution in a composite medium with a resistance at the interface. Numbers on curves are values of $D_1 t$; $D_1 = D_2$; $h = \frac{1}{10}D_1$.

3.5. The semi-infinite composite medium

This is the case of a semi-infinite medium which has a skin or surface layer having diffusion properties different from those of the rest of the medium. Thus, suppose in the semi-infinite region $-l < x < \infty$, the diffusion coefficient is D_1 in the region $-l < x < 0$, and that the concentration is denoted by c_1 there, while the corresponding quantities in $x > 0$ are D_2 and c_2. If we assume the conditions at the interface to be

$$c_1 = c_2, \qquad x = 0, \tag{3.52}$$

$$D_1 \, \partial c_1/\partial x = D_2 \, \partial c_2/\partial x, \qquad x = 0, \tag{3.53}$$

the solution to the problem of zero initial concentration and the surface $x = -l$ maintained at constant concentration C_0 is given by Carslaw and Jaeger (1959, p. 321), and is

$$c_1 = C_0 \sum_{n=0}^{\infty} \alpha^n \left\{ \operatorname{erfc} \frac{(2n+1)l+x}{2\sqrt{(D_1 t)}} - \alpha \operatorname{erfc} \frac{(2n+1)l-x}{2\sqrt{(D_1 t)}} \right\}, \tag{3.54}$$

$$c_2 = \frac{2kC_0}{k+1} \sum_{n=0}^{\infty} \alpha^n \operatorname{erfc} \frac{(2n+1)l+kx}{2\sqrt{(D_1 t)}}, \tag{3.55}$$

where

$$k = (D_1/D_2)^{\frac{1}{2}}, \qquad \alpha = \frac{1-k}{1+k}. \tag{3.56}$$

The total quantity entering the medium through unit area of the surface $x = -l$ in time t is M_t, where M_t is given by

$$\frac{M_t}{lC_0} = 2\left(\frac{D_1 t}{\pi l^2}\right)^{\frac{1}{2}} \left\{ 1 + 2 \sum_{n=1}^{\infty} \alpha^n \exp\left(-n^2 l^2/D_1 t\right) \right\} - 4 \sum_{n=1}^{\infty} n\alpha^n \operatorname{erfc} \frac{nl}{\sqrt{(D_1 t)}}. \tag{3.57}$$

Following Carslaw and Jaeger (1959, p. 322), for very large times the exponentials in (3.57) may all be replaced by unity. This is true also of the error-function complements in (3.57) and so for large times we have approximately

$$\frac{M_t}{lC_0} = 2\left(\frac{D_1 t}{\pi l^2}\right)^{\frac{1}{2}} \left(1 + \frac{2\alpha}{1-\alpha}\right) - \frac{4\alpha}{(1-\alpha)^2}, \tag{3.58}$$

provided $\alpha^2 < 1$. Fig. 3.8 shows M_t/lC_0 as a function of $(D_1 t/l^2)^{\frac{1}{2}}$.

Whipple (1954) has given formulae for the concentration in a semi-infinite region of low diffusion coefficient bisected by a thin well-diffusing slab, at different times after the boundary of the semi-infinite region has been raised suddenly from zero to unit concentration. This is of interest in grain-boundary diffusion.

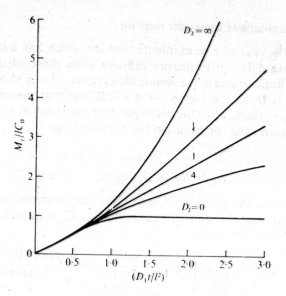

FIG. 3.8. Sorption curves for a composite semi-infinite medium. Numbers on curves are values of D_1/D_2.

3.6. Weber's disc

This is a classical problem of the field due to an electrified disc. More recent interest relates to the diffusion current at a circular electrode. Several authors (Tranter 1951; Grigull 1961; Saito 1968) have developed the same solution in different ways. When the diameter of the electrode is $2a$ we require solutions of

$$\frac{\partial^2 C}{\partial r^2} + \frac{1}{r}\frac{\partial C}{\partial r} + \frac{\partial^2 C}{\partial z^2} = 0, \qquad (3.59)$$

where C is the concentration say of oxygen in the solution, the axis of z passes perpendicularly through the centre of the disc and r is the radial distance from the z-axis. The boundary conditions are

$$C = 0, \qquad z = 0, \qquad r \leqslant a,$$
$$\partial C/\partial z = 0, \qquad z = 0, \qquad r > a,$$
$$C = C_0, \qquad r \geqslant 0; \qquad z = \infty,$$
$$C = C_0, \qquad z \geqslant 0, \qquad r = \infty. \qquad (3.60)$$

where C_0 is the concentration in the bulk of the solution. The concentration is given by

$$C_0 - C = \frac{2C_0}{\pi} \int_0^\infty \frac{\sin(ma)}{m} J_0(mr) e^{-mz} \, dm$$

$$= \frac{2C_0}{\pi} \tan^{-1} \frac{a}{\sqrt{(\frac{1}{2}[(r^2+z^2-a^2)+\sqrt{\{(r^2+z^2-a^2)^2+4z^2a^2\}}])}}, \quad (3.61)$$

$$z > 0.$$

The concentration gradient at the disc surface is

$$\left(\frac{\partial C}{\partial z}\right)_{z=0} = -\frac{2C_0}{\pi} \int_0^\infty \sin(ma) J_0(mr) \, dm$$

$$= \frac{2C_0}{\pi} \frac{1}{\sqrt{(a^2-r^2)}}, \quad 0 \leqslant r \leqslant a. \quad (3.62)$$

Saito (1968) gives an approximate analytical solution for a narrow band electrode and sketches the concentration distribution for each electrode.

4

DIFFUSION IN A PLANE SHEET

4.1. Introduction

IN this chapter we consider various cases of one-dimensional diffusion in a medium bounded by two parallel planes, e.g. the planes at $x = 0$, $x = l$. These will apply in practice to diffusion into a plane sheet of material so thin that effectively all the diffusing substance enters through the plane faces and a negligible amount through the edges.

4.2. Steady state

Consider the case of diffusion through a plane sheet or membrane of thickness l and diffusion coefficient D, whose surfaces, $x = 0$, $x = l$, are maintained at constant concentrations C_1, C_2 respectively. After a time, a steady state is reached in which the concentration remains constant at all points of the sheet. The diffusion equation in one dimension then reduces to

$$d^2C/dx^2 = 0, \tag{4.1}$$

provided the diffusion coefficient D is constant. On integrating with respect to x we have

$$dC/dx = \text{constant}, \tag{4.2}$$

and by a further integration we have, on introducing the conditions at $x = 0$, $x = l$,

$$\frac{C - C_1}{C_2 - C_1} = \frac{x}{l}. \tag{4.3}$$

Both (4.2) and (4.3) show that the concentration changes linearly from C_1 to C_2 through the sheet. Also, the rate of transfer of diffusing substance is the same across all sections of the membrane and is given by

$$F = -D\,dC/dx = D(C_1 - C_2)/l. \tag{4.4}$$

If the thickness l and the surface concentrations C_1, C_2 are known, D can be deduced from an observed value of F by using (4.4). Experimental arrangements for measuring D in this and other ways have been reviewed by Newns (1950).

If the surface $x = 0$ is maintained at a constant concentration C_1 and at $x = l$ there is evaporation into an atmosphere for which the equilibrium concentration just within the surface is C_2, so that

$$\partial C/\partial x + h(C - C_2) = 0, \qquad x = l,$$

then we find

$$\frac{C - C_1}{C_2 - C_1} = \frac{hx}{1 + hl}, \tag{4.3a}$$

and

$$F = \frac{Dh(C_1 - C_2)}{1 + hl}. \tag{4.4a}$$

If the surface conditions are

$$\partial C/\partial x + h_1(C_1 - C) = 0, \quad x = 0; \qquad \partial C/\partial x + h_2(C - C_2) = 0, \quad x = l,$$

then

$$C = \frac{h_1 C_1\{1 + h_2(l - x)\} + h_2 C_2(1 + h_1 x)}{h_1 + h_2 + h_1 h_2 l}, \tag{4.3b}$$

and

$$F = \frac{Dh_1 h_2(C_1 - C_2)}{h_1 + h_2 + h_1 h_2 l}. \tag{4.4b}$$

Corresponding solutions for a funnel-shaped region are easily obtained (Jacobs 1967).

4.2.1. *Permeability constant*

In some practical systems, the surface concentrations C_1, C_2 may not be known but only the gas or vapour pressures p_1, p_2 on the two sides of the membrane. The rate of transfer in the steady state is then sometimes written

$$F = P(p_1 - p_2)/l, \tag{4.5}$$

and the constant P is referred to as the permeability constant. Here P is expressed, for example, as cm^3 gas at some standard temperature and pressure passing per second through $1 cm^2$ of the surface of a membrane 1 cm thick when the pressure difference across the membrane is 1 cm of mercury. The permeability constant is a much less fundamental constant than the diffusion coefficient which is expressed in units such as $cm^2 s^{-1}$, particularly as different investigators use different units and even different definitions of P.

If the diffusion coefficient is constant, and if the sorption isotherm is linear, i.e. if there is a linear relationship between the external vapour pressure and the corresponding equilibrium concentration within the membrane, then eqns (4.4) and (4.5) are equivalent, but not otherwise. The linear isotherm may be written

$$C = Sp, \tag{4.6}$$

where C is the concentration within the material of the membrane in equilibrium with an external vapour pressure p, and S is the solubility. Since C_1, p_1 and C_2, p_2 in (4.4) and (4.5) are connected by (4.6) it follows that, with due regard to units,

$$P = DS. \tag{4.7}$$

4.2.2. *Concentration-dependent diffusion coefficient*

If the diffusion coefficient varies with concentration it is clear that the simple value of D deduced from a measurement of the steady rate of flow is some kind of mean value over the range of concentration involved. Thus, if D is a function of C (4.1) is to be replaced by

$$\frac{d}{dx}\left(D\frac{dC}{dx}\right) = 0, \tag{4.8}$$

and hence the relationship

$$F = -D\, dC/dx = \text{constant} \tag{4.9}$$

still holds, as of course it must in the steady state. Integrating between C_1 and C_2, the two surface concentrations, we have

$$F = -\frac{1}{l}\int_{C_1}^{C_2} D\, dC = D_l(C_1 - C_2)/l, \tag{4.10}$$

where

$$D_l = \frac{1}{C_1 - C_2}\int_{C_2}^{C_1} D\, dC, \tag{4.11}$$

and this is the mean value deduced from a measurement of F. It follows from (4.9) that if D depends on C the concentration no longer depends linearly on distance through the membrane. Concentration distributions for D depending on C in a number of ways are given in Chapter 9.

4.2.3. *Composite membrane*

If we have a composite membrane composed of n sheets of thicknesses $l_1, l_2, \ldots, l_n$, and diffusion coefficients $D_1, D_2, \ldots, D_n$, the fall in concentration through the whole membrane is the sum of the falls through the component sheets. Since the rate of transfer F is the same across each section, the total drop in concentration is

$$\frac{Fl_1}{D_1} + \frac{Fl_2}{D_2} + \; ... \; + \frac{Fl_n}{D_n} = (R_1 + R_2 + \; ... \; + R_n)F,\tag{4.12}$$

where $R_1 = l_1/D_1$, etc. may be termed formally the resistance to diffusion of each sheet. Thus the resistance to diffusion of the whole membrane is simply the sum of the resistances of the separate layers, assuming that there are no barriers to diffusion between them. This subject is treated more generally in § 12.2(i) (p. 266).

4.3. Non-steady state

All the solutions presented here can be obtained either by the method of separation of the variables or by the Laplace transform as described in § 2.4. Many of the results are quoted by Barrer (1951), Carslaw and Jaeger (1959), Jacobs (1967), Jost (1952) and others. The emphasis here is on numerical evaluation.

4.3.1. *Surface concentrations constant. Initial distribution $f(x)$*

If

$$C = C_1, \qquad x = 0, \qquad t \geqslant 0,\tag{4.13}$$

$$C = C_2, \qquad x = l, \qquad t \geqslant 0,\tag{4.14}$$

$$C = f(x), \qquad 0 < x < l, \qquad t = 0,\tag{4.15}$$

the solution in the form of a trigonometrical series is

$$C = C_1 + (C_2 - C_1)\frac{x}{l} + \frac{2}{\pi}\sum_1^\infty \frac{C_2 \cos n\pi - C_1}{n} \sin \frac{n\pi x}{l} \exp\left(-Dn^2\pi^2 t/l^2\right)$$

$$+ \frac{2}{l}\sum_1^\infty \sin \frac{n\pi x}{\pi} \exp\left(-D\pi^2 n^2 t/l^2\right) \int_0^l f(x') \sin \frac{n\pi x'}{l} \, dx'.\tag{4.16}$$

In the cases of most common occurrence $f(x)$ is either zero or constant so that the integral in (4.16) is readily evaluated. Very often the problem is symmetrical about the central plane of the sheet, and the formulae are then most convenient if this is taken as $x = 0$ and the surfaces at $x = \pm l$.

4.3.2. *Uniform initial distribution. Surface concentrations equal*

This is the case of sorption and desorption by a membrane. If the region $-l < x < l$ is initially at a uniform concentration C_0, and the surfaces are kept at a constant concentration C_1, the solution (4.16) becomes

$$\frac{C - C_0}{C_1 - C_0} = 1 - \frac{4}{\pi}\sum_{n=0}^\infty \frac{(-1)^n}{2n+1} \exp\left\{-D(2n+1)^2\pi^2 t/4l^2\right\} \cos \frac{(2n+1)\pi x}{2l}.\tag{4.17}$$

If M_t denotes the total amount of diffusing substance which has entered the sheet at time t, and M_∞ the corresponding quantity after infinite time, then

$$\frac{M_t}{M_\infty} = 1 - \sum_{n=0}^{\infty} \frac{8}{(2n+1)^2\pi^2} \exp\left\{-D(2n+1)^2\pi^2 t/4l^2\right\}. \tag{4.18}$$

The corresponding solutions useful for small times are

$$\frac{C - C_0}{C_1 - C_0} = \sum_{n=0}^{\infty} (-1)^n \operatorname{erfc} \frac{(2n+1)l - x}{2\sqrt{(Dt)}}$$

$$+ \sum_{n=0}^{\infty} (-1)^n \operatorname{erfc} \frac{(2n+1)l + x}{2\sqrt{(Dt)}}, \tag{4.19}$$

and

$$\frac{M_t}{M_\infty} = 2\left(\frac{Dt}{l^2}\right)^{\frac{1}{2}}\left\{\pi^{-\frac{1}{2}} + 2\sum_{n=1}^{\infty} (-1)^n \operatorname{ierfc} \frac{nl}{\sqrt{(Dt)}}\right\}. \tag{4.20}$$

The modifications to these expressions for $C_0 = 0$ or $C_1 = 0$ are obvious.

Jason and Peters (1973) analyse the bimodal diffusion of water in fish muscle by combining two expressions of the type (4.18), one for each mode and each having its own diffusion coefficient.

Eqn (4.18) can be solved graphically for Dt/l^2 and hence D obtained from measured sorption or desorption time curves. An alternative suggested by Talbot and Kitchener (1956) is the approximate formula

$$\theta = -\ln x + \frac{q}{p} + \frac{rq^2}{2p^3}$$

where $\theta = \pi^2 Dt/4l^2$, $x = \frac{1}{8}\pi^2 M_t/M_\infty$ and $p = 1 + x^8 + x^{24}$, $q = \frac{1}{9}x^8 + \frac{1}{25}x^{24}$, and, $r = 1 + 9x^8 + 25x^{24}$.

The solution is correct to four significant figures when $M_t/M_\infty < \frac{2}{3}$ for desorption. The solution (4.18) also applies to diffusion along a cylindrical rod or tube of length l, with one end and its surface sealed and the other end maintained at a constant concentration.

Talbot and Kitchener also obtained a solution for a slightly tapering tube. If d_1 and d_2 are the diameters of the two ends, the degree of taper is specified by $\mu = (d_2 - d_1)/d_1$, where d_1 is the closed end, and may be positive or negative. For small μ, e.g. around 0·01, we have effectively radial diffusion and a solution is

$$\frac{M_t}{M_\infty} = \frac{6(1+\mu)^2}{3 + 3\mu + \mu^2} \sum_{n=1}^{\infty} \frac{\alpha_n^2 + \mu^2}{\alpha_n^2(\alpha_n^2 + \mu^2 + \mu)} \exp(-D\alpha_n^2 t/l^2),$$

where the α_n are the roots of $\tan \alpha = -\alpha/\mu$, and, to the first order in μ, $\alpha_n^2 = (n - \frac{1}{2})^2\pi^2 + 2\mu$. The solution becomes, with this approximation,

$$\frac{M_t}{M_\infty} = \sum_{n=1}^{\infty} \frac{8}{(2n-1)^2\pi^2}\{1 + (k_n - \lambda\theta)\mu \exp(-(2n-1)^2\theta)\},$$

where $\theta = \pi^2 Dt/4l^2$, $\lambda = 8/\pi^2$, $k_n = 1 - \frac{3}{2}\lambda/(2n-1)^2$, i.e.

$$\lambda = 0.81057, \qquad k_1 = -0.21585, \qquad k_2 = 0.86491, \qquad k_3 = 0.95137.$$

Thus putting $x = \frac{1}{8}\pi^2 M_t/M_\infty$ as before, θ must satisfy

$$x = \{1 + (k_1 - \lambda\theta)\mu\}\, e^{-\theta} + \tfrac{1}{9}\{1 + (k_2 - \lambda\theta)\mu\}\, e^{-9\theta}$$
$$+ \tfrac{1}{25}\{1 + (k_3 - \lambda\theta)\mu\}\, e^{-25\theta} + \dots .$$

If the solution (4.18) is applied to the tapered tube to calculate θ, the approximate value, say θ_0, obtained can be corrected by using

$$\theta = \theta_0 \left\{ 1 - \mu \frac{\lambda(1+q) - s/\theta_0}{p} \right\} = \theta_0 (1 - \kappa\mu)$$

where $s = k_1 + \frac{1}{9}k_2 x^8 + \frac{1}{25}k_3 x^{24}$, p and q are defined above, and $\kappa\mu$ is the correction term. Talbot and Kitchener (1956) quote

$$\kappa = 0.807 - 0.60x^8 + (0.212 - 0.25x^8)/\theta_0$$

approximately and discuss in more detail the capillary tube method (Anderson and Saddington 1949) for measuring diffusion coefficients.

It is clear that expressions (4.17), (4.18), (4.19), (4.20) can be written in terms of the dimensionless parameters

$$T = Dt/l^2, \qquad X = x/l, \tag{4.21}$$

so that the solutions for all values of D, l, t, and x can be obtained from graphs or tabulated values covering these two parameters. Graphs of $(C - C_0)/(C_1 - C_0)$ are shown for various times in Fig. 4.1. These are reproduced with change of nomenclature from Carslaw and Jaeger's book (1959, p. 101). Tabulated values of $(C - C_0)/(C_1 - C_0)$ and of M_t/M_∞ are given by Henry (1939). Values of M_t/M_∞ have also been tabulated by McKay (1930) and extensive numerical values for the concentration at the centre of the sheet, $x = 0$, are given by Olson and Schulz (1942). The curve labelled zero fractional uptake in Fig. 4.6 shows how M_t/M_∞ varies with the square root of time in a sheet of thickness $2a$ when the concentration at each surface remains constant.

4.3.3. *Uniform initial distribution. Surface concentrations different*

This is the case of flow through a membrane. If one face $x = 0$ of a membrane is kept at a constant concentration C_1 and the other $x = l$ at C_2, and the membrane is initially at a uniform concentration C_0, there is a finite interval of time during which the steady-state condition previously discussed

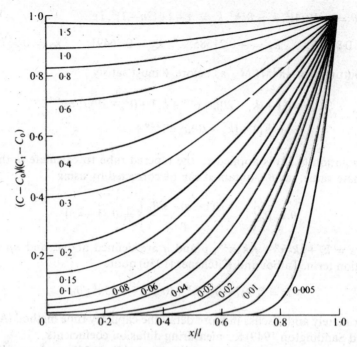

FIG. 4.1. Concentration distributions at various times in the sheet $-l < x < l$ with initial uniform concentration C_0 and surface concentration C_1. Numbers on curves are values of Dt/l^2.

in § 4.2 (p. 44) is set up. During this time the concentration changes according to

$$C = C_1 + (C_2 - C_1)\frac{x}{l} + \frac{2}{\pi}\sum_{n=1}^{\infty}\frac{C_2\cos n\pi - C_1}{n}\sin\frac{n\pi x}{l}\exp(-Dn^2\pi^2 t/l^2)$$

$$+ \frac{4C_0}{\pi}\sum_{m=0}^{\infty}\frac{1}{2m+1}\sin\frac{(2m+1)\pi x}{l}\exp\{-D(2m+1)^2\pi^2 t/l^2\}. \quad (4.22)$$

As t approaches infinity the terms involving the exponentials vanish and we have simply the linear concentration distribution obtained in § 4.2. Barnes (1934) examined the errors introduced by assuming a linear gradient to exist across the membrane during the whole course of diffusion.

If M_t denotes the total amount of diffusing substance which enters the sheet during time t, and M_∞ the corresponding amount during infinite time, then

$$\frac{M_t}{M_\infty} = 1 - \frac{8}{\pi^2}\sum_{n=0}^{\infty}\frac{1}{(2n+1)^2}\exp\{-D(2n+1)^2\pi^2 t/l^2\}. \quad (4.23)$$

In this case $M_\infty = l\{\frac{1}{2}(C_1 + C_2) - C_0\}$ and the total content of the membrane at time t is given by $M_t + lC_0$. The expression (4.23) is similar to (4.18) and is readily evaluated from the curve labelled zero fractional uptake in Fig. 4.6, with the proviso that in (4.23) l signifies the whole thickness of the membrane but in (4.18) it denotes the half-thickness.

The rate at which the gas or other diffusing substance emerges from unit area of the face $x = l$ of the membrane is given by $-D(\partial C/\partial x)_{x=l}$ which is easily deduced from (4.22). By integrating then with respect to t, we obtain the total amount of diffusing substance Q_t which has passed through the membrane in time t.

$$Q_t = D(C_1 - C_2)\frac{t}{l} + \frac{2l}{\pi^2}\sum_1^\infty \frac{C_1 \cos n\pi - C_2}{n^2}\{1 - \exp(-Dn^2\pi^2 t/l^2)\}$$

$$+ \frac{4C_0 l}{\pi^2}\sum_{m=0}^\infty \frac{1}{(2m+1)^2}\{1 - \exp(-D(2m+1)^2\pi^2 t/l^2)\}. \tag{4.24}$$

In the commonest experimental arrangement both C_0 and C_2 are zero, i.e. the membrane is initially at zero concentration and the concentration at the face through which diffusing substance emerges is maintained effectively at zero concentration. In this case we find

$$\frac{Q_t}{lC_1} = \frac{Dt}{l^2} - \frac{1}{6} - \frac{2}{\pi^2}\sum_1^\infty \frac{(-1)^n}{n^2}\exp(-Dn^2\pi^2 t/l^2), \tag{4.24a}$$

which, as $t \to \infty$, approaches the line

$$Q_t = \frac{DC_1}{l}\left(t - \frac{l^2}{6D}\right). \tag{4.25}$$

This has an intercept L on the t-axis given by

$$L = l^2/6D \tag{4.26}$$

Following Daynes (1920), Barrer (1951) has used (4.26) as the basis of a method for obtaining the diffusion constant, the permeability constant, and the solubility of a gas by analysing stationary and non-stationary flow through a membrane. Thus from an observation of the intercept, L, D is deduced by (4.26); from the steady-state flow rate the permeability constant P is deduced by using (4.5), and S follows from (4.7). The intercept L is referred to as the 'time lag'.

A graph of Q_t/lC_1 as a function Dt/l^2 is shown for the case $C_0 = C_2 = 0$ in Fig. 4.2. To within the accuracy of plotting the steady state is achieved when $Dt/l^2 = 0.45$ approximately

An alternative form of solution useful for small times, usually attributed to Holstein (Rogers, Buritz, and Alpert 1954) is easily derived by using Laplace transforms as in § 2.4.3(i). Still keeping $C_0 = C_2 = 0$, we find the rate of

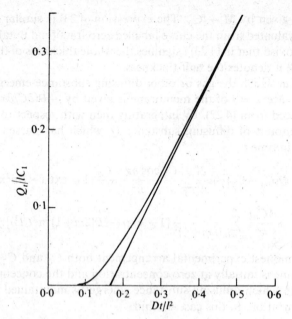

FIG. 4.2. Approach to steady-state flow through a plane sheet.

flow $F(t)$ per unit area of the surface $x = l$ to be

$$F(t) = -\left(\frac{D\partial C}{\partial x}\right)_{x=l} = 2C_1 \sum_{1}^{\infty} \left(\frac{D}{\pi t}\right)^{\frac{1}{2}} \exp\left\{-(2m+1)^2 l^2/(4Dt)\right\} \quad (4.27)$$

This series converges most rapidly for small t. Rogers *et al.* take only the leading term and obtain essentially

$$\ln(t^{\frac{1}{2}}F) = \ln\left\{2C_1\left(\frac{D}{\pi}\right)^{\frac{1}{2}}\right\} - \frac{l^2}{4Dt}. \quad (4.27a)$$

From the slope and intercept of this line experimental data yield D and C_1, and hence solubility. They discuss the advantages of their method compared with the use of the time lag given by eqn (4.26).

Other applications of Holstein's solution are discussed in § 10.6 and § 10.6.2 (pp. 216 and 224). Jenkins, Nelson, and Spirer (1970) examined the more general problems of deducing both the diffusion coefficient and the solubility coefficient from experimental data and mathematical solutions when the outflow volume is finite so that the concentration varies with time at the outgoing face. The necessary solution is given by Carslaw and Jaeger (1959). Jenkins *et al.* tabulate some useful calculated data and also consider varying concentrations at the ingoing face. In particular, they examined the assumption, frequently made in time-lag measurements of the Daynes type, that steady flow is established after a period of about 3 times the time lag. They concluded that the time lag is underestimated by about 4 per cent by making this assumption.

Paul and Dibenedetto (1965) also obtained solutions for finite outflow volumes. Špaček and Kubin (1967) allowed the concentrations on both sides of the membrane to vary with time.

4.3.4. *Variable surface concentration*

The solution to the general problem of diffusion in the region $0 < x < l$ with the surfaces at concentrations $\phi_1(t)$ and $\phi_2(t)$ and the initial concentration $f(x)$ is given by Carslaw and Jaeger (1959, p. 102). For empirical values of $\phi_1(t)$, $\phi_2(t)$, and $f(x)$, three integrals arise which have to be evaluated graphically or numerically. In certain cases, however, where the surface concentration can be represented by a mathematical expression, the solution can be considerably simplified.

(i) One case of practical interest is that of a sheet in which the concentration is initially zero and each surface of which approaches an equilibrium concentration C_0, exponentially, i.e.

$$\phi_1(t) = \phi_2(t) = C_0\{1 - \exp(-\beta t)\}. \tag{4.28}$$

This can represent a surface concentration which is changed rapidly but not instantaneously, a situation which usually arises when an instantaneous change is attempted in an experiment. For the sheet whose surfaces are at $\pm l$ the solution is

$$\frac{C}{C_0} = 1 - \exp(-\beta t)\frac{\cos x(\beta/D)^{\frac{1}{2}}}{\cos l(\beta/D)^{\frac{1}{2}}}$$

$$-\frac{16\beta l^2}{\pi}\sum_{n=0}^{\infty}\frac{(-1)^n \exp(-D(2n+1)^2\pi^2 t/4l^2)}{(2n+1)\{4\beta l^2 - D\pi^2(2n+1)^2\}}\cos\frac{(2n+1)\pi x}{2l}, \tag{4.29}$$

provided β is not equal to any of the values $D(2n+1)^2\pi^2/4l^2$. The sorption-time curve, i.e. the curve showing the total amount M_t of diffusing substance in the sheet as a function of time t, is obtained by integrating (4.29) with respect to x between the limits $-l$ and $+l$ and is

$$\frac{M_t}{2lC_0} = 1 - \exp(-\beta t)(D/\beta l^2)^{\frac{1}{2}}\tan(\beta l^2/D)^{\frac{1}{2}}$$

$$-\frac{8}{\pi^2}\sum_{n=0}^{\infty}\frac{\exp\{-(2n+1)^2\pi^2 Dt/4l^2\}}{(2n+1)^2[1-(2n+1)^2\{D\pi^2/(4\beta l^2)\}]}. \tag{4.30}$$

Fig. 4.3 shows uptake curves for different values of the parameter $\beta l^2/D$ plotted against $(Dt/l^2)^{\frac{1}{2}}$. When $\beta = \infty$, the surface concentration rises instantaneously to C_0 and the curve of Fig. 4.3 has the characteristic initial

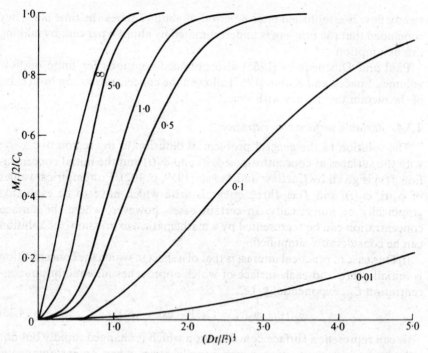

FIG. 4.3. Calculated sorption curves for surface concentration given by $C_0(1 - e^{-\beta t})$. Numbers on curves are values of $\beta l^2/D$.

linear portion followed by the approach to the equilibrium value $M_t = 2lC_0$. The uptake curves for finite values of $\beta l^2/D$, for which the surface concentration rises at a finite rate, all show points of inflexion. At first the rate of uptake increases as sorption proceeds but later decreases as the final equilibrium is approached. Curves of this kind are often referred to as sigmoid sorption curves. They may arise in practice because surface equilibrium conditions are not established instantaneously, but they may result from other causes (see § 11.2, p. 255; § 14.4.6, p. 347).

(ii) If the surface concentrations vary linearly with time, i.e.

$$\phi_1(t) = \phi_2(t) = kt, \tag{4.31}$$

the solution is

$$\frac{DC}{kl^2} = \frac{Dt}{l^2} + \frac{1}{2}\left(\frac{x^2}{l^2} - 1\right)$$

$$+ \frac{16}{\pi^3} \sum_{n=0}^{\infty} \frac{(-1)^n}{(2n+1)^3} \exp\left\{-D(2n+1)^2\pi^2 t/4l^2\right\}\cos\frac{(2n+1)\pi x}{2l}. \tag{4.32}$$

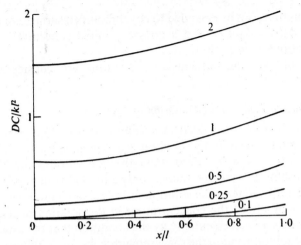

FIG. 4.4. Concentration distributions in a plane sheet for surface concentration kt. Numbers on curves are values of Dt/l^2.

The corresponding expression for M_t is

$$\frac{DM_t}{kl^3} = \frac{2Dt}{l^2} - \frac{2}{3} + \frac{64}{\pi^4} \sum_{n=0}^{\infty} \frac{\exp\{-D(2n+1)^2\pi^2t/4l^2\}}{(2n+1)^4}. \quad (4.33)$$

Some numerical results are given by Williamson and Adams (1919) and Gurney and Lurie (1923). Fig. 4.4 shows DC/kl^2 plotted as a function of x/l for various values of Dt/l^2. Fig. 4.5 shows DM_t/kl^3 as a function of the single variable Dt/l^2.

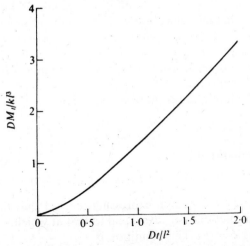

FIG. 4.5. Sorption curve for plane sheet with surface concentration kt.

These solutions can be extended to cover modified surface conditions such as $\phi_1(t) = \phi_2(t) = C_0 + kt$, and a non-zero initial concentration by superposing solutions as in § 3.3(v).

Sinusoidally varying surface concentrations are considered in § 10.6 (p. 217).

4.3.5. *Diffusion from a stirred solution of limited volume*

If a plane sheet is suspended in a volume of solution so large that the amount of solute taken up by the sheet is a negligible fraction of the whole, and the solution is well stirred, then the concentration in the solution remains constant. If, however, there is only a limited volume of solution, the concentration of solute in the solution falls as solute enters the sheet. If the solution is well stirred the concentration in the solution depends only on time, and is determined essentially by the condition that the total amount of solute in the solution and in the sheet remains constant as diffusion proceeds. It is useful from an experimental point of view to have only a limited amount of solution since the rate of uptake of solute by the sheet can be deduced from observations of the uniform concentration in the solution. It is often simpler to do this than to observe directly the amount in the sheet. This has been stressed by Carman and Haul (1954), who have written mathematical solutions in forms most appropriate for the measurement of diffusion coefficients by this method.

The general problem can be stated mathematically in terms of a solute diffusing from a well-stirred solution. The modifications necessary for corresponding alternative problems, such as that of a sheet suspended in a vapour, are obvious.

Suppose that an infinite sheet of uniform material of thickness $2l$ is placed in a solution and that the solute is allowed to diffuse into the sheet. The sheet occupies the space $-l \leqslant x \leqslant l$, while the the solution is of limited extent and occupies the spaces $-l-a \leqslant x \leqslant -l$, $l \leqslant x \leqslant l+a$. The concentration of the solute in the solution is always uniform and is initially C_0, while initially the sheet is free from solute.

We require a solution of the diffusion equation

$$\frac{\partial C}{\partial t} = D \frac{\partial^2 C}{\partial x^2}, \tag{4.34}$$

with the initial condition

$$C = 0, \quad -l < x < l, \quad t = 0, \tag{4.35}$$

and with a boundary condition expressing the fact that the rate at which solute leaves the solution is always equal to that at which it enters the sheet over the surfaces $x = \pm l$. This condition is

$$a \partial C / \partial t = \mp D \partial C / \partial x, \quad x = \pm l, \quad t > 0. \tag{4.36}$$

We assume here that the concentration of solute just within the surface of the sheet is the same as that in the solution. This may not be so but there may be a partition factor K which is not unity, such that the concentration just within the sheet is K times that in the solution. This can clearly be allowed for by using a modified length of solution a/K in place of a in (4.36) and elsewhere.

A solution of this problem by March and Weaver (1928), based on the use of an integral equation, was cumbersome for numerical evaluation. More convenient forms of solution have been obtained by Carslaw and Jaeger (1959, p. 128), Wilson (1948), Berthier (1952), while Crank (1948c) has developed solutions particularly suitable for small values of the time. The solution is most readily obtained by the use of the Laplace transform. In a form expressing the total amount of solute M_t in the sheet at time t as a fraction of M_∞, the corresponding quantity after infinite time, the solution is

$$\frac{M_t}{M_\infty} = 1 - \sum_{n=1}^{\infty} \frac{2\alpha(1+\alpha)}{1+\alpha+\alpha^2 q_n^2} \exp(-Dq_n^2 t/l^2), \qquad (4.37)$$

where the q_ns are the non-zero positive roots of

$$\tan q_n = -\alpha q_n, \qquad (4.38)$$

and $\alpha = a/l$, the ratio of the volumes of solution and sheet, or if there is a partition factor K then $\alpha = a/Kl$. Some roots of (4.38) are given in Table 4.1 for values of α corresponding to several values of final fractional uptake. Roots for other values of α are given by Carslaw and Jaeger (1959, p. 492) and by Carman and Haul (1954). It is sometimes convenient to express α in terms of the fraction of total solute finally taken up by the sheet. Thus, in the final equilibrium state, since the total amount of solute in solution and sheet was originally contained in the solution of concentration C_0, we have

$$\frac{aC_\infty}{K} + lC_\infty = aC_0, \qquad (4.39)$$

where C_∞ is the uniform concentration in the sheet finally. The content M_∞ of the sheet finally is given by

$$M_\infty = 2lC_\infty = \frac{2aC_0}{1+a/(lK)} = \frac{2aC_0}{1+\alpha}. \qquad (4.40)$$

The fractional uptake of the sheet finally is therefore given by

$$\frac{M_\infty}{2aC_0} = \frac{1}{1+\alpha}. \qquad (4.41)$$

If, for example, 50 per cent of the solute initially in the solution is finally in the sheet, $\alpha = 1$. In the particular case of an infinite amount of solute ($\alpha = \infty$)

the roots of (4.38) are $q_n = (n+\tfrac{1}{2})\pi$, and we have

$$\frac{M_t}{M_\infty} = 1 - \sum_{n=0}^{\infty} \frac{8}{(2n+1)^2\pi^2} \exp\{-D(n+\tfrac{1}{2})^2\pi^2t/l^2\}, \qquad (4.42)$$

which is expression (4.18) for the case of a constant concentration C_0 at the surface of a sheet. The smaller Dt/l^2 is, the more terms in the series in (4.37) are needed for a given accuracy. When more than three or four terms are needed it is better to use an alternative form of solution. For most values of α the simplest expression is

$$\frac{M_t}{M_\infty} = (1+\alpha)\{1 - \exp(T/\alpha^2)\,\mathrm{erfc}\,(T/\alpha^2)^{\frac{1}{2}}\}, \qquad (4.43)$$

where $T = Dt/l^2$. If very small values of α are required, corresponding to very high fractional uptakes of solute by the sheet, there may be a range of Dt/l^2 in which neither (4.37) nor (4.43) is convenient but where the following is useful:

$$\frac{M_t}{M_\infty} = (1+\alpha)\left\{1 - \frac{\alpha}{\pi^{\frac{1}{2}}T^{\frac{1}{2}}} + \frac{\alpha^3}{2\pi^{\frac{1}{2}}T^{\frac{1}{2}}} - \frac{3\alpha^5}{4\pi^{\frac{1}{2}}T^{\frac{1}{2}}} + \cdots\right\}. \qquad (4.44)$$

This is obtained from (4.43) by substituting the asymptotic expansion for $\exp(T/\alpha^2)\,\mathrm{erfc}\,(T/\alpha^2)^{\frac{1}{2}}$ when T/α^2 is large.

Fig. 4.6 gives curves showing M_t/M_∞ against $(Dt/l^2)^{\frac{1}{2}}$ for five values of final fractional uptake. Fig. 4.6 shows that the greater the final fractional uptake of the sheet the faster is the solute removed from the solution. Clearly by comparing the rate of fall of concentration in the solution observed experimentally, with the corresponding calculated curve showing M_t/M_∞ as a function of Dt/l^2, the diffusion coefficient D can be deduced. This has been suggested by Berthier (1952) as a method for measuring self-diffusion using radioactive isotopes. He gives a table of M_t/M_∞ for values of $1/\alpha$ between 0 and 1·0 at intervals of 0·1. For precision measurements it is advisable to check Berthier's values as in some instances not enough terms of the series solutions have been retained to obtain the accuracy quoted. The concentration within the sheet is given by the expression

$$C = C_\infty\left\{1 + \sum_{n=1}^{\infty} \frac{2(1+\alpha)\exp(-Dq_n^2t/l^2)}{1+\alpha+\alpha^2q_n^2} \frac{\cos(q_nx/l)}{\cos q_n}\right\}. \qquad (4.45)$$

We have considered diffusion into a plane sheet initially free of solute. There is the complementary problem in which all the solute is initially uniformly distributed through the sheet and subsequently diffuses out into a well-stirred solution. It is easily seen that the mathematical solutions presented above for sorption by the sheet also describe desorption, provided M_t is taken to mean the amount of solute leaving the sheet up to time t,

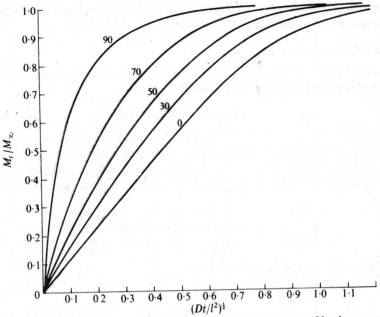

FIG. 4.6. Uptake by a plane sheet from a stirred solution of limited volume. Numbers on curves show the percentage of total solute finally taken up by the sheet.

and M_∞ the corresponding amount after infinite time. For the problem of desorption from the sheet we require a solution of (4.34) satisfying (4.36) but with the initial condition (4.35) replaced by

$$C = C_0, \qquad -l < x < l, \qquad t = 0. \tag{4.46}$$

On writing

$$C_1 = C_0 - C, \tag{4.47}$$

(4.46) and the other equations for desorption are identical with (4.34), (4.35), (4.36), with C_1 written for C. Hence the equations and solutions for desorption are identical with those for sorption provided M_t, M_∞ are suitably interpreted and $C_1, (C_1)_\infty$ replace C, C_∞ in expression (4.45). The parameter α is equal to a/Kl as before, but its relation to the final uptake of the sheet expressed by (4.40) and (4.41) no longer holds. Instead we have that the fractional uptake of the solution is given by

$$\frac{M_\infty}{2lC_0} = \frac{1}{1 + 1/\alpha}. \tag{4.48}$$

Jaeger and Clarke (1947) have presented solutions of a number of other problems in diffusion from a well-stirred solution in terms of certain fundamental functions. Accurately drawn graphs of these functions, from which solutions of limited accuracy are readily constructed, are given in their paper. Permeation into finite volumes is discussed in § 4.3.3 (pp. 52, 53).

4.3.6. *Surface evaporation*

In § 3.3.1 (p. 35) the rate of loss of diffusing substance by evaporation from the surface of a sheet was represented by

$$-D \, \partial C/\partial x = \alpha(C_0 - C_s), \tag{4.49}$$

where C_s is the actual concentration just within the sheet and C_0 is the concentration required to maintain equilibrium with the surrounding atmosphere. If the sheet $-l < x < l$ is initially at a uniform concentration C_2, and the law of exchange of the type (4.49) holds on both surfaces, the solution is

$$\frac{C - C_2}{C_0 - C_2} = 1 - \sum_{n=1}^{\infty} \frac{2L \cos (\beta_n x/l) \exp (-\beta_n^2 Dt/l^2)}{(\beta_n^2 + L^2 + L) \cos \beta_n}, \tag{4.50}$$

where the β_ns are the positive roots of

$$\beta \tan \beta = L \tag{4.51}$$

and

$$L = l\alpha/D, \tag{4.52}$$

a dimensionless parameter. Roots of (4.51) are given in Table 4.2 for several values of L. Roots for other values of L are given by Carslaw and Jaeger (1959, p. 491). The total amount of diffusing substance M_t entering or leaving the sheet up to time t, depending on whether C_0 is greater or less than C_2, is expressed as a fraction of M_∞, the corresponding quantity after infinite time, by

$$\frac{M_t}{M_\infty} = 1 - \sum_{n=1}^{\infty} \frac{2L^2 \exp (-\beta_n^2 Dt/l^2)}{\beta_n^2(\beta_n^2 + L^2 + L)}. \tag{4.53}$$

A solution suitable for small values of time may be obtained in the usual way by expanding the expression for the Laplace transform in a series of negative exponentials (Carslaw and Jaeger 1959, p. 310). The terms in the series expression for concentration very soon become cumbersome for numerical evaluation, however. In practice, it is usually sufficient to use only the leading terms corresponding to the interval during which the sheet is effectively semi-infinite, when the concentration is given by expression (3.35), and (3.37) gives the value of M_t for half the sheet.

Graphs showing M_t/M_∞ for several values of L are plotted in Fig. 4.7 from numerical values given by Newman (1931). Carslaw and Jaeger (1959, p. 123) give corresponding curves as well as others showing how the concentrations at the surfaces and the centre of the sheet vary with time. Newman

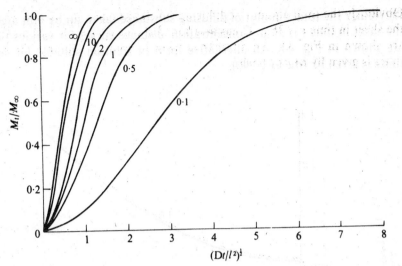

FIG. 4.7. Sorption or desorption curves for the surface condition (4.49). Numbers on curves are values of $L = l\alpha/D$.

also gives a table of values from which M_t/M_∞ can easily be deduced for a parabolic initial distribution instead of a uniform initial concentration. All these equations and solutions have a practical application in the drying of porous solids.

Carslaw and Jaeger (1959, p. 114) give solutions of more general problems in which there is an initial, non-uniform concentration distribution, or in which the vapour pressure is different on the two surfaces of the sheet, or in which evaporation occurs from one surface only, the other being maintained at a constant concentration. Jaeger and Clarke (1947) have also given in graphical form the solutions of a number of problems with an evaporation type of boundary condition.

The more complicated case in which the rate of transfer on the surface is proportional to some power of the surface concentration was discussed by Jaeger (1950a).

4.3.7. Constant flux F_0 at the surfaces

If the sheet $-l < x < l$ is initially at a constant concentration C_0, and diffusing substance enters at a constant rate F_0 over unit area of each surface, i.e.

$$D\, \partial C/\partial x = F_0, \qquad x = l, \tag{4.54}$$

then

$$C - C_0 = \frac{F_0 l}{D}\left\{\frac{Dt}{l^2} + \frac{3x^2 - l^2}{6l^2} - \frac{2}{\pi^2}\sum_{n=1}^{\infty}\frac{(-1)^n}{n^2}\exp\left(-Dn^2\pi^2 t/l^2\right)\cos\frac{n\pi x}{l}\right\}. \tag{4.55}$$

Obviously the total amount of diffusing substance taken up by unit area of the sheet in time t is $2F_0 t$. Concentration–distance curves for various times are shown in Fig. 4.8. An alternative form of solution suitable for small times is given by Macey (1940).

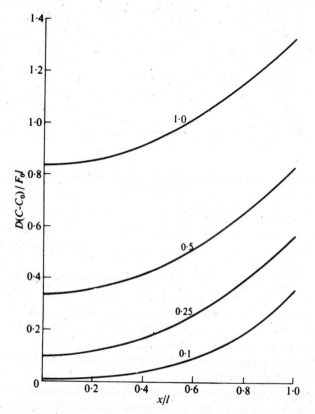

FIG. 4.8. Concentration distributions in a plane sheet for constant flux F_0 at the surface. Numbers on curves are values of Dt/l^2.

4.3.8. Impermeable surfaces

An impermeable surface is one at which the concentration gradient is zero. This condition holds at the central plane of a sheet provided the initial and boundary conditions are symmetrical about that plane. It follows therefore that the symmetrical solutions already given for the plane sheet occupying the region $-l < x < l$ apply also to the sheet $0 < x < l$ when the face $x = 0$ is impermeable. If, on the other hand, both surfaces $x = 0$ and

$x = l$ are impermeable and the initial distribution is $f(x)$, the solution is

$$C = \frac{1}{l}\int_0^l f(x')\,dx' + \frac{2}{l}\sum_{n=1}^{\infty}\exp\left(-Dn^2\pi^2 t/l^2\right)\cos\frac{n\pi x}{l}\int_0^l f(x')\cos\frac{n\pi x'}{l}\,dx'.$$

$$(4.56)$$

Barrer (1951) suggests that diffusion from one layer to another as discussed in § 2.2.4 may be treated by regarding the system as a single layer with impermeable boundaries and applying (4.56), where the initial distribution is

$$\begin{aligned}f(x) &= C_0, && 0 < x < h\\ f(x) &= 0, && h < x < l\end{aligned}\Bigg\}.$$

$$(4.57)$$

The solution (4.56) becomes

$$C = C_0\left\{\frac{h}{l} + \frac{2}{\pi}\sum_{n=1}^{\infty}\frac{1}{n}\sin\frac{n\pi h}{l}\exp\left(-Dn^2\pi^2 t/l^2\right)\cos\frac{n\pi x}{l}\right\}. \qquad (4.58)$$

Jackson, Oldland, and Pajaczkowski (1968) made a similar use of (4.56) in a radiotracer method of measuring diffusion coefficients (§ 10.8, p. 252). The solution (4.58) is complementary to expression (2.17) which is convenient for small times. Numerical values based on (2.17) and (4.58) are available in the well-known tables by Stefan (1879) and Kawalki (1894), some of which are reproduced by Jost (1952).

Another special case of (4.56) has been evaluated by Crank and Henry (1949c) in an investigation of different methods of conditioning a sheet to a required uniform concentration. They consider the problem in which a sheet, initially at zero concentration throughout, has its surfaces maintained at a constant concentration C_0 for a time t_0, after which they are rendered impermeable. The subsequent change in concentration is described by (4.56) with $f(x)$ given by

$$f(x) = C_0\left[1 - \frac{4}{\pi}\sum_{m=0}^{\infty}\frac{1}{2m+1}\sin\frac{(2m+1)\pi x}{l}\exp\left\{-D(2m+1)^2\pi^2 t_0/l^2\right\}\right].$$

$$(4.59)$$

In this case the two integrals in (4.56) reduce to

$$\int_0^l f(x)\,dx = C_0 l\left[1 - \frac{8}{\pi^2}\sum_{m=0}^{\infty}\frac{1}{(2m+1)^2}\exp\left\{-D(2m+1)^2\pi^2 t_0/l^2\right\}\right], \quad (4.60)$$

and

$$\int_0^l f(x)\cos\frac{2p\pi x}{l}\,dx = \frac{8C_0 l}{\pi^2}\sum_{m=0}^{\infty}\frac{1}{4p^2-(2m+1)^2}\exp\left\{-D(2m+1)^2\pi^2 t_0/l^2\right\},$$

$$(4.61)$$

where $2p$ is substituted for n since only the terms involving even n are non-zero.

4.3.9. *Composite sheet*

Various problems of diffusion into a composite sheet comprised of two layers for which the diffusion coefficients are different have been solved (see, for example, Carslaw and Jaeger (1959, p. 319)). The solutions are similar in form to those presented in this chapter but obviously more complicated. In view of the additional number of parameters involved, no attempt is made to give numerical results here. Studies of the time-lag involved in establishing the steady-state flow through a composite sheet of several layers are discussed in § 12.2(ii) (p. 268).

4.4. Edge effects in membranes

We have treated flow through a membrane as a one-dimensional phenomenon. In the usual experimental arrangements an appreciable portion of the membrane is clamped between impermeable annular plates of outer radius b. At the ingoing and outgoing faces, flow only occurs through the circular aperture, radius a, where $a < b$. But inside the membrane the flow lines spread into the clamped region. This 'edge effect' means that the usual assumption of one-dimensional diffusion is not strictly correct. Barrer, Barrie, and Rogers (1962), using a solution developed by Jaeger and Beck (1955), examined the importance of the edge effect. They replaced the usual experimental condition of constant concentrations maintained on the two faces of the membrane by conditions of uniform constant flux. The mathematical solution yields a mean concentration difference between the two faces of the membrane which can be equated to the prescribed uniform difference in the original experiment. Barrer *et al.* (1962) obtain the relationship

$$F_0/F = 1 - 16S/\pi^2, \tag{4.62}$$

with

$$S = \sum_{q=1}^{\infty} \frac{1}{q^2} \frac{I_1(q\alpha)}{I_1(q\beta)} \{I_1(q\beta)K_1(q\alpha) - K_1(q\beta)I_1(q\alpha)\}, \tag{4.63}$$

where F is the steady-state flux per unit area with edge effect, F_0 without edge effect, $\beta = \pi b/l$, $\alpha = \pi a/l$, with l the membrane thickness, and I_1, K_1 are modified Bessel functions. Fig. 4.9, taken from Barrer *et al.*'s paper, shows that the relative difference in the flux $(F - F_0)/F_0$ for a given overlap $(b - a)/a$ is greater the thicker the membrane and for an l/a ratio of 0·2 or less is approaching the limits of experimental error.

The conclusion is that with thin membranes ($l/a \leqslant 0·2$) edge effects may safely be neglected but otherwise appreciable errors creep in. Barrer *et al.*

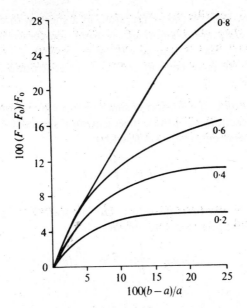

FIG. 4.9. Relative difference in flux $(F - F_0)/F_0$ as a function of the overlap $(b - a)/a$ for fixed l/a ratios. (From Barrer *et al.* 1962).

show that the errors can lead to apparent differences between diffusion coefficients deduced from steady-state and time-lag measurements.

4.5. Approximate two-dimensional solutions

Problems in which the diffusion is predominately in one direction occur frequently, perhaps because of the boundary conditions or the shape of the medium. Crank and Parker (1966) obtained approximate solutions to two-dimensional problems in biased heat flow. Previously it was widely accepted that for a thin sheet a good, one-dimensional approximation is obtained by assuming the temperature to be uniform across the thickness of the sheet. It appears, however, that a 'quadratic profile' provides a better approximation than the usual 'constant profile'. Crank and Parker considered the problem defined by the equations

$$v_{xx} + v_{yy} = v_t \tag{4.64}$$

$$v_x = 0, \quad x = 0; \quad v_x = -hv, \quad x = b; \quad t \geqslant 0, \tag{4.65}$$

$$v = 0, \quad y = 0; \quad v = 1, \quad y = 1, \quad t \geqslant 0,$$

$$v = 0, \quad 0 \leqslant y < 1, \quad 0 \leqslant x \leqslant b, \quad t = 0. \tag{4.66}$$

These equations describe the temperature in a sheet of thickness b and unit length, whose ends are maintained at fixed temperatures and from the surfaces of which heat is lost according to Newton's law of cooling. In diffusion terms, we have ends at constant concentration and evaporation from the surface $x = b$.

(i) *Constant profile.* If diffusion is small in the x-direction compared with that in the y-direction, then the concentration v should vary slowly with x. Thus the mean concentration u defined by

$$u = \frac{1}{b} \int_0^b v \, dx \tag{4.67}$$

should be a good approximation to v. On integrating (4.64) with respect to x from 0 to b and using (4.67) we obtain

$$u_t - u_{yy} = \frac{1}{b}[v_x]_0^b = \frac{-h}{b}(v)_{x=b},$$

i.e.

$$u_t - u_{yy} = -pu, \tag{4.68}$$

where

$$p = \{h(v)_{x=b}\}/bu. \tag{4.69}$$

The 'constant profile' solution (Fox 1934) assumes $v \equiv u$, i.e. $p = h/b$. Crank and Parker discuss two ways of improving this crude approximation.

(ii) *Partial separation of variables.* We replace the assumption $v \equiv u$ by $v = Xu$, where X is a function of x only and u is defined by (4.68) for some value of p. Then from (4.64)

$$uX_{xx} = X(u_t - u_{yy}) = -Xpu$$

from (4.68), so that $X_{xx} = -pX$ and finally

$$X = A \sin (x\sqrt{p}) + B \cos (x\sqrt{p}),$$

where A, B are arbitrary constants. In order to satisfy (4.65) with $v = Xu$ we have $A = 0$ and

$$\sqrt{p} \tan (b\sqrt{p}) = h. \tag{4.70}$$

If hb is small enough we get a good approximation from the first root only p_1 of (4.70). Then (4.68) becomes

$$u_t - u_{yy} = -p_1 u. \tag{4.71}$$

The boundary conditions on u are found from (4.66) and (4.67) and are

$$u = 0, \quad y = 0; \quad u = 1, \quad y = 1; \quad u = 0, \quad t = 0, \quad 0 \leqslant y < 1. \tag{4.72}$$

The solution of (4.71) subject to (4.72) is

$$u_1 = \frac{\sinh(y\sqrt{p_1})}{\sinh\sqrt{p_1}}$$

$$+ 2\pi \exp(-p_1 t) \sum_{r=1}^{\infty} \frac{(-1)^r r}{p_1 + r^2\pi^2} \exp(-r^2\pi^2 t) \sin r\pi y. \qquad (4.73)$$

Thus, an approximation to the solution of (4.64), (4.65), and (4.66) is

$$v_1 = B u_1 \cos x\sqrt{p_1}. \qquad (4.74)$$

The arbitrary constant B may be chosen so that v_1 is as close as possible to v at $t = 0$. Since u_1 is identically zero at $t = 0$ except when $y = 1$, the only deviation of v_1 from v to be minimized is along $y = 1$. A least-squares fitting of v_1 gives

$$B = 2h/[\{h + b(h^2 + p_1)\} \cos(b\sqrt{p_1})]. \qquad (4.75)$$

A Chebyshev criterion (Fox and Parker 1968) leads to

$$B = 2/\{1 + \cos(b\sqrt{p})\}. \qquad (4.76)$$

Crank and Parker quote error bounds on $v - v_1$.

(iii) *Quadratic profile.* An alternative approach starts again from (4.68) but with p regarded simply as a parameter, for the time being unknown. We define an approximation $w(x, y, t)$ to v such that

$$w_{xx} = u_t - u_{yy} = -pu. \qquad (4.77)$$

Since u and p are independent of x we can integrate (4.77) to give

$$w = (A + Bx - \tfrac{1}{2}px^2)u, \qquad (4.78)$$

where A and B are arbitrary functions of y and t. They are obtained by making w obey the conditions (4.65) on v and we have

$$B = 0, \quad A = pb(1 + \tfrac{1}{2}hb)/h,$$

so that

$$w = (1 + \tfrac{1}{2}hb - \tfrac{1}{2}hx^2/b)\frac{pbu}{h}. \qquad (4.79)$$

We can easily see that w satisfies eqns (4.64)–(4.66) to within $O(hb)$ if $p = h\{1 + O(hb)\}/b$. Thus by using w for v in (4.69) to obtain p, we have finally

$$w = \frac{k(1 + \tfrac{1}{2}hb - \tfrac{1}{2}hx^2/b)u}{1 + \tfrac{1}{3}hb}, \qquad (4.80)$$

referred to as the 'quadratic-profile' solution. The constant k is arbitrary but must not differ from unity by more than $O(hb)$. Least-squares or Chebyshev criteria can again be used but we choose $k = 1$. Crank and Parker (1966) tabulated their approximate solutions and compared them with the full analytical solution which is

$$v = \sum_{s=1}^{\infty} \frac{2h}{h + b(h^2 + \alpha_s^2)} \frac{\cos \alpha_s x}{\cos \alpha_s b}$$

$$\times \left\{ \frac{\sinh \alpha_s y}{\sinh \alpha_s} + 2\pi \exp(-\alpha_s^2 t) \sum_{r=1}^{\infty} \frac{(-1)^r \exp(-r^2\pi^2 t)}{r^2\pi^2 + \alpha_s^2} \sin r\pi y \right\}, \quad (4.81)$$

where α_s, $(s = 1, 2, \ldots)$ are the positive roots of

$$\alpha \tan \alpha b = h, \quad (4.82)$$

for a range of values of h and b. They concluded that the 'constant-profile' solution which is usually used as an approximation for a thin sheet or rod is, in fact, poor for a thin sheet with a high surface evaporation. But it provides a good approximation for a thick sheet with low surface loss. Both the quadratic profile' and the 'partial separation of variables' yield much better approximations over all. Concentration profiles across the sheet support these general conclusions.

As $hb \to 0$, the roots of (4.82) approach

$$b\alpha_1 \sim \sqrt{(hb)}, b\alpha_{n+1} \sim n\pi, n \geqslant 1.$$

Also the constant terms outside the curly brackets in (4.81) tend towards unity for $s = 1$ and the other terms either stay of the same order of magnitude or decrease, as s increases. It is easy to see that as $hb \to 0$ the analytical solution (4.81) approaches the approximate solution given by (4.73), (4.74), and (4.79). Crank and Parker show, however, that the approximate methods can readily be extended to cover some non-linear situations.

5

DIFFUSION IN A CYLINDER

5.1. Introduction

WE consider a long circular cylinder in which diffusion is everywhere radial. Concentration is then a function of radius r and time t only, and the diffusion equation (1.7) becomes

$$\frac{\partial C}{\partial t} = \frac{1}{r} \frac{\partial}{\partial r}\left(rD\frac{\partial C}{\partial r}\right). \tag{5.1}$$

5.2. Steady state

If the medium is a hollow cylinder whose inner and outer radii are a and b respectively, and if the diffusion coefficient is constant, the equation describing the steady-state condition is

$$\frac{d}{dr}\left(r\frac{dC}{dr}\right) = 0, \qquad a < r < b. \tag{5.2}$$

The general solution of this is

$$C = A + B \ln r, \tag{5.3}$$

where A and B are constants to be determined from the boundary conditions at $r = a$, $r = b$. If the surface $r = a$ is kept at a constant concentration C_1, and $r = b$ at C_2, then

$$C = \frac{C_1 \ln (b/r) + C_2 \ln (r/a)}{\ln (b/a)}. \tag{5.4}$$

The quantity of diffusing substance Q_t which diffuses through unit length of the cylinder in time t is given by

$$Q_t = \frac{2\pi Dt(C_2 - C_1)}{\ln (b/a)}. \tag{5.5}$$

If Q_t is measured in a concentration-dependent system, the mean value of the diffusion coefficient obtained from (5.5) is $(\int_{C_1}^{C_2} D \, dC)/(C_2 - C_1)$ as for the plane sheet (see (4.11)). The concentration distribution defined by (5.4) is not linear, as it is for the plane sheet. Typical distributions are shown in Fig. 5.1 for the cases $C_2 = 0$, $b/a = 2, 5, 10$.

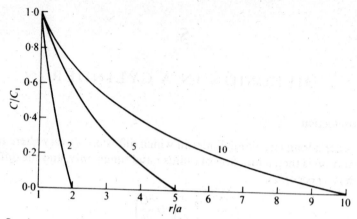

FIG. 5.1. Steady-state concentration distributions through cylinder wall. Numbers on curves are values of b/a.

Another steady-state problem leading to an interesting result is that of the hollow cylinder whose surface $r = a$ is kept at a constant concentration C_1, and at $r = b$ there is evaporation into an atmosphere for which the equilibrium concentration just within the surface is C_2. The boundary condition, with the constant of proportionality denoted by h, is

$$\frac{dC}{dr} + h(C - C_2) = 0, \qquad r = b, \tag{5.6}$$

and we find

$$C = \frac{C_1\{1 + hb \ln (b/r)\} + hbC_2 \ln (r/a)}{1 + hb \ln (r/a)}. \tag{5.7}$$

The outward rate of diffusion per unit length of the cylinder is Q_t, where

$$Q_t = 2\pi D(C_1 - C_2)\frac{hb}{1 + hb \ln (b/a)}. \tag{5.8}$$

By differentiating this expression with respect to b, it is easily seen that if $ah > 1$ the rate of diffusion decreases steadily as b increases from a, but if $ah < 1$ the rate first increases and later decreases, passing through a maximum when $b = 1/h$. This is due to the two opposing changes associated with an increase in b. On the one hand, the rate of evaporation is increased because of the increase in area of the surface, $r = b$, as b increases, but on the other hand, the gradient of concentration through the cylinder decreases as b is increased. In certain circumstances, therefore, the rate of diffusion through the wall of a pipe may be increased by making the wall thicker (Porter and Martin 1910). This is illustrated in Fig. 5.2 for $ah = \frac{1}{2}$.

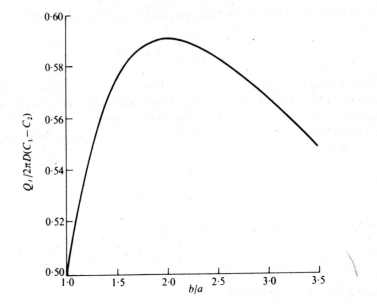

FIG. 5.2. Effect of thickness of cylinder wall on steady-state rate of flow.

If the surface conditions are

$$\partial C/\partial r + h_1(C_1 - C) = 0; \qquad \partial C/\partial r + h_2(C - C_2) = 0, \qquad (5.6a)$$

we find

$$C = \frac{ah_1 C_1\{1 + bh_2 \ln(b/r)\} + bh_2 C_2\{1 + ah_1 \ln(r/a)\}}{ah_1 + bh_2 + abh_1 h_2 \ln(b/a)}, \qquad (5.7a)$$

and

$$Q_t = \frac{2\pi Dtabh_1 h_2(C_2 - C_1)}{ah_1 + bh_2 + abh_1 h_2 \ln(b/a)}. \qquad (5.8a)$$

Other problems on diffusion in regions bounded by surfaces of the cylindrical coordinate system and in which the flow is not necessarily radial are treated by Carslaw and Jaeger (1959, p. 214).

5.3. Non-steady state: solid cylinder

Following essentially the method of separating the variables described in § 2.3 (p. 17), we see that

$$C = \hat{u} \exp(-D\alpha^2 t) \qquad (5.9)$$

is a solution of (5.1) for D constant provided u is a function of r only, satisfying

$$\frac{d^2u}{dr^2} + \frac{1}{r}\frac{du}{dr} + \alpha^2 u = 0, \tag{5.10}$$

which is Bessel's equation of order zero. Solutions of (5.10) may be obtained in terms of Bessel functions, suitably chosen so that the initial and boundary conditions are satisfied. Thus if the initial concentration distribution is $f(r)$ and the surface $r = a$ is maintained at zero concentration, a solution of (5.1) is wanted satisfying

$$C = 0, \qquad r = a, \qquad t \geqslant 0, \tag{5.11}$$

$$C = f(r), \qquad 0 < r < a, \qquad t = 0. \tag{5.12}$$

The boundary condition (5.11) is satisfied by

$$C = \sum_{n=1}^{\infty} A_n J_0(\alpha_n r) \exp(-D\alpha_n^2 t), \tag{5.13}$$

provided the α_ns are roots of

$$J_0(a\alpha_n) = 0, \tag{5.14}$$

where $J_0(x)$ is the Bessel function of the first kind of order zero. Roots of (5.14) are tabulated in tables of Bessel functions. For this function C is finite at $r = 0$. The initial condition (5.12) becomes

$$f(r) = \sum_{n=1}^{\infty} A_n J_0(r\alpha_n), \tag{5.15}$$

it being assumed that $f(r)$ can be expanded in a series of Bessel functions of order zero. The A_ns are determined by multiplying both sides of (5.15) by $rJ_0(\alpha_n r)$ and integrating from 0 to a using the results,

$$\int_0^a rJ_0(\alpha r)J_0(\beta r)\,dr = 0, \tag{5.16}$$

when α and β are different roots of (5.14), and

$$\int_0^a r\{J_0(\alpha r)\}^2\,dr = \tfrac{1}{2}a^2 J_1^2(a\alpha_n), \tag{5.17}$$

where $J_1(x)$ is the Bessel function of the first order and α is a root of (5.14). The derivation of the relationships (5.16) and (5.17), and of corresponding expressions which hold when α is a root not of (5.14) but of alternative equations which commonly arise in diffusion problems, is given by Carslaw and Jaeger (1959, p. 196). Finally the solution satisfying (5.11) and (5.12) is

$$C = \frac{2}{a^2} \sum_{n=1}^{\infty} \exp\left(-D\alpha_n^2 t\right) \frac{J_0(r\alpha_n)}{J_1^2(a\alpha_n)} \int_0^a r f(r) J_0(r\alpha_n)\, dr. \tag{5.18}$$

Alternatively, solutions for both large and small times can be obtained by use of the Laplace transform.

5.3.1. Surface concentration constant: initial distribution $f(r)$

If in the cylinder of radius a the conditions are

$$C = C_0, \qquad r = a, \qquad t \geqslant 0, \tag{5.19}$$

$$C = f(r), \qquad 0 < r < a, \qquad t = 0, \tag{5.20}$$

the solution is

$$C = C_0 \left\{ 1 - \frac{2}{a} \sum_{n=1}^{\infty} \frac{1}{\alpha_n} \frac{J_0(r\alpha_n)}{J_1(a\alpha_n)} \exp\left(-D\alpha_n^2 t\right) \right\}$$

$$+ \frac{2}{a^2} \sum_{n=1}^{\infty} \exp\left(-D\alpha_n^2 t\right) \frac{J_0(r\alpha_n)}{J_1^2(a\alpha_n)} \int r f(r) J_0(r\alpha_n)\, dr, \tag{5.21}$$

where the α_ns are the positive roots of (5.14).

If the concentration is initially uniform throughout the cylinder $f(r) = C_1$ and (5.21) reduces to

$$\frac{C - C_1}{C_0 - C_1} = 1 - \frac{2}{a} \sum_{n=1}^{\infty} \frac{\exp\left(-D\alpha_n^2 t\right) J_0(r\alpha_n)}{\alpha_n J_1(a\alpha_n)}. \tag{5.22}$$

If M_t denotes the quantity of diffusing substance which has entered or left the cylinder in time t and M_∞ the corresponding quantity after infinite time, then

$$\frac{M_t}{M_\infty} = 1 - \sum_{n=1}^{\infty} \frac{4}{a^2 \alpha_n^2} \exp\left(-D\alpha_n^2 t\right). \tag{5.23}$$

The corresponding solution useful for small times is

$$\frac{C - C_1}{C_0 - C_1} = \frac{a^{\frac{1}{2}}}{r^{\frac{1}{2}}} \operatorname{erfc} \frac{a - r}{2\sqrt{(Dt)}} + \frac{(a - r)(Dta)^{\frac{1}{2}}}{4ar^{\frac{3}{2}}} \operatorname{ierfc} \frac{a - r}{2\sqrt{(Dt)}}$$

$$+ \frac{(9a^2 - 7r^2 - 2ar)Dt}{32a^{\frac{3}{2}}r^{\frac{5}{2}}} \, i^2\operatorname{erfc} \frac{a - r}{2\sqrt{(Dt)}} + \dots , \tag{5.24}$$

which holds provided r/a is not small. The case of r/a small is discussed by Carsten and McKerrow (1944). They give a series solution involving modified Bessel functions of order $n \pm \frac{1}{4}$. The necessary functions are tabulated in their paper and numerical calculation is straightforward.

Also for small times we have

$$\frac{M_t}{M_\infty} = \frac{4}{\pi^{\frac{1}{2}}}\left(\frac{Dt}{a^2}\right)^{\frac{1}{2}} - \frac{Dt}{a^2} - \frac{1}{3\pi^{\frac{1}{2}}}\left(\frac{Dt}{a^2}\right)^{\frac{3}{2}} + \dots \ . \tag{5.25}$$

Clearly these solutions are not as valuable as the corresponding ones for the plane sheet. In practice the range in t over which they are convenient for evaluation is less than in the plane case.

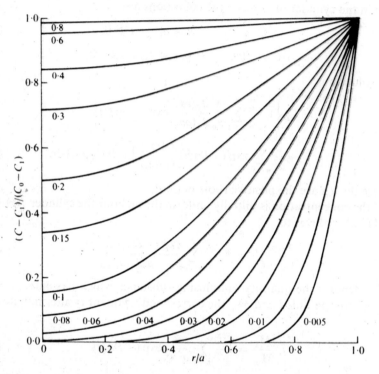

FIG. 5.3. Concentration distributions at various times with initial concentration C_1 and surface concentration C_0. Numbers on curves are values of Dt/a^2.

The solutions for the cylinder can be written in terms of the two dimensionless parameters Dt/a^2 and r/a. Curves showing $(C-C_1)/(C_0-C_1)$ as a function of r/a for different values of Dt/a^2 drawn by Carslaw and Jaeger (1959, p. 200) are reproduced in Fig. 5.3. The curve of Fig. 5.7 for zero fractional uptake shows how M_t/M_∞ depends on Dt/a^2 when the concentration at the surface of the cylinder remains constant.

5.3.2. *Variable surface concentration*

If the initial concentration in the cylinder is zero and that at the surface is $\phi(t)$, the solution is

$$C = \frac{2D}{a} \sum_{n=1}^{\infty} \exp\left(-D\alpha_n^2 t\right) \frac{\alpha_n J_0(r\alpha_n)}{J_1(a\alpha_n)} \int_0^t \exp\left(D\alpha_n^2 \lambda\right)\phi(\lambda)\,d\lambda, \qquad (5.26)$$

where the α_ns are the roots of (5.14).

(i) As for the plane sheet, a case of practical interest is when

$$\phi(t) = C_0\{1 - \exp\left(-\beta t\right)\}, \qquad (5.27)$$

representing a surface concentration which approaches a steady value C_0, but not instantaneously. The solution (5.26) then becomes

$$\frac{C}{C_0} = 1 - \frac{J_0\{(\beta r^2/D)^{\frac{1}{2}}\}}{J_0\{(\beta a^2/D)^{\frac{1}{2}}\}} \exp\left(-\beta t\right) + \frac{2\beta}{aD} \sum_{n=1}^{\infty} \frac{J_0(r\alpha_n)}{\alpha_n J_1(a\alpha_n)} \frac{\exp\left(-D\alpha_n^2 t\right)}{(\alpha_n^2 - \beta/D)}, \qquad (5.28)$$

and the sorption–time curve is given by

$$\frac{M_t}{\pi a^2 C_0} = 1 - \frac{2J_1\{(\beta a^2/D)^{\frac{1}{2}}\} \exp\left(-\beta t\right)}{(\beta a^2/D)^{\frac{1}{2}} J_0\{(\beta a^2/D)^{\frac{1}{2}}\}} + \frac{4}{a^2} \sum_{n=1}^{\infty} \frac{\exp\left(-D\alpha_n^2 t\right)}{\alpha_n^2 \{\alpha_n^2/(\beta/D) - 1\}}. \qquad (5.29)$$

Fig. 5.4 shows uptake curves for different values of the parameter $\beta a^2/D$.

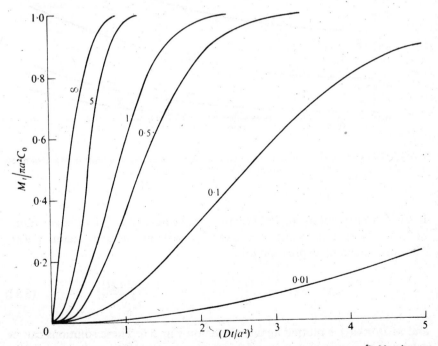

FIG. 5.4. Calculated sorption curves for surface concentration given by $C_0(1 - e^{-\beta t})$. Numbers on curves are values of $\beta a^2/D$.

(ii) If the surface concentration varies linearly with time, i.e.

$$\phi(t) = kt, \tag{5.30}$$

the solution is

$$C = k\left(t - \frac{a^2 - r^2}{4D}\right) + \frac{2k}{aD}\sum_{n=1}^{\infty} \exp\left(-D\alpha_n^2 t\right)\frac{J_0(r\alpha_n)}{\alpha_n^3 J_1(a\alpha_n)}. \tag{5.31}$$

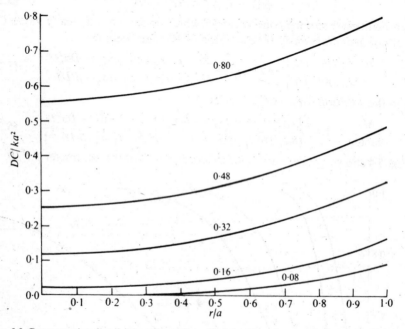

FIG. 5.5. Concentration distributions through a cylinder with surface concentration kt. Numbers on curves are values of Dt/a^2.

In Fig. 5.5, curves showing $DC/(ka^2)$ are drawn against r/a for different values of Dt/a^2. Numerical values are given by Williamson and Adams (1919). The corresponding expression for M_t is

$$M_t = \pi ka^2 t - \frac{\pi ka^4}{8D} + \frac{4\pi k}{D}\sum_{n=1}^{\infty}\frac{\exp\left(-D\alpha_n^2 t\right)}{\alpha_n^4}, \tag{5.32}$$

and $M_t D/(\pi ka^4)$ is plotted against Dt/a^2 on Fig. 5.6. These solutions can be extended by superposition as in § 3.3(v), to cover modified surface conditions such as $\phi(t) = C_0 + kt$ and a non-zero initial concentration.

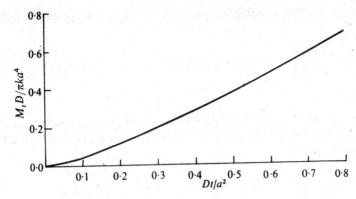

Fig. 5.6. Sorption curve for cylinder with surface concentration kt.

5.3.3. *Diffusion from a stirred solution of limited volume*

The problem differs only in detail from the corresponding problem considered for the plane sheet in § 4.3.5 (p. 56) and the results can be written down without explanation.

Suppose that the cylinder occupies the space $r < a$ while the cross-section of the bath of solution in which it is immersed is A (excluding the space occupied by the cylinder). The concentration of solute in the solution is always uniform and is initially C_0. The cylinder is initially free from solute. The total amount of solute M_t in the cylinder after time t is expressed as a fraction of the corresponding amount M_∞ after infinite time by the relation (Wilson 1948).

$$\frac{M_t}{M_\infty} = 1 - \sum_{n=1}^{\infty} \frac{4\alpha(1+\alpha)}{4+4\alpha+\alpha^2 q_n^2} \exp\left(-Dq_n^2 t/a^2\right), \qquad (5.33)$$

where the q_ns are the positive, non-zero roots of

$$\alpha q_n J_0(q_n) + 2J_1(q_n) = 0, \qquad (5.34)$$

and $\alpha = A/\pi a^2$, the ratio of the volumes of solution and cylinder. If there is a partition factor K between solute in equilibrium in the cylinder and in the solution, $\alpha = A/(\pi a^2 K)$. The parameter α is expressed in terms of the final fractional uptake of solute by the cylinder by the expression

$$\frac{M_\infty}{AC_0} = \frac{1}{1+\alpha}. \qquad (5.35)$$

The roots of (5.34) are given in Table 5.1 for several values of α in order to assist the evaluation of (5.33). The convergence of the series in (5.33) becomes inconveniently slow for numerical evaluation when Dt/a^2 is small. An alternative solution suitable for small Dt/a^2 when α is moderate is (Crank 1948c).

$$\frac{M_t}{M_\infty} = \frac{1+\alpha}{1+\frac{1}{4}\alpha}\left[1 - \exp\left\{4(1+\tfrac{1}{4}\alpha)^2 Dt/(a^2\alpha^2)\right\} \operatorname{erfc}\left\{2(1+\tfrac{1}{4}\alpha)\frac{(Dt/a^2)^{\frac{1}{2}}}{\alpha}\right\}\right].$$

$$(5.36)$$

Carman and Haul (1954) have derived an alternative equation which is less easy to use but which is accurate up to considerably higher values of M_t/M_∞. Their equation is

$$\frac{M_t}{M_\infty} = \frac{\gamma_3}{\gamma_3+\gamma_4}\exp\left\{4\gamma_3^2 Dt/(a^2\alpha^2)\right\}\operatorname{erfc}\left\{\frac{2\gamma_3}{\alpha}\left(\frac{Dt}{a^2}\right)^{\frac{1}{2}}\right\}$$

$$+\frac{\gamma_4}{\gamma_3+\gamma_4}\exp\left\{4\gamma_4^2 Dt/(a^2\alpha^2)\right\}\operatorname{erfc}\left\{-\frac{2\gamma_4}{\alpha}\left(\frac{Dt}{a^2}\right)^{\frac{1}{2}}\right\}, \quad (5.37)$$

where

$$\gamma_3 = \tfrac{1}{2}\{(1+\alpha)^{\frac{1}{2}}+1\}, \qquad \gamma_4 = \gamma_3 - 1. \quad (5.38)$$

For α very small it is convenient to use the asymptotic expansion for erfc in (5.36) and hence to write

$$\frac{M_t}{M_\infty} = \frac{1+\alpha}{1+\frac{1}{4}\alpha}\left\{1 - \frac{\alpha(Dt/a^2)^{-\frac{1}{2}}}{2\pi^{\frac{1}{2}}(1+\frac{1}{4}\alpha)} + \frac{\alpha^3(Dt/a^2)^{-\frac{3}{2}}}{16\pi^{\frac{1}{2}}(1+\frac{1}{4}\alpha)^3} - \frac{3\alpha^5(Dt/a^2)^{-\frac{5}{2}}}{128\pi^{\frac{1}{2}}(1+\frac{1}{4}\alpha)^5} + \cdots\right\}. \quad (5.39)$$

If α is very large, the following expression is more convenient,

$$\frac{M_t}{M_\infty} = \frac{2}{\alpha}(1+\alpha)\left\{\frac{2}{\pi^{\frac{1}{2}}}\left(\frac{Dt}{a^2}\right)^{\frac{1}{2}} - \left(\frac{1}{2}+\frac{2}{\alpha}\right)\frac{Dt}{a^2} - \frac{4}{3\pi^{\frac{1}{2}}}\left(\frac{1}{8}-\frac{2}{\alpha}-\frac{4}{\alpha^2}\right)\left(\frac{Dt}{a^2}\right)^{\frac{3}{2}} + \cdots\right\}, \quad (5.40)$$

which for the special case of $\alpha = \infty$, becomes

$$\frac{M_t}{M_\infty} = 2\left\{\frac{2}{\pi^{\frac{1}{2}}}\left(\frac{Dt}{a^2}\right)^{\frac{1}{2}} - \frac{1}{2}\frac{Dt}{a^2} - \frac{1}{6\pi^{\frac{1}{2}}}\left(\frac{Dt}{a^2}\right)^{\frac{3}{2}} + \cdots\right\}. \quad (5.41)$$

The derivation of these solutions for small times is given by Crank (1948c). Fig. 5.7 shows curves of M_t/M_∞ against $(Dt/a^2)^{\frac{1}{2}}$ for five values of the final fractional uptake. Berthier (1952) gives a table of M_t/M_∞ for values of $1/\alpha$ between 0 and 1 at intervals of 0·1. As in the plane case his values are not always reliable to three decimal places.

The concentration of solute C within the cylinder is given by

$$C = C_\infty\left\{1 + \sum_{n=1}^{\infty}\frac{4(\alpha+1)\exp(-Dq_n^2 t/a^2)}{(4+4\alpha+\alpha^2 q_n^2)}\frac{J_0(q_n r/a)}{J_0(q_n)}\right\}. \quad (5.42)$$

As for the plane sheet (§ 4.3.5, p. 58) the above equations also describe the course of desorption into a well-stirred solution, initially free from solute,

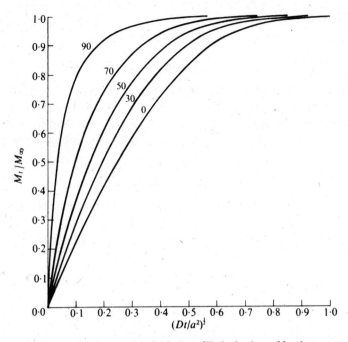

FIG. 5.7. Uptake by a cylinder from a stirred solution of limited volume. Numbers on curves show percentages of total solute finally taken up by cylinder.

from a cylinder in which the concentration is initially uniform and equal to C_0. The only modifications are that (5.35) is to be replaced by

$$\frac{M_\infty}{\pi a^2 C_0} = \frac{1}{1+1/\alpha},$$ (5.43)

and C_1, $(C_1)_\infty$ replace C, C_∞ in (5.42), where

$$C_1 = C_0 - C, \qquad (C_1)_\infty = C_0 - C_\infty.$$ (5.44)

5.3.4. *Surface evaporation*

If the cylinder is initially at a uniform concentration C_2, and there is a surface condition

$$-D\, \partial C/\partial r = \alpha(C_s - C_0),$$ (5.45)

where C_s is the actual concentration just within the cylinder and C_0 is the concentration required to maintain equilibrium with the surrounding atmosphere, the required solution is

$$\frac{C - C_2}{C_0 - C_2} = 1 - \sum_{n=1}^{\infty} \frac{2L J_0(r\beta_n/a)}{(\beta_n^2 + L^2) J_0(\beta_n)} \exp\left(-\beta_n^2 Dt/a^2\right).$$ (5.46)

The β_ns are the roots of

$$\beta J_1(\beta) - L J_0(\beta) = 0, \tag{5.47}$$

and

$$L = a\alpha/D, \tag{5.48}$$

a dimensionless parameter. Roots of (5.47) are given in Table 5.2 for several values of L. The total amount of diffusing substance M_t entering or leaving the cylinder, depending on whether C_0 is greater or less than C_2, is expressed as a fraction of M_∞, the corresponding quantity after infinite time, by

$$\frac{M_t}{M_\infty} = 1 - \sum_{n=1}^{\infty} \frac{4L^2 \exp(-\beta_n^2 Dt/a^2)}{\beta_n^2(\beta_n^2 + L^2)}. \tag{5.49}$$

The solutions suitable for small values of time, provided r/a is not small, are

$$\frac{C - C_2}{C_0 - C_2} = \frac{2h(Dta)^{\frac{1}{2}}}{r^{\frac{1}{2}}} \text{ierfc} \frac{a-r}{2\sqrt{(Dt)}} + \frac{4ha^{\frac{1}{2}}Dt}{r^{\frac{1}{2}}} \left\{ \frac{1}{8r} + \frac{3}{8a} - h \right\} i^2\text{erfc} \frac{a-r}{2\sqrt{(Dt)}} + \cdots,$$
$$\tag{5.50}$$

where $h = \alpha/D$, and

$$\frac{M_t}{M_\infty} = \frac{2DtL}{a^2} - \frac{8L^2}{3\pi^{\frac{1}{2}}} \left(\frac{Dt}{a^2}\right)^{\frac{3}{2}} - L^2 \left(\frac{Dt}{a^2}\right)^2 (\tfrac{1}{2} - L) - \cdots. \tag{5.51}$$

Tabulated values of M_t/M_∞ are given by Newman (1931) from which the graphs of Fig. 5.8 are drawn. Newman also gives values of a second function from which M_t/M_∞ can easily be deduced for a parabolic initial distribution.

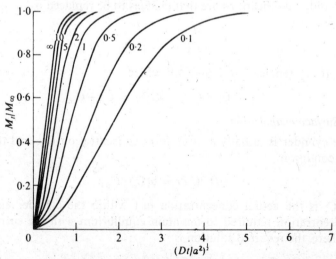

FIG. 5.8. Sorption and desorption curves for the surface condition (5.45). Numbers on curves are values of $L = a\alpha/D$.

5.3.5. *Constant flux F_0 at the surface*

If the cylinder is initially at a uniform concentration C_0, and there is a constant rate of transfer of diffusing substance F_0 per unit area of the surface, i.e.

$$-D\,\partial C/\partial r = F_0, \qquad r = a, \tag{5.52}$$

then we have

$$C - C_0 = -\frac{F_0 a}{D}\left\{\frac{2Dt}{a^2} + \frac{r^2}{2a^2} - \frac{1}{4}\right.$$

$$\left. -2\sum_{n=1}^{\infty} \exp\left(-D\alpha_n^2 t/a^2\right)\frac{J_0(r\alpha_n/a)}{\alpha_n^2 J_0(\alpha_n)}\right\}, \tag{5.53}$$

where the α_ns are the positive roots of

$$J_1(\alpha) = 0. \tag{5.54}$$

Roots of (5.54) are given in tables of Bessel functions, and the first five roots are to be found in Table 5.2 when $L = 0$. Obviously the amount of diffusing substance lost by unit length of the cylinder in time t is $2\pi a F_0 t$. This is a problem which has been discussed in connexion with the drying of clay by Macey (1940, 1942), Jaeger (1944), and others. A solution useful for small values of time is

$$C - C_0 = -\frac{F_0}{D}\left\{2\left(\frac{Dat}{r}\right)^{\frac{1}{2}}\text{ierfc}\,\frac{a-r}{2\sqrt{(Dt)}}\right.$$

$$\left. +\frac{Dt(a+3r)}{2a^{\frac{1}{2}}r^{\frac{3}{2}}}\,\text{i}^2\text{erfc}\,\frac{a-r}{2\sqrt{(Dt)}} + \dots\right\}. \tag{5.55}$$

Concentration–distance curves, plotted from (5.53) are shown in Fig. 5.9.

5.3.6. *Impermeable surface*

If the surface of the cylinder is impermeable and there is an initial concentration distribution $f(r)$ then

$$C = \frac{2}{a^2}\left\{\int_0^a r'f(r')\,dr' + \sum_{n=1}^{\infty} \exp\left(-D\alpha_n^2 t\right)\right.$$

$$\left. \times \frac{J_0(r\alpha_n)}{J_0^2(a\alpha_n)}\int_0^a r'f(r')J_0(\alpha_n r')\,dr'\right\}, \tag{5.56}$$

where the α_ns are roots of

$$J_1(a\alpha_n) = 0, \tag{5.57}$$

which are given in standard tables.

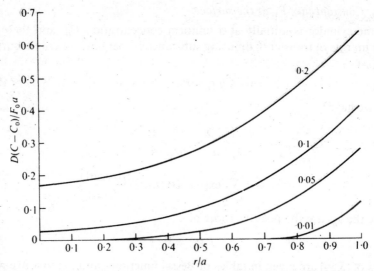

FIG. 5.9. Concentration distributions in a cylinder for constant flux F_0 at the surface. Numbers on curves are values of Dt/a^2.

5.3.7. *Composite cylinder*

Various problems of radial diffusion into a composite cylinder comprised of two coaxial cylinders having different diffusion coefficients have been solved (see e.g. Carslaw and Jaeger (1959, p. 345). The extra parameters involved make any attempt at general numerical evaluation too formidable to be attempted here. Some solutions relating to diffusion accompanied by non-linear absorption are discussed in Chapter 9.

Ölcer (1968) summarized the available solutions and solved the composite hollow cylinder problem in a very general form for combined radial and axial diffusion with internal sources included. Stevenson (1974) obtains a solution and gives numerical values typical of a small tubular membrane related to permeability measurements.

5.4. The hollow cylinder

Carslaw and Jaeger (1959, p. 207) give the general solution to the problem of the hollow cylinder with the surface $r = a$ maintained at a constant concentration C_1, and $r = b$ at C_2, when the initial distribution is $f(r)$, in the region $a \leqslant r \leqslant b$. In the special case of a constant initial concentration, $f(r) = C_0$, and when $C_1 = C_2$, the solution is

$$\frac{C - C_0}{C_1 - C_0} = 1 - \pi \sum_{n=1}^{\infty} \frac{J_0(a\alpha_n) U_0(r\alpha_n)}{J_0(a\alpha_n) + J_0(b\alpha_n)} \exp(-D\alpha_n^2 t). \tag{5.58}$$

where
$$U_0(r\alpha_n) = J_0(r\alpha_n) Y_0(b\alpha_n) - J_0(b\alpha_n) Y_0(r\alpha_n), \tag{5.59}$$

and the α_ns are the positive roots of

$$U_0(a\alpha_n) = 0. \tag{5.60}$$

Roots of (5.60) are given in Table 5.3 for different values of b/a. This table is reproduced from Carslaw and Jaeger's book (1959, p. 493). In (5.58) and (5.59) J_0 and Y_0 are Bessel functions of the first and second kind respectively, of order zero. They are both listed in standard tables. The expression for the amount of diffusing substance entering or leaving the region $a \leqslant r \leqslant b$ in time t is given by

$$\frac{M_t}{M_\infty} = 1 - \frac{4}{b^2 - a^2} \sum_{n=1}^{\infty} \frac{J_0(a\alpha_n) - J_0(b\alpha_n)}{\alpha_n^2 \{J_0(a\alpha_n) + J_0(b\alpha_n)\}} \exp(-D\alpha_n^2 t). \tag{5.61}$$

In Fig. 5.10, curves of M_t/M_∞ are plotted against $\{Dt/(b-a)^2\}^{\frac{1}{2}}$ for different values of b/a.

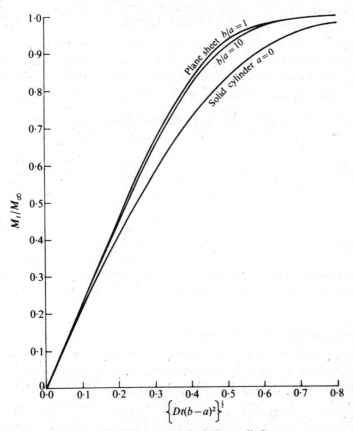

FIG. 5.10. Uptake curves for hollow cylinder.

5.4.1. *Flow through cylinder wall*

If the surface $r = a$ is maintained at C_1, $r = b$ at C_2, and the region $a \leqslant r \leqslant b$ is initially at C_0, the concentration approaches the steady-state distribution discussed in § 5.2 (p. 69) according to the expression

$$C = \frac{C_1 \ln (b/r) + C_2 \ln (r/a)}{\ln (b/a)} + \pi C_0 \sum_{n=1}^{\infty} \frac{J_0(a\alpha_n) U_0(r\alpha_n) \exp (-D\alpha_n^2 t)}{J_0(a\alpha_n) + J_0(b\alpha_n)}$$

$$- \pi \sum_{n=1}^{\infty} \frac{\{C_2 J_0(a\alpha_n) - C_1 J_0(b\alpha_n)\} J_0(a\alpha_n) U_0(r\alpha_n)}{J_0^2(a\alpha_n) - J_0^2(b\alpha_n)} \exp (-D\alpha_n^2 t), \quad (5.62)$$

where the α_ns are roots of (5.60). The amount of diffusing substance entering or leaving the cylinder wall in time t is given by M_t, where

$$M_t = \tfrac{1}{2}\pi \frac{(C_2 - C_1)(a^2 - b^2)}{\ln (b/a)} + \pi\{b^2(C_2 - C_0) - a^2(C_1 - C_0)\}$$

$$+ 4\pi \sum_{n=1}^{\infty} \frac{\{(C_1 - C_0)J_0(b\alpha_n) - (C_2 - C_0)J_0(a\alpha_n)\}}{\alpha_n^2 \{J_0(a\alpha_n) + J_0(b\alpha_n)\}} \exp (-D\alpha_n^2 t). \quad (5.63)$$

Normally a quantity of greater practical interest is the amount Q_t escaping from unit length of the outer surface $r = b$. This is readily deduced from (5.62) and is given by obtaining $-2\pi D(r \, \partial c/\partial r)_{r=b}$ and integrating with respect to time t. In the most commonly occurring case $C_0 = C_2 = 0$, and we then find

$$\frac{Q_t}{\pi C_1} = \frac{2(Dt - L)}{\ln (b/a)} - 4 \sum_{n=1}^{\infty} \frac{J_0(a\alpha_n) J_0(b\alpha_n) \exp (-D\alpha_n^2 t)}{\alpha_n^2 \{J_0^2(a\alpha_n) - J_0^2(b\alpha_n)\}}. \quad (5.64)$$

For a given b/a the graph of $Q_t/(\pi b^2 C_1)$ against t approaches, at large t, a straight line which makes an intercept L on the t-axis given by

$$\frac{a^2 - b^2 + (a^2 + b^2) \ln (b/a)}{4 \ln (b/a)}. \quad (5.65)$$

As Barrer (1951) suggests, this intercept provides a means of measuring the diffusion coefficient D for a material in the form of a hollow cylinder. Fig. 5.11 shows graphs of $Q_t/(\pi b^2 C_1)$, the amount evaporated from unit area of the outer surface $r = b$, as a function of $Dt/(b-a)^2$ for different values of b/a.

5.4.2. *A general boundary condition*

Jaeger (1940) has given the solution to the problem of diffusion into a hollow cylinder in which the concentration is initially zero and the boundary conditions on the two surfaces are

$$k_1 \frac{\partial C}{\partial t} + k_2 \frac{\partial C}{\partial r} + k_3 C = k_4, \quad r = a, \quad (5.66)$$

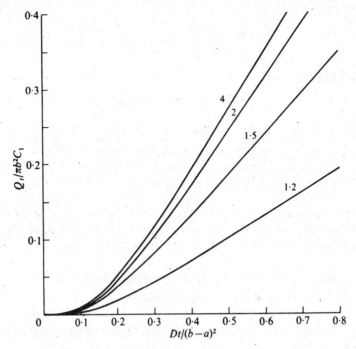

FIG. 5.11. Approach to steady-state flow through wall of cylinder. Numbers on curves are values of b/a.

$$k_1' \frac{\partial C}{\partial t} + k_2' \frac{\partial C}{\partial r} + k_3' C = k_4', \qquad r = b. \qquad (5.67)$$

These conditions include as special cases.

(i) Constant concentrations, C_1 on $r = a$ and C_2 on $r = b$, when $k_1 = k_2 = k_1' = k_2' = 0$, $k_4/k_3 = C_1$, $k_4'/k_3' = C_2$.

(ii) Evaporation conditions on the surfaces

$$-D\, \partial C/\partial r = \gamma_1(C_1 - C), \qquad r = a, \qquad (5.68)$$

$$-D\, \partial C/\partial r = \gamma_2(C - C_2), \qquad r = b, \qquad (5.69)$$

when $k_1 = k_1' = 0$, $k_2 = k_2' = D$, $k_3 = -\gamma_1$, $k_4 = -\gamma_1 C_1$, $k_3' = \gamma_2$, $k_4' = \gamma_2 C_2$. This includes the obvious modification for an impermeable surface.

(iii) Diffusion proceeding from a well-stirred solution occurring in the region $0 \leqslant r \leqslant a$, the concentration at $r = a$ being always the same as that throughout the solution; the surface $r = b$ impermeable. At $r = a$ the condition is

$$\pi a^2\, \partial C/\partial t = 2\pi a D\, \partial C/\partial r, \qquad r = a,$$

i.e.

$$\frac{\partial C}{\partial t} - \frac{2D}{a}\frac{\partial C}{\partial r} = 0, \qquad r = a, \tag{5.70}$$

and we have also

$$\partial C/\partial r = 0, \qquad r = b, \tag{5.71}$$

so that $k_1 = k'_2 = 1$, $k_2 = -2D/a$, $k'_1 = k'_3 = k'_4 = k_3 = k_4 = 0$. Other cases, such as that of diffusion from a well-stirred solution in the region $0 \leqslant r \leqslant a$ with the surface $r = b$ maintained at a constant concentration, or the surface $r = a$ maintained at a constant concentration while there is loss by evaporation from the surface $r = b$, and other combinations of these boundary conditions are all deducible from the general solution. The derivation of the solution by the use of Laplace transforms is given by Jaeger (1940) (Carslaw and Jaeger 1959, p. 332). The final result is

$$C = \frac{ak_4\{k'_2 - bk'_3 \ln (r/b)\} - bk'_4\{k_2 - ak_3 \ln (r/a)\}}{ak_3k'_2 - bk_2k'_3 - abk_3k'_3 \ln (a/b)}$$

$$- \pi \sum_{n=1}^{\infty} \exp(-D\alpha_n^2 t)F(\alpha_n)C_0(r;\alpha_n)[k_4\{A'_n J_0(b\alpha_n) - k'_2\alpha_n J_1(b\alpha_n)\}$$

$$- k'_4\{A_n J_0(a\alpha_n) - k_2\alpha_n J_1(a\alpha_n)\}], \tag{5.72}$$

where

$$\left.\begin{array}{ll} A_n = k_3 - Dk_1\alpha_n^2; & A'_n = k'_3 - Dk'_1\alpha_n^2 \\ B = k_2 + 2Dk_1/a; & B' = k'_2 + 2Dk'_1/b \end{array}\right\}, \tag{5.73}$$

$$C_0(r;\alpha_n) = J_0(r\alpha_n)\{A_n Y_0(a\alpha_n) - k_2\alpha_n Y_1(a\alpha_n)\}$$

$$- Y_0(r\alpha_n)\{A_n J_0(a\alpha_n) - k_2\alpha_n J_1(a\alpha_n)\}, \tag{5.74}$$

$$F(\alpha_n) = \frac{A'_n J_0(b\alpha_n) - k'_2\alpha_n J_1(b\alpha_n)}{\{A'_n J_0(b\alpha_n) - k'_2\alpha_n J_1(b\alpha_n)\}^2(A_n^2 + k_2 B\alpha_n^2)}{- \{A_n J_0(a\alpha_n) - k_2\alpha_n J_1(a\alpha_n)\}^2(A_n'^2 + k'_2 B'\alpha_n^2)}; \tag{5.75}$$

and where the α_ns are the positive roots of

$$\{(k_3 - k_1 D\alpha^2)J_0(a\alpha) - k_2\alpha J_1(a\alpha)\}\{(k'_3 - k'_1 D\alpha^2)Y_0(b\alpha) - k'_2\alpha Y_1(b\alpha)\}$$

$$- \{(k'_3 - k'_1 D\alpha^2)J_0(b\alpha) - k'_2\alpha J_1(b\alpha)][(k_3 - k_1 D\alpha^2)Y_0(a\alpha) - k_2\alpha Y_1(a\alpha)\} = 0. \tag{5.76}$$

5.5. The region bounded internally by the cylinder $r = a$

(i) If the initial concentration throughout the region $r > a$ is C_0, and the surface $r = a$ is maintained at C_1, then

$$\frac{C - C_0}{C_1 - C_0} = 1 + \frac{2}{\pi} \int_0^\infty \exp(-Du^2 t) \frac{J_0(ur) Y_0(ua) - Y_0(ur) J_0(ua)}{J_0^2(ua) + Y_0^2(ua)} \frac{du}{u}. \quad (5.77)$$

A solution useful for small times is

$$\frac{C - C_0}{C_1 - C_0} = \left(\frac{a}{r}\right)^{\frac{1}{2}} \operatorname{erfc} \frac{r - a}{2\sqrt{(Dt)}} + \frac{(r - a)(Dt)^{\frac{1}{2}}}{4 a^{\frac{1}{2}} r^{\frac{3}{2}}} \operatorname{ierfc} \frac{r - a}{2\sqrt{(Dt)}}$$

$$+ \frac{Dt(9a^2 - 2ar - 7r^2)}{32 a^{\frac{3}{2}} r^{\frac{5}{2}}} i^2 \operatorname{erfc} \frac{r - a}{2\sqrt{(Dt)}} + \dots . \quad (5.78)$$

Fig. 5.12 shows how concentration depends on radius at successive times. The expression for the amount of diffusing substance F crossing unit area of the surface $r = a$ in unit time is

$$F = -D \left(\frac{\partial C}{\partial r}\right)_{r=a} = \frac{4(C_1 - C_0)D}{\pi^2 a} \int_0^\infty \exp(-Du^2 t) \frac{du}{u\{J_0^2(au) + Y_0^2(au)\}}. \quad (5.79)$$

Numerical values of the integral in (5.79) have been tabulated by Jaeger and Clarke (1942) and are shown graphically by Carslaw and Jaeger (1959, p. 388). For small times we have

$$F = \frac{D(C_1 - C_0)}{a} \{(\pi T)^{-\frac{1}{2}} + \tfrac{1}{2} - \tfrac{1}{4}(T/\pi)^{\frac{1}{2}} + \tfrac{1}{8} T - \dots\}, \quad (5.80)$$

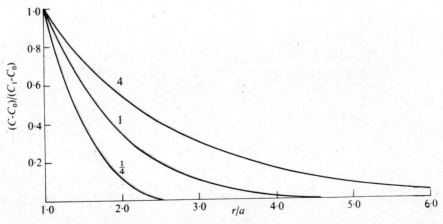

FIG. 5.12. Concentration–distance curves in the region $r > a$. Numbers on curves are values of Dt/a^2

and for large times

$$F = \frac{2D}{a}(C_1 - C_0)\left[\frac{1}{\ln(4T) - 2\gamma} - \frac{\gamma}{\{\ln(4T) - 2\gamma\}^2} - \cdots\right], \quad (5.81)$$

where $T = Dt/a^2$ and $\gamma = 0.57722$ is Euler's constant.

(ii) If the region $r > a$ is initially at a uniform concentration C_0, and there is transfer of diffusing substance across $r = a$ according to

$$-D\,\partial C/\partial r = \alpha(C - C_1), \qquad r = a, \quad (5.82)$$

then we have

$$\frac{C - C_1}{C_0 - C_1} = -\frac{2h}{\pi}\int_0^\infty \exp(-Du^2 t)$$

$$\times \frac{J_0(ur)\{uY_1(ua) + hY_0(ua)\} - Y_0(ur)\{uJ_1(ua) + hJ_0(ua)\}}{\{uJ_1(ua) + hJ_0(ua)\}^2 + \{uY_1(ua) + hY_0(ua)\}^2}\frac{du}{u}, \quad (5.83)$$

where $h = \alpha/D$. Graphs showing how $(C - C_1)/(C_0 - C_1)$ at the surface $r = a$ varies with time have been drawn by Carslaw and Jaeger, (1959 p. 338), for several values of ah. Clearly because of (5.82) these curves also show the rate of transfer across unit area of the surface $r = a$. Carslaw and Jaeger (1959, p. 338) also give the solution of the corresponding problem when there is a constant rate of flow of diffusing substance across the surface $r = a$.

6

DIFFUSION IN A SPHERE

6.1. Introduction

IF we restrict ourselves to cases in which the diffusion is radial, the diffusion equation for a constant diffusion coefficient takes the form

$$\frac{\partial C}{\partial t} = D\left(\frac{\partial^2 C}{\partial r^2} + \frac{2}{r}\frac{\partial C}{\partial r}\right). \tag{6.1}$$

On putting

$$u = Cr, \tag{6.2}$$

(6.1) becomes

$$\frac{\partial u}{\partial t} = D\frac{\partial^2 u}{\partial r^2}. \tag{6.3}$$

Since this is the equation for linear flow in one dimension, the solutions of many problems in radial flow in a sphere can be deduced immediately from those of the corresponding linear problems.

6.2. Steady state

In this case the equation is

$$\frac{d}{dr}\left(r^2\frac{dC}{dr}\right) = 0, \tag{6.4}$$

of which the general solution is

$$C = B + A/r, \tag{6.5}$$

where A and B are constants to be determined from the boundary conditions. If in the hollow sphere, $a \leqslant r \leqslant b$, the surface $r = a$ is kept at a constant concentration C_1, and $r = b$ at C_2, then

$$C = \frac{aC_1(b-r) + bC_2(r-a)}{r(b-a)}. \tag{6.6}$$

The quantity of diffusing substance Q_t which passes through the spherical wall in time t is given by

$$Q_t = 4\pi Dt\frac{ab}{b-a}(C_2 - C_1). \tag{6.7}$$

If Q_t is measured in a concentration-dependent system, the mean value of the diffusion coefficient obtained from (6.7) is

$$\left(\int_{C_2}^{C_1} D \, dC \right) \Big/ (C_1 - C_2)$$

as for the plane sheet (see eqn. (4.11)).

If the surface $r = a$ is maintained at a concentration C_1 and at $r = b$ there is evaporation according to the condition

$$\frac{dC}{dr} + h(C - C_2) = 0, \qquad r = b, \tag{6.8}$$

we find

$$C = \frac{aC_1\{hb^2 + r(1 - hb)\} + hb^2 C_2(r - a)}{r\{hb^2 + a(1 - hb)\}}. \tag{6.9}$$

The amount Q_t passing through the spherical wall in time t is now given by

$$Q_t = \frac{4\pi D t h a b^2 (C_1 - C_2)}{hb^2 + a(1 - hb)}. \tag{6.10}$$

If $ah > 2$, the rate of diffusion decreases steadily as b increases, but if $ah < 2$ the rate first increases and later decreases, passing through a maximum when $b = 2/h$. As in the case of the cylinder, this maximum is due to the combination of a decreasing gradient and an increasing surface area as b is increased.

If the surface conditions are

$$\partial C / \partial r + h_1(C_1 - C) = 0, \quad r = a; \qquad \partial C / \partial r + h_2(C - C_2) = 0, \quad r = b, \tag{6.11}$$

the solutions are

$$C = \frac{C_1 a^2 h_1\{b^2 h_2 - r(bh_2 - 1)\} + C_2 b^2 h_2\{r(ah_1 + 1) - a^2 h_1\}}{r\{b^2 h_2(ah_1 + 1) - a^2 h_1(bh_2 - 1)\}}, \tag{6.12}$$

and

$$Q_t = \frac{4\pi a^2 b^2 h_1 h_2 D t (C_1 - C_2)}{b^2 h_2(ah_1 + 1) - a^2 h_1(bh_2 - 1)}. \tag{6.13}$$

6.3. Non-steady state

6.3.1. *Surface concentration constant: initial distribution $f(r)$*

If we make the substitution $u = Cr$, suggested above, the equations for u are

$$\frac{\partial u}{\partial t} = D \frac{\partial^2 u}{\partial r^2}, \tag{6.14}$$

$$u = 0, \qquad r = 0, \qquad t > 0, \tag{6.15}$$

$$\dot{u} = aC_0, \qquad r = a, \qquad t > 0, \tag{6.16}$$

$$u = rf(r), \qquad t = 0, \qquad 0 < r < a, \tag{6.17}$$

where C_0 is the constant concentration at the surface of the sphere. These are the equations of diffusion in a plane sheet of thickness a, with its ends, $r = 0$ and $r = a$, kept at zero and aC_0 respectively, and with the initial distribution $rf(r)$. This problem has been considered in §4.3.1 (p. 47) and the solution follows immediately by making the appropriate substitutions in eqn (4.16). If the sphere is initially at a uniform concentration C_1 and the surface concentration is maintained constant at C_0, the solution becomes

$$\frac{C-C_1}{C_0-C_1} = 1 + \frac{2a}{\pi r} \sum_{n=1}^{\infty} \frac{(-1)^n}{n} \sin\frac{n\pi r}{a} \exp(-Dn^2\pi^2 t/a^2). \tag{6.18}$$

The concentration at the centre is given by the limit as $r \to 0$, that is by

$$\frac{C-C_1}{C_0-C_1} = 1 + 2\sum_{n=1}^{\infty} (-1)^n \exp(-Dn^2\pi^2 t/a^2). \tag{6.19}$$

The total amount of diffusing substance entering or leaving the sphere is given by

$$\frac{M_t}{M_\infty} = 1 - \frac{6}{\pi^2} \sum_{n=1}^{\infty} \frac{1}{n^2} \exp(-Dn^2\pi^2 t/a^2). \tag{6.20}$$

The corresponding solutions for small times are

$$\frac{C-C_1}{C_0-C_1} = \frac{a}{r} \sum_{n=0}^{\infty} \left\{ \text{erfc}\frac{(2n+1)a-r}{2\sqrt{(Dt)}} - \text{erfc}\frac{(2n+1)a+r}{2\sqrt{(Dt)}} \right\}, \tag{6.21}$$

and

$$\frac{M_t}{M_\infty} = 6\left(\frac{Dt}{a^2}\right)^{\frac{1}{2}} \left\{ \pi^{-\frac{1}{2}} + 2\sum_{n=1}^{\infty} \text{ierfc}\frac{na}{\sqrt{(Dt)}} \right\} - 3\frac{Dt}{a^2}. \tag{6.22}$$

These solutions can be written in terms of the two dimensionless parameters Dt/a^2 and r/a. Curves showing $(C-C_1)/(C_0-C_1)$ as a function of r/a for different values of Dt/a^2, drawn by Carslaw and Jaeger (1959, p. 235) are reproduced in Fig. 6.1. The curve for zero fractional exhaustion on Fig. 6.4 shows M_t/M_∞ as a function of Dt/a^2.

6.3.2. Variable surface concentration

If the initial concentration in the sphere is zero and that at the surface is $\phi(t)$ the solution is

$$C = -\frac{2D}{ra} \sum_{n=1}^{\infty} (-1)^n \exp(-Dn^2\pi^2 t/a^2) n\pi \sin\frac{n\pi r}{a} \int_0^t \exp(Dn^2\pi^2\lambda/a^2)\phi(\lambda) \, d\lambda. \tag{6.23}$$

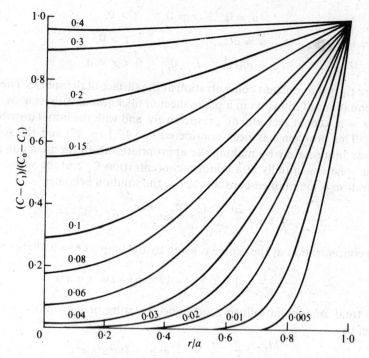

FIG. 6.1. Concentration distributions at various times in a sphere with initial concentration C_1 and surface concentration C_0. Numbers on curves are values of Dt/a^2.

(i) When

$$\phi(t) = C_0\{1 - \exp(-\beta t)\}, \tag{6.24}$$

(6.23) becomes

$$\frac{C}{C_0} = 1 - \frac{a}{r}\exp(-\beta t)\frac{\sin\{(\beta a^2/D)^{\frac{1}{2}}r/a\}}{\sin\{\beta a^2/D\}^{\frac{1}{2}}}$$

$$- \frac{2\beta a^3}{\pi Dr}\sum_{n=1}^{\infty}(-1)^n\frac{\exp(-Dn^2\pi^2 t/a^2)}{n(n^2\pi^2 - \beta a^2/D)}\sin\frac{n\pi r}{a}, \tag{6.25}$$

and the sorption-time curve is given by

$$\frac{3M_t}{4\pi a^3 C_0} = 1 - \frac{3D}{\beta a^2}\exp(-\beta t)\left\{1 - \left(\frac{\beta a^2}{D}\right)^{\frac{1}{2}}\cot\left(\frac{\beta a^2}{D}\right)^{\frac{1}{2}}\right\}$$

$$+ \frac{6\beta a^2}{\pi^2 D}\sum_{n=1}^{\infty}\frac{\exp(-Dn^2\pi^2 t/a^2)}{n^2(n^2\pi^2 - \beta a^2/D)}. \tag{6.26}$$

Fig. 6.2 shows uptake curves for different values of the parameter $\beta a^2/D$.

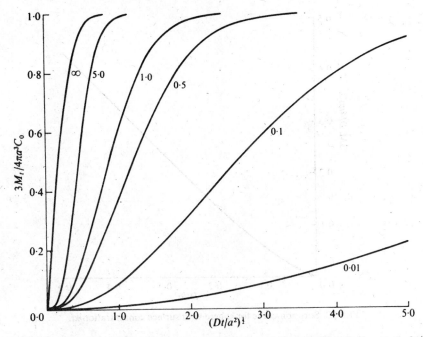

FIG. 6.2. Calculated sorption curves for surface concentration given by $C_0\{1 - \exp(-\beta t)\}$. Numbers on curves are values of $\beta a^2/D$.

(ii) If the surface concentration varies linearly with time, i.e.

$$\phi(t) = kt. \tag{6.27}$$

the solutions are

$$C = k\left(t - \frac{a^2 - r^2}{6D}\right) - \frac{2ka^3}{D\pi^3 r} \sum_{n=1}^{\infty} \frac{(-1)^n}{n^3} \exp(-Dn^2\pi^2 t/a^2) \sin\frac{n\pi r}{a}, \tag{6.28}$$

and

$$\frac{M_t}{\frac{4}{3}\pi a^3 k} = \left(t - \frac{a^2}{15D}\right) + \frac{6a^2}{\pi^4 D} \sum_{n=1}^{\infty} \frac{1}{n^4} \exp(-Dn^2\pi^2 t/a^2). \tag{6.29}$$

$M_t D/(\frac{4}{3}\pi a^5 k)$ is plotted against Dt/a^2 in Fig. 6.3.

6.3.3. Diffusion from a well-stirred solution of limited volume

The problem and method of solution are very similar to those of the plane sheet and the results can be given without explanation. Suppose that the sphere occupies the space $r < a$, while the volume of the bath of solution (excluding the space occupied by the sphere) is V. The concentration of solute in the solution is always uniform and is initially C_0. The sphere is initially

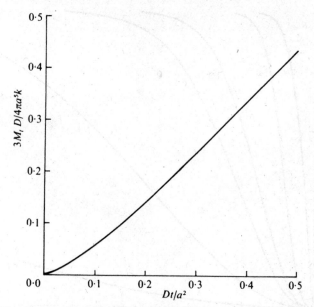

FIG. 6.3. Sorption curve for sphere with surface concentration kt.

free from solute. The total amount of solute M_t in the sphere after time t is expressed as a fraction of the corresponding quantity after infinite time by the relation

$$\frac{M_t}{M_\infty} = 1 - \sum_{n=1}^{\infty} \frac{6\alpha(\alpha+1)\exp(-Dq_n^2 t/a^2)}{9+9\alpha+q_n^2\alpha^2},$$

(6.30)

where the q_ns are the non-zero roots of

$$\tan q_n = \frac{3q_n}{3+\alpha q_n^2},$$

(6.31)

and $\alpha = 3V/(4\pi a^3)$, the ratio of the volumes of solution and sphere, or if there is a partition factor K between solute in equilibrium in the sphere and the solution, $\alpha = 3V/(4\pi a^3 K)$. The parameter α is expressed in terms of the final fractional uptake of solute by the sphere by the relation

$$\frac{M_\infty}{V C_0} = \frac{1}{1+\alpha}.$$

(6.32)

The roots of (6.31) are given in Table 6.1 for several values of α. An alternative solution suitable for small times given by Carman and Haul (1954) is

$$\frac{M_t}{M_\infty} = (1+\alpha)\left[1 - \frac{\gamma_1}{\gamma_1+\gamma_2}\mathrm{e}\operatorname{erfc}\left\{\frac{3\gamma_1}{\alpha}\left(\frac{Dt}{a^2}\right)^{\frac{1}{2}}\right\} - \frac{\gamma_2}{\gamma_1+\gamma_2}\mathrm{e}\operatorname{erfc}\left\{-\frac{3\gamma_2}{\alpha}\left(\frac{Dt}{a^2}\right)^{\frac{1}{2}}\right\}\right]$$

$$+ \text{higher terms},$$

(6.33)

where in their notation

$$\gamma_1 = \tfrac{1}{2}\{(1 + \tfrac{4}{3}\alpha)^{\frac{1}{2}} + 1\}, \qquad \gamma_2 = \gamma_1 - 1, \tag{6.34}$$

and

$$\text{e erfc } z \equiv \exp z^2 \text{ erfc } z. \tag{6.35}$$

In Fig. 6.4, M_t/M_∞ against $(Dt/a^2)^{\frac{1}{2}}$ is plotted for five final fractional uptakes. Berthier (1952) gives a table of M_t/M_∞ for values of $1/\alpha$ between 0 and 1 at intervals of 0·1. His values are not always reliable to three decimal places.

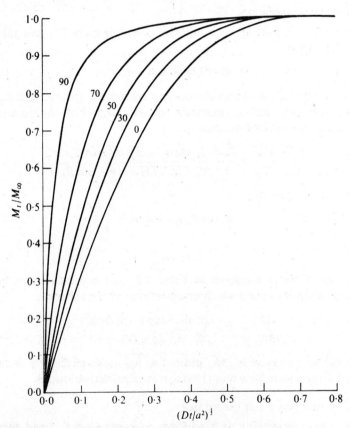

FIG. 6.4. Uptake by a sphere from a stirred solution of limited volume. Numbers on curves show percentage of solute finally taken up by sphere.

The concentration of solute C within the sphere is given by

$$C = C_\infty \left\{ 1 + \sum_{n=1}^{\infty} \frac{6(1+\alpha)\exp(-Dq_n^2 t/a^2)}{9 + 9\alpha + q_n^2\alpha^2} \frac{a}{r} \frac{\sin(q_n r/a)}{\sin q_n} \right\}. \tag{6.36}$$

The above equations also describe the course of desorption into a well-stirred solution, initially free from solute, from a sphere in which the concentration is uniform and equal to C_0. The only modifications are that (6.32) is to be replaced by

$$\frac{3M_\infty}{4\pi a^3 C_0} = \frac{1}{1+1/\alpha}, \tag{6.37}$$

and C_1, $(C_1)_\infty$ replace C, C_∞ in (6.36), where

$$C_1 = C_0 - C, \qquad (C_1)_\infty = C_0 - C_\infty \tag{6.38}$$

6.3.4. *Surface evaporation*

If the sphere is initially at a uniform concentration C_1, and there is a surface condition

$$-D\, \partial C/\partial r = \alpha(C_s - C_0), \tag{6.39}$$

where C_s is the actual concentration just within the sphere, and C_0 is the concentration required to maintain equilibrium with the surrounding atmosphere, the required solution is

$$\frac{C - C_0}{C_1 - C_0} = \frac{2La}{r} \sum_{n=1}^{\infty} \frac{\exp(-D\beta_n^2 t/a^2)}{\{\beta_n^2 + L(L-1)\}} \frac{\sin \beta_n r/a}{\sin \beta_n}. \tag{6.40}$$

The β_ns are the roots of

$$\beta_n \cot \beta_n + L - 1 = 0, \tag{6.41}$$

and

$$L = a\alpha/D. \tag{6.42}$$

Some roots of (6.41) are given in Table 6.2. The expression for the total amount of diffusing substance entering or leaving the sphere is

$$\frac{M_t}{M_\infty} = 1 - \sum_{n=1}^{\infty} \frac{6L^2 \exp(-\beta_n^2 Dt/a^2)}{\beta_n^2 \{\beta_n^2 + L(L-1)\}}. \tag{6.43}$$

Fig. 6.5 shows curves of M_t/M_∞ plotted as functions of $(Dt/a^2)^{\frac{1}{2}}$ for several values of L for which Newman (1931) gives tabulated solutions.

6.3.5. *Constant flux F_0 at the surface*

If the sphere is initially at a uniform concentration C_0, and there is a constant rate of transfer F_0 per unit of surface, i.e.

$$-D\, \partial C/\partial r = F_0, \qquad r = a, \tag{6.44}$$

then we have

$$C_0 - C = \frac{F_0 a}{D}\left\{\frac{3Dt}{a^2} + \frac{1}{2}\frac{r^2}{a^2} - \frac{3}{10} - 2\frac{a}{r}\sum_{n=1}^{\infty}\frac{\sin(\alpha_n r)}{\alpha_n^2 a^2 \sin(\alpha_n a)}\exp(-D\alpha_n^2 t)\right\}, \tag{6.45}$$

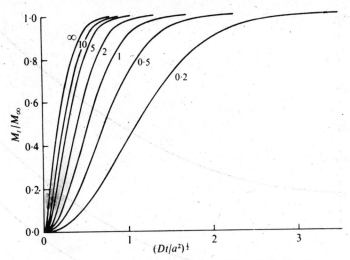

FIG. 6.5. Sorption or desorption curves for the surface condition (6.39). Numbers on curves are values of $L = a\alpha/D$.

where the $a\alpha_n$s are the positive roots of

$$a\alpha_n \cot a\alpha_n = 1. \qquad (6.46)$$

The amount of diffusing substance lost by the sphere in time t is $4\pi a^2 F_0 t$. Some roots of (6.46) are given in Table 6.2 when $L = 0$. Fig. 6.6 shows curves of $D(C_0 - C)/(F_0 a)$ plotted against r/a for different values of Dt/a^2.

6.3.6. Impermeable surface

If the surface of the sphere is impermeable and there is an initial concentration $f(r)$ then

$$C = \frac{3}{a^3} \int_0^a r^2 f(r)\, dr + \frac{2}{ar} \sum_{n=1}^{\infty} \exp(-D\alpha_n^2 t) \frac{\sin \alpha_n r}{\sin^2 \alpha_n a} \int_0^a r' f(r') \sin \alpha_n r'\, dr', \qquad (6.47)$$

where the α_ns are the positive roots of (6.46).

6.3.7. Composite sphere

Problems of diffusion into a composite sphere comprised of an inner core and an outer shell for which the diffusion coefficients are different have been considered by Carslaw and Jaeger (1959, p. 351), Bromwich (1921), Carslaw (1921), Carslaw and Jaeger (1939), Bell (1945), and others. The extra parameters involved make any attempt at general numerical evaluation too formidable to be attempted here.

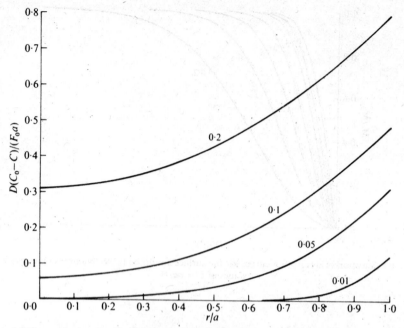

FIG. 6.6. Concentration distributions in a sphere for constant flux F_0 at the surface. Numbers on curves are values of Dt/a^2.

6.4. Hollow sphere

Carslaw and Jaeger (1959, p. 246) give the general solution to the problem of the hollow sphere with the surface $r = a$ maintained at a constant concentration C_1, and $r = b$ at C_2, when the initial distribution is $f(r)$ in the region $a \leqslant r \leqslant b$. Some special cases have been considered by Barrer (1944) who also suggests a number of practical systems to which his solutions might be applied. In the special case of a constant initial concentration, $f(r) = C_0$, and when $C_1 = C_2$, the solution is

$$\frac{C - C_0}{C_1 - C_0} = 1 + \frac{2}{\pi r} \sum_{n=1}^{\infty} \left(\frac{b \cos n\pi - a}{n} \right) \sin \frac{n\pi(r-a)}{b-a} \exp \left\{ -Dn^2\pi^2 t/(b-a)^2 \right\}.$$

(6.48)

The total amount of diffusing substance entering or leaving the hollow sphere $a \leqslant r \leqslant b$ in time t is given by

$$\frac{M_t}{M_\infty} = 1 - \frac{6}{\pi^2(a^2 + ab + b^2)} \sum_{n=1}^{\infty} \left(\frac{b \cos n\pi - a}{n} \right)^2 \exp \left\{ -Dn^2\pi^2 t/(b-a)^2 \right\}.$$

(6.49)

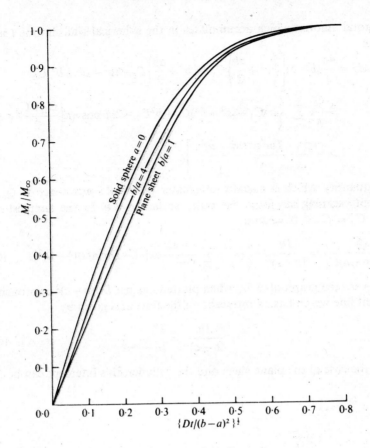

FIG. 6.7. Uptake curves for hollow sphere.

Graphs of M_t/M_∞ against $\{Dt/(b-a)^2\}^{\frac{1}{2}}$ are shown in Fig. 6.7 for different values of b/a.

6.4.1. *Flow through spherical wall*

If the surface $r = a$ is maintained at C_1, and $r = b$ at C_2, and the region $a \leqslant r \leqslant b$ is initially at C_0, the concentration approaches the steady-state distribution discussed in § 6.2, according to the expression

$$C \quad \frac{aC_1}{r} + \frac{(bC_2 - aC_1)(r-a)}{r(b-a)} + \frac{2}{r\pi} \sum_{n=1}^{\infty} \frac{b(C_2 - C_0)\cos n\pi - a(C_1 - C_0)}{n}$$

$$\times \sin\frac{n\pi(r-a)}{b-a}\exp\{-Dn^2\pi^2t/(b-a)^2\}. \tag{6.50}$$

The total amount which accumulates in the spherical wall in time t is M_t, where

$$M_t = \frac{4\pi}{3}(b-a)\left[\left(a^2 + \frac{ab}{2}\right)C_1 + \left(b^2 + \frac{ab}{2}\right)C_2 - (a^2 + ab + b^2)C_0\right.$$

$$-\frac{6}{\pi^2}\sum_{n=1}^{\infty}\left\{a^2C_1 - (a^2 + b^2)C_0 + b^2C_2 - 2ab\cos n\pi\left(\frac{C_1 + C_2}{2} - C_0\right)\right\}$$

$$\left.\times\frac{\exp\{-Dn^2\pi^2t/(b-a)^2\}}{n^2}\right]. \tag{6.51}$$

The quantity which is usually of greater practical importance is Q_t, the amount escaping say from the outer surface $r = b$. In the simplest case, when $C_0 = C_2 = 0$, we find

$$\frac{Q_t}{4\pi ab(b-a)C_1} = \frac{Dt}{(b-a)^2} - \frac{1}{6} - \frac{2}{\pi^2}\sum_{n=1}^{\infty}\frac{(-1)^n}{n^2}\exp\{-Dn^2\pi^2t/(b-a)^2\}. \tag{6.52}$$

As $t \to \infty$, the graph of (6.52) when plotted against $Dt/(b-a)^2$ approaches a straight line which has an intercept on the time axis given by

$$\frac{Dt}{(b-a)^2} = \frac{1}{6}. \tag{6.53}$$

As in the case of the plane sheet and the cylinder, this intercept can be used

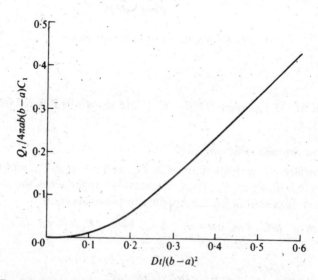

FIG. 6.8. Approach to steady-state flow through the wall of a sphere.

to determine the diffusion coefficient D experimentally (Barrer 1951). Fig. 6.8 shows the way in which the graph of (6.52) approaches the straight line.

6.4.2. Surface evaporation

Carslaw and Jaeger (1959, p. 246) give a solution to the problem in which evaporation into an atmosphere of zero vapour pressure occurs at $r = a$ and $r = b$ according to the expressions

$$k_1 \frac{\partial C}{\partial r} - h_1 C = 0, \qquad r = a; \tag{6.54}$$

$$k_2 \frac{\partial C}{\partial r} + h_2 C = 0, \qquad r = b. \tag{6.55}$$

If the initial concentration distribution in the spherical wall is $f(r)$, the solution is

$$C = \frac{2}{r} \sum_{n=1}^{\infty} \exp\{-D\alpha_n^2 t R_n(r)\} \int_a^b r' R_n(r') f(r')\, dr', \tag{6.56}$$

where

$$G = ah_1 + k_1, \qquad H = bh_2 - k_2, \tag{6.57}$$

$R_n(r) =$

$$\frac{(H^2 + b^2 k_2^2 \alpha_n^2)^{\frac{1}{2}}\{G \sin(r-a)\alpha_n + ak_1\alpha_n \cos(r-a)\alpha_n\}}{\{(b-a)(a^2 k_1^2 \alpha_n^2 + G^2)(b^2 k_2^2 \alpha_n^2 + H^2) + (Hak_1 + Gbk_2)(GH + abk_1 k_2 \alpha_n^2)\}^{\frac{1}{2}}}, \tag{6.58}$$

and $\pm\alpha_n$, $n = 1, 2, \ldots$ are the roots of

$$(GH - abk_1 k_2 \alpha^2) \sin(b-a)\alpha + \alpha(ak_1 H + bk_2 G) \cos(b-a)\alpha = 0. \tag{6.59}$$

By putting either or both k_1 and k_2 zero we obtain the solutions for one or both surfaces maintained at zero concentration, and similarly h_1 or h_2 zero corresponds to an impermeable surface. If both h_1 and h_2 are zero, so that both surfaces are impermeable, a term

$$\frac{3}{(b^3 - a^3)} \int_a^b r^2 f(r)\, dr$$

is to be added to (6.56).

If evaporation takes place into atmospheres of different vapour pressures the solution can be deduced from the above by following the method used by Carslaw and Jaeger (1959, p. 118) for the plane sheet. Thus we write

$C = u + w$, where u is a function of r only, satisfying

$$\frac{d}{dr}\left(r^2 \frac{du}{dr}\right) = 0, \qquad a < r < b,$$

$$k_1 \frac{du}{dr} + h_1(C_1 - u) = 0, \qquad r = a,$$

$$k_2 \frac{du}{dr} + h_2(u - C_2) = 0, \qquad r = b,$$

so that

$$u = \frac{C_1 a^2 h_1\{b^2 h_2 - r(bh_2 - k_2)\} + C_2 b^2 h_2\{r(ah_1 + k_1) - a^2 h_1\}}{r\{b^2 h_2(ah_1 + k_1) - a^2 h_1(bh_2 - k_2)\}},$$

and w is a function of r and t such that

$$\frac{\partial w}{\partial t} = \frac{D}{r^2} \frac{\partial}{\partial r}\left(r^2 \frac{\partial w}{\partial r}\right), \qquad a < r < b,$$

$$k_1 \frac{\partial w}{\partial r} - h_1 w = 0, \qquad r = a,$$

$$k_2 \frac{\partial w}{\partial r} + h_2 w = 0, \qquad r = b,$$

$$w = f(r) - u, \qquad t = 0,$$

and hence w is given by (6.56) on using $f(r) - u$ for $f(r)$.

6.5. The region bounded internally by the sphere $r = a$

Solutions of this problem follow readily from those of the corresponding problems of the semi-infinite sheet by using the transformation $u = Cr$.

(i) If the region $r > a$ is initially at a uniform concentration C_0, and the surface $r = a$ is maintained at C_1, the solution is

$$\frac{C - C_0}{C_1 - C_0} = \frac{a}{r} \operatorname{erfc} \frac{r - a}{2\sqrt{(Dt)}}, \tag{6.60}$$

which is readily evaluated using Table 2.1.

(ii) If the initial concentration is C_0, and there is a boundary condition

$$-\partial C/\partial r = h(C_1 - C), \qquad r = a, \tag{6.61}$$

then we find

$$\frac{C-C_0}{C_1-C_0} = \frac{ha^2}{r(1+ah)}\left[\text{erfc}\frac{r-a}{2\sqrt{(Dt)}} - \exp\{h'(r-a)+h'^2Dt\}\text{erfc}\left\{\frac{r-a}{2\sqrt{(Dt)}}\right.\right.$$

$$\left.\left.+h'\sqrt{(Dt)}\right\}\right], \tag{6.62}$$

where $h' = h+(1/a)$. The ratio $(C-C_0)/(C_1-C_0)$ on the surface, $r = a$, is

$$\frac{ha}{ha+1}\{1-\exp(h'^2Dt)\,\text{erfc}\,h'\sqrt{(Dt)}\}, \tag{6.63}$$

which again is readily evaluated from Table 2.1.

CONCENTRATION-DEPENDENT DIFFUSION: METHODS OF SOLUTION

7.1. Time–dependent diffusion coefficients

IN Chapters 2–6 mathematical solutions were presented for various problems in which the diffusion coefficient was taken to be constant. These solutions can also be used if the diffusion coefficient D depends on the time t for which diffusion has been proceeding but is independent of other variables, i.e. if D is a function of t only. In this case the equation for diffusion in one dimension becomes

$$\frac{\partial C}{\partial t} = D(t)\frac{\partial^2 C}{\partial x^2},\tag{7.1}$$

and on writing

$$dT = D(t)\,dt,\tag{7.2}$$

i.e.

$$T = \int_0^t D(t')\,dt',\tag{7.3}$$

equation (7.1) reduces to

$$\frac{\partial C}{\partial T} = \frac{\partial^2 C}{\partial x^2}.\tag{7.4}$$

The solutions for constant D can therefore be used to give C as a function of x and T, and T is then converted into t using the relationship (7.3). If the integral in (7.3) cannot be evaluated formally, the relationship between T and t has to be obtained by graphical or numerical integration. If the boundary conditions involve time explicitly, e.g. if the surface concentration is a given function of t, this function must be rewritten in terms of T in order to obtain the appropriate solution of (7.4) in x and T. The transformation (7.2) of course, can be used for all forms of the diffusion equation, e.g. for diffusion in a plane sheet, cylinder, or sphere.

Dodson (1973) used (7.3) in a problem of diffusion in a cooling solid presented by the calculation of the age of a rock or mineral from its accumulated products of radioactive decay. His equation takes the form

$$\partial c/\partial t = D(0)\,e^{-t}\nabla^2 c + \alpha\,e^{-\alpha t},$$

which in terms of the variables

$$\cdot q = 1 - e^{-\alpha t} - c, \qquad u = \int_0^t D(0)\, e^{-t}\, dt = D(0)(1 - e^{-t})$$

becomes

$$\partial q / \partial u = \nabla^2 q.$$

Solutions for a time-dependent surface condition on q are deduced by Dodson from general forms by Carslaw and Jaeger (1959).

7.2. Concentration-dependent diffusion: infinite and semi-infinite media

A case of great practical interest is that in which the diffusion coefficient depends only on the concentration of diffusing substance. Such a concentration-dependence exists in most systems, but often, e.g. in dilute solutions, the dependence is slight and the diffusion coefficient can be assumed constant for practical purposes. In other cases, however, such as the diffusion of vapours in high-polymer substances, the concentration dependence is a very marked, characteristic feature. A number of methods have been used to obtain numerical solutions, some applicable to any type of concentration-dependent diffusion coefficient, and others restricted to particular types, e.g. exponential or linear dependence. In other cases, algebraic solutions have been expressed in terms of a single integral and these will be referred to as formal solutions even though the integral has to be evaluated numerically.

7.2.1. *Boltzmann's transformation*

The equation for one-dimensional diffusion when the diffusion coefficient D is a function of concentration C is

$$\frac{\partial C}{\partial t} = \frac{\partial}{\partial x}\left(D \frac{\partial C}{\partial x} \right). \tag{7.5}$$

Boltzmann (1894) showed that for certain boundary conditions, provided D is a function of C only, C may be expressed in terms of a single variable $x/2t^{\frac{1}{2}}$ and that (7.5) may therefore be reduced to an ordinary differential equation by the introduction of a new variable η, where

$$\eta = \tfrac{1}{2} x / t^{\frac{1}{2}}. \tag{7.6}$$

Thus we have

$$\frac{\partial C}{\partial x} = \frac{1}{2t^{\frac{1}{2}}} \frac{dC}{d\eta}, \tag{7.7}$$

and

$$\frac{\partial C}{\partial t} = -\frac{x}{4t^{\frac{3}{2}}} \frac{dC}{d\eta}, \tag{7.8}$$

and hence

$$\frac{\partial}{\partial x}\left(D\frac{\partial C}{\partial x}\right) = \frac{\partial}{\partial x}\left(\frac{D}{2t^{\frac{1}{2}}}\frac{\mathrm{d}C}{\mathrm{d}\eta}\right) = \frac{1}{4t}\frac{\mathrm{d}}{\mathrm{d}\eta}\left(D\frac{\mathrm{d}C}{\mathrm{d}\eta}\right), \qquad (7.9)$$

so that finally (7.5) becomes

$$-2\eta\frac{\mathrm{d}C}{\mathrm{d}\eta} = \frac{\mathrm{d}}{\mathrm{d}\eta}\left(D\frac{\mathrm{d}C}{\mathrm{d}\eta}\right), \qquad (7.10)$$

an ordinary differential equation in C and η. The transformation (7.6) can be used when diffusion takes place in infinite or semi-infinite media provided the concentration is initially constant in the regions $x \gtrless 0$ or $x > 0$ as the case may be. For the problem defined by

$$C = C_1, \qquad x < 0, \qquad t = 0, \qquad (7.11)$$

$$C = C_2, \qquad x > 0, \qquad t = 0, \qquad (7.12)$$

eqn (7.10) is to be solved for the boundary conditions

$$C = C_1, \qquad \eta = -\infty, \qquad (7.13)$$

$$C = C_2, \qquad \eta = +\infty. \qquad (7.14)$$

Similarly, the conditions for a semi-infinite medium

$$C = C_0, \qquad x = 0, \qquad t > 0, \qquad (7.15)$$

$$C = C_1, \qquad x > 0, \qquad t = 0, \qquad (7.16)$$

become

$$C = C_0, \qquad \eta = 0, \qquad (7.17)$$

$$C = C_1, \qquad \eta = \infty. \qquad (7.18)$$

It is only when the initial and boundary conditions are expressible in terms of η alone, and x and t are not involved separately, that the transformation (7.6) and the eqn (7.10) can be used. They cannot be used, for example, when diffusion occurs in a finite sheet of thickness l, and the boundary conditions are

$$C = C_0, \qquad x = 0, \qquad x = l, \qquad (7.19)$$

because the second condition becomes

$$C = C_0, \qquad \eta = \tfrac{1}{2}l/t^{\frac{1}{2}}, \qquad (7.20)$$

which is not expressible in terms of η only but involves t explicitly. In general the transformation can be used for diffusion and semi-infinite media when the initial concentrations are uniform and may be zero.

7.2.2. Numerical iterative methods

(i) Crank and Henry (1949a,b) gave a method of solving (7.10) subject to (7.17) and (7.18) based on iterative quadrature. Eqn (7.10) can be written

$$\frac{-2\eta}{D} D \frac{dc}{d\eta} = \frac{d}{d\eta}\left(D\frac{dc}{d\eta}\right), \tag{7.21}$$

where $c = C/C_0$, a dimensionless concentration, and $c = 1$ on $\eta = 0$. We illustrate the method for the case $c = 0$, $\eta = \infty$. By integrating (7.21) twice, we obtain

$$c = 1 - A\int_0^\eta \frac{1}{D}\exp\left(-\int_0^\eta \frac{2\eta'}{D}d\eta'\right)d\eta', \tag{7.22}$$

where A is a constant of integration to be chosen such that

$$A\int_0^\infty \frac{1}{D}\exp\left(-\int_0^\eta \frac{2\eta'}{D}d\eta'\right)d\eta = 1, \tag{7.23}$$

in order that (7.18) shall be satisfied. The condition (7.17) is automatically satisfied by (7.22). The diffusion coefficient D is a known function $f(c)$ of c, e.g. $D = \exp(kc)$. If we have a first approximation to the function $c(\eta)$ we deduce the function $D(\eta)$ using $D = f(c)$ and carry out the integrations in (7.22) numerically using (7.23) to obtain a second approximation, and so on. The convergence of this iteration is poor when D varies markedly with c. Crank and Henry obtained some improvement by rewriting the procedure in terms of the variable s defined by

$$s = \int_0^c D\,dc \bigg/ \int_0^1 D\,dc.\dagger \tag{7.24}$$

(ii) Philip (1969a) reviewed his extensive contributions to the theory of infiltration which include one-dimensional diffusion as a special case. One method (Philip 1955) which has better convergence and accuracy than the one just described starts by rewriting (7.10) as

$$-2\eta = \frac{d}{dc}\left(D\frac{dc}{d\eta}\right),$$

---

† This transformation seems to have been used first by Kirchhoff (1894) and more recently by Eyres, Hartree, Ingham, Jackson, Sarjant, and Wagstaff (1946). Philip (1973a) has shown that for periodic non-linear diffusion, the time average of s satisfies Laplace's equation and is therefore readily evaluated for appropriate boundary conditions.

which on integration yields

$$2 \int_0^c \eta \, dc = -D \, dc/d\eta, \tag{7.25}$$

subject to the condition (7.17). The lower limit of the integral is 0 because (7.18) implies $dc/d\eta \to 0$ as $c \to 0$. Philip's method uses a finite-difference form of (7.25) in an iterative procedure starting from an estimate of $\int_0^1 \eta \, dc$. An analytical solution is used in the region of $c \to 0$. He compares his numerical solutions with the analytical solutions obtained by Neumann's method § 13.2.4 (p. 296) for two examples in which D is a two-step function of c.

(iii) Lee (1969) put the problem on a sounder mathematical footing by examining the analytical properties of the initial-value system comprising eqn (7.10) written in terms of the variable s defined in (7.24), i.e.

$$\frac{d^2s}{d\eta^2} = -2\eta F(s)\frac{ds}{d\eta} \tag{7.26}$$

subject to the conditions

$$s = 1, \qquad \eta = 0 \tag{7.27}$$

and

$$ds/d\eta = -g, \qquad \eta = 0, \tag{7.28}$$

where $F(s) = 1/D(s)$ and the parameter $g > 0$. He showed that the unique solutions are monotonic decreasing, and either reach $s = 0$ at a finite value of η or tend asymptotically as $\eta \to \infty$, to $s = s_\infty$, $1 > s_\infty > 0$. He refers to the former solutions as Class I type and the latter of Class ∞. He established that a solution of (7.26) subject to (7.28) and

$$s = 0, \qquad \eta = +\infty \tag{7.29}$$

exists, which is the solution of the initial-value system with $s_\infty = 0$. Subsequently, Lee (1971a, 1972) developed iterative methods for obtaining numerical solutions of the initial-value problem for successive approximations to the parameter g in (7.28) until the solution for which $s_\infty = 0$ is obtained within prescribed error estimates. The methods can conveniently be programmed for a digital computer and one solution takes only two or three minutes of computer time.

First, Lee considers a solution $s = s(\eta, g)$ of the initial-value system. From (7.26) it is evident that for $\eta \ll 1$, $d^2s/d\eta^2 \ll 1$. Thus $ds/d\eta$ changes very slowly for $\eta \ll 1$, and its value at a point is very close to the slope of the chord joining that point and the initial point $\eta = 0$, $s = 1$.

The term $ds/d\eta$ in (7.26) may be replaced by $(1-s)/(1-\eta)$, and so we approximate to (7.26) for $\eta \ll 1$ by

$$\frac{d^2 s}{d\eta^2} = 2(1-s)F(s).\qquad(7.30)$$

Lee (1971a) used the solution of (7.30) as a first approximation in his iterative process. The solution is easily obtained and can be of some intrinsic interest as an approximate solution which may be accurate enough for some purposes. The reader is referred to Lee's paper (1971a) for iterations based on the solution of (7.30).

We turn instead to the method described in the later paper (1972) which, by avoiding the use of (7.30), is simpler and requires less computing time. Lee established some relevant theorems so that we can consider two positive numbers g_U and g_L

$$g_L = \tfrac{1}{2}(L/\pi)^{\frac{1}{2}} \quad \text{and} \quad g_U = \tfrac{1}{2}(U/\pi)^{\frac{1}{2}},\qquad(7.31)$$

where

$$U = \sup_{0 \leqslant s \leqslant 1} F(s)\qquad(7.32)$$

and

$$L = \inf_{0 \leqslant s \leqslant 1} F(s)\qquad(7.33)$$

such that

(i) for all $g > g_U$ the solutions $s(\eta, g)$ are of Class I;
(ii) for all $g < g_L$ the solutions $s(\eta, g)$ are of Class ∞ with $s_\infty > 0$.

The gradient G we require to give the solution of (7.26) lies in the range $g_L < G < g_U$.

We now take as a first approximation of G an interpolated value g_1, where

$$g_1 = \tfrac{1}{2}(g_U + g_L).\qquad(7.34)$$

Depending on $s(\eta, g_1)$ being of Class I or Class ∞, the second approximation is obtained by a similar interpolation between the derivative at $\eta = 0$ of $s(\eta, g_1)$ and that of $s(\eta, g_L)$ or $s(\eta, g_U)$.

In general, if the last iterative solution is of Class ∞ with an asymptotic value other than zero, the next approximation is obtained by interpolating between the derivative at $\eta = 0$ of the last solution and that of the last Class I solution; if, on the other hand, the last iteration is of Class I then the next approximation follows by interpolating between the derivative of the last solution and that of the last Class ∞ solution. Lee used a fourth-order Runge–Kutta method of numerical integration to obtain the successive approximate solutions.

Since the derivatives of these solutions are negative and monotonic increasing to zero, an error estimate $0 < e_d \ll 1$ can always be prescribed so that the asymptotic value of a solution can be assumed as soon as the absolute value of the derivative becomes less than e_d, and hence the solution is asymptotic. Otherwise, the value of the solution reaches zero before the value of its derivative exceeds $-e_d$, and the solution is of Class I.

For a solution of Class ∞ the asymptotic value assumed above indicates the degree of approximation of the solution to the required one which approaches zero asymptotically. Thus, for a prescribed error estimate $0 < e_a \ll 1$, iteration stops as soon as the assumed asymptotic value of the last iterative solution is less than e_a. The last solution is taken as the required solution $s(\eta, s_0')$.

Lee (1969) showed that the derivatives are monotonic increasing to zero approximately at an exponential rate $\exp(-\eta^2 m)$, where

$$m = \inf_{0 < s < 1} F(s).$$

It follows that the increment of the asymptotic value due to $e_d \neq 0$ is always less than

$$-e_d \int_0^\infty \exp(-\eta^2 m)\, d\eta = -e_d(\pi/4m)^{\frac{1}{2}},$$

which can be made very much less than e_a if we choose $e_d \ll m^{\frac{1}{2}} e_a$. Thus we can choose consistent error estimates.

Lee (1971a) tabulated results for a practical example relating to the uptake of excess calcium by calcium chloride (Wagner 1968). The diffusion coefficient D takes the form

$$D = D_0/(1 + Ac).$$

The variable s defined by (7.24) becomes

$$s(c) = \ln(1 + Ac)/\ln(1 + A),$$

and in the eqn (7.26) we have

$$F(s) = \exp\{s \ln(1 + A)\}.$$

Lee evaluated numerical solutions satisfying (7.27) and (7.29) which are reproduced in Table 7.1 for a wide range of values of A. The prescribed error estimate e_a is 0·01 and e_d is 10^{-4}. The table shows the integration step sizes and the number of iterations required. A GE-225 digital computer took two to three minutes from data input to print final results for one solution.

By his second method Lee (1972) tabulated solutions of the equation

$$\frac{d^2 p}{d\eta^2} = -2\eta \left\{ 1 + \frac{1}{(a_0 p + b_0)^2} \right\} \frac{dp}{d\eta},$$

subject to the conditions

$$p = 1, \quad \eta = 0; \qquad p = 0, \quad \eta = \infty,$$

where $\eta \geqslant 0, 1 \geqslant p \geqslant 0, a_0, b_0$ are constants.

7.2.3. Two- and three-dimensional series solutions

Philip (1966) developed series solutions for radially symmetrical diffusion in infinite media surrounding cylinders and spheres. The equation is

$$\frac{\partial C}{\partial \tau} = \rho^{1-m} \frac{\partial}{\partial \rho} \left(D \rho^{m-1} \frac{\partial C}{\partial \rho} \right), \tag{7.35}$$

subject to the conditions

$$C = C_0, \qquad \rho > 1, \qquad \tau = 0, \tag{7.36}$$

$$C = C_1, \qquad \rho = 1, \qquad \tau \geqslant 0, \tag{7.37}$$

where

$$\rho = r/r_0, \qquad \tau = t/r_0^2 \tag{7.38}$$

and $r = r_0$ is the surface of the cylindrical $(m = 2)$ or spherical $(m = 3)$ cavity from which diffusant is supplied. We seek solutions of (7.35) of the form

$$\rho(C, \tau) = 1 + \phi_1 \tau^{\frac{1}{2}} + \phi_2 \tau + \phi_3 \tau^{\frac{3}{2}} + \dots, \tag{7.39}$$

where each ϕ is a function of C. Philip (1957) gave an *a priori* justification of this form of series in a related study of infiltration. By using

$$(\partial C/\partial \tau)_\rho (\partial \rho/\partial C)_\tau = -(\partial \rho/\partial \tau)_C \tag{7.40}$$

we convert (7.35) into

$$-\frac{1}{m} \frac{\partial \rho^m}{\partial \tau} = \frac{\partial}{\partial C} \left(D \rho^{m-1} \frac{\partial C}{\partial \rho} \right), \tag{7.41}$$

and integration with respect to C yields

$$-\frac{1}{m} \frac{\partial}{\partial \tau} \int_{C_0}^{C} \rho^m \, dC = D \rho^{m-1} \frac{\partial C}{\partial \rho}. \tag{7.42}$$

We have assumed that $\lim_{C \to C_0} D \rho^{m-1} \partial C/\partial \rho = 0$, i.e. the flux is zero at infinity for all finite τ. By substituting (7.39) in (7.42) and assuming that the expansion on each side is convergent we obtain the following set of relations on equating coefficients:

$$\int_{C_0}^{C} \phi_1 \, dC = -2(D/\phi_1'), \tag{7.43}$$

$$\int_{C_0}^{C} \phi_2 \, dC = \{D\phi_2'/(\phi_1')^2\} - (m-1) \int_{C_0}^{C} D \, dC, \tag{7.44}$$

$$\int_{C_0}^{C} \phi_3 \, dC = -\frac{2}{3}\left[\frac{D\{(\phi_2')^2 - \phi_1'\phi_3'\}}{(\phi_1')^3} - (m-1)\int_{C_0}^{C} D\phi_1 \, dC\right], \qquad (7.45)$$

and so on. Here ϕ' signifies differentiation with respect to C, and because of (7.37) we have the conditions

$$C = C_1, \qquad \phi_n = 0, \qquad n = 1, 2, 3, \dots. \qquad (7.46)$$

The first equation (7.43) is clearly of the same form as (7.25). Detailed comparison allowing for the different symbolism shows them to be identical. Thus, the leading term in the series for m-dimensional diffusion is the solution for one-dimensional diffusion. This is to be anticipated as long as the depth of penetration is small compared with r_0.

Once equation (7.43) has been solved, say by the methods of § 7.2.2 (i), (ii), (p. 107), then (7.44), (7.45) etc. may be solved in turn. They are all linear equations and are amenable to numerical solution. One workable procedure is analogous that that outlined in § 7.2.2 (ii) (p. 107). Philip (1969a) gives a sample set of moisture profiles in soil, computed from the first three terms of the series solutions for the cylindrical and spherical cases.

7.2.4. *Some special numerical solutions*

Wagner (1952) for exponential and Stokes (1952) for linear concentration dependent diffusion each transformed eqn (7.10), so that the relationship between D and C is removed from the differential equation and appears instead in one of the boundary conditions. Methods of numerical forward integration are used to build up families of solutions for different exponential and linear relationships.

(i) *Exponential diffusion coefficients. Infinite medium.* Wagner (1952) has given the following method of dealing with a diffusion coefficient which varies exponentially with concentration, according to the expression

$$D = D_a \exp[\beta\{C - \tfrac{1}{2}(C_1 + C_2)\}], \qquad (7.47)$$

where D_a is the diffusion coefficient for the average concentration $C_a = \tfrac{1}{2}(C_1 + C_2)$, and β is a constant given by

$$\beta = d \ln D/dC. \qquad (7.48)$$

Wagner deals first with the problem in an infinite medium for which the diffusion equation is

$$\frac{\partial C}{\partial t} = \frac{\partial}{\partial x}\left(D\frac{\partial C}{\partial x}\right) \qquad (7.49)$$

and the initial conditions are

$$C = C_1, \qquad x < 0, \qquad t = 0, \qquad (7.50)$$

$$C = C_2, \qquad x > 0, \qquad t = 0. \qquad (7.51)$$

For purposes of tabulating the final results it is convenient to take new variables

$$y = \tfrac{1}{2}x/(D_0 t)^{\frac{1}{2}}, \tag{7.52}$$

where D_0 is the diffusion coefficient for the concentration C_0 at $x = 0$ and

$$\psi = \frac{C - \tfrac{1}{2}(C_1 + C_2)}{\tfrac{1}{2}(C_2 - C_1)}, \tag{7.53}$$

so that

$$C = \tfrac{1}{2}(C_1 + C_2) + \tfrac{1}{2}(C_2 - C_1)\psi. \tag{7.54}$$

To compute numerical values of the function ψ, we introduce another dimensionless variable

$$\gamma = \beta(C - C_0), \tag{7.55}$$

whereupon (7.47) becomes

$$D = D_0 \, e^\gamma, \tag{7.56}$$

with D_0 as diffusion coefficient for $C = C_0$. Substitution of (7.52), (7.55), and (7.56) in (7.49) gives

$$\frac{d}{dy}\left(e^\gamma \frac{d\gamma}{dy}\right) + 2y\frac{d\gamma}{dy} = 0. \tag{7.57}$$

Introducing the auxiliary variable

$$u = e^\gamma \tag{7.58}$$

into (7.57), we have

$$u\frac{d^2u}{dy^2} + 2y\frac{du}{dy} = 0. \tag{7.59}$$

Solutions of (7.59) from $y = -\infty$ to $y = +\infty$ for the initial conditions

$$\gamma = 0, \quad u = 1, \quad y = 0, \tag{7.60}$$

were obtained for different values of the parameter g given by

$$g \equiv (d\gamma/dy)_{y=0} \equiv (du/dy)_{y=0} = 0{\cdot}2, 0{\cdot}4, \ldots, 2{\cdot}4. \tag{7.61}$$

For each value of g there are two limiting values of γ, γ_1 at $y = -\infty$ and γ_2 at $y = +\infty$, with opposite signs. In view of (7.55) these values are related to the initial concentrations C_1 and C_2 by

$$\gamma_1 = \beta(C_1 - C_0), \tag{7.62}$$

$$\gamma_2 = \beta(C_2 - C_0). \tag{7.63}$$

On subtracting corresponding sides of (7.62) and (7.63) we have

$$\gamma_2 - \gamma_1 = \beta(C_2 - C_1), \qquad (7.64)$$

which according to (7.48) is the natural logarithm of the ratio of the diffusion coefficients for the concentrations C_2 and C_1. Fig. 7.1 shows the relation between $\gamma_2 - \gamma_1$ and the parameter g.

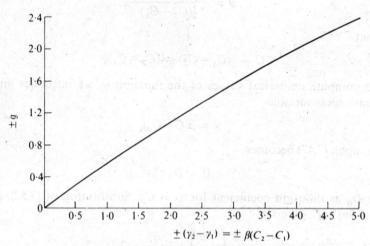

FIG. 7.1. Auxiliary parameter g as a function of $(\gamma_2 - \gamma_1) = \beta(C_2 - C_1)$.

Moreover, it follows from (7.62) and (7.63) that

$$C_0 = \tfrac{1}{2}(C_1 + C_2) - \tfrac{1}{2}(\gamma_1 + \gamma_2)/\beta. \qquad (7.65)$$

Substitution of (7.65) in (7.47) gives

$$D_0 = D_a \exp\{-\tfrac{1}{2}(\gamma_1 + \gamma_2)\}. \qquad (7.66)$$

To facilitate the computation of D_0, a graph of $\tfrac{1}{2}(\gamma_1 + \gamma_2)$ as a function of g is shown in Fig. 7.2.

According to (7.55) and (7.53) we obtain the values of ψ from the relationship

$$\psi = \frac{\gamma - \tfrac{1}{2}(\gamma_1 + \gamma_2)}{\tfrac{1}{2}(\gamma_2 - \gamma_1)}. \qquad (7.67)$$

Numerical values of the function $\psi(y, g)$ are compiled in Table 7.2. For negative values of the parameter g the relationship

$$\psi(y, -g) = -\psi(y, g) \qquad (7.68)$$

can be used and the sign of y must be reversed.

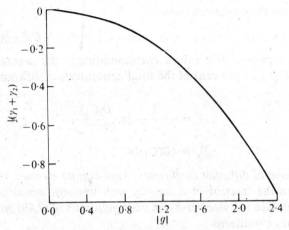

FIG. 7.2. $\frac{1}{2}(\gamma_1 + \gamma_2)$ as a function of g.

In view of the regularity of the differences between the ψ values for different g values but equal y values, it is believed that the error in ψ is in general not greater than one figure in the last decimal place, except for $g = -1.6$ to $g = -2.4$ for which the accuracy is less because of rather sudden changes in the values of the first derivative of ψ as a function of y.

In order to obtain the concentration distribution for a diffusion coefficient with given initial concentrations C_1 and C_2 and known values of D_a and β, we proceed as follows:

(1) read the value of the auxiliary parameter g for the known value of $(\gamma_2 - \gamma_1) = \beta(C_2 - C_1)$ from Fig. 7.1;
(2) read the value of $\frac{1}{2}(\gamma_1 + \gamma_2)$ for the auxiliary parameter g from Fig. 7.2;
(3) calculate the value of D_0 from (7.66);
(4) plot ψ for the auxiliary parameter g as a function of y with the aid of Table 7.2.

In view of (7.52) and (7.54) the plot of ψ versus y gives the concentration C at any point x at time t.

Instead of using Figs. 7.1 and 7.2, the values of g and $\frac{1}{2}(\gamma_1 + \gamma_2)$ may be calculated from the empirical interpolation formulae

$$g = 0.564(\gamma_2 - \gamma_1) - 5(\gamma_2 - \gamma_1)^3 10^{-3} + 64(\gamma_2 - \gamma_1)^5 10^{-5}, \tag{7.69}$$

$$\tfrac{1}{2}(\gamma_1 + \gamma_2) = -0.144g^2 - 0.0038g^4. \tag{7.70}$$

Thus far it has been assumed that the diffusion coefficient is an exponential function of the concentration in accordance with (7.47). If this assumption does not hold strictly, we may use the foregoing analysis as an approximation. Wagner (1952) recommends that the analysis be based on the diffusion coefficients for the concentrations

$$C_{-\frac{1}{4}} = \tfrac{1}{2}(C_1 - C_2) - \tfrac{1}{4}(C_2 - C_1) \quad \text{and} \quad C_{\frac{1}{4}} = \tfrac{1}{2}(C_1 + C_2) + \tfrac{1}{4}(C_2 - C_1)$$

as the most representative values, corresponding to the average concentration $\tfrac{1}{2}(C_1 + C_2) \pm 25$ per cent of the total concentration difference $(C_2 - C_1)$. Thus

$$\beta = \frac{2}{C_2 - C_1} \ln \frac{D(C_{\frac{1}{4}})}{D(C_{-\frac{1}{4}})}, \tag{7.71}$$

$$D_a = \{D(C_{\frac{1}{4}})D(C_{-\frac{1}{4}})\}^{\frac{1}{2}}. \tag{7.72}$$

(ii) *Exponential diffusion coefficients: semi-infinite medium.* If a substance diffuses from the interior of a sample, with uniform initial concentration C_0, to the surface (or vice versa) we need solutions of (7.49) with the initial and boundary conditions

$$C = C_0, \quad x > 0, \quad t = 0, \tag{7.73}$$

$$C = C_s, \quad x = 0, \quad t > 0, \tag{7.74}$$

where x is the distance from the surface and C_s denotes the surface concentration for $t > 0$.

In this case we consider an exponential dependence of the diffusion coefficient D on concentration given by

$$D = D_s \exp \{\beta(C - C_s)\}, \tag{7.75}$$

where D_s is the diffusion coefficient for the surface concentration C_s.

Upon introduction of (7.75) and the auxiliary variables

$$y_s = \tfrac{1}{2}x/(D_s t)^{\frac{1}{2}}, \tag{7.76}$$

$$c = (C - C_s)/(C_0 - C_s), \tag{7.77}$$

and

$$\gamma = \beta(C - C_s) \tag{7.78}$$

into (7.49) we obtain (7.57) with y_s instead of y. In view of (7.74), (7.76), (7.78) the integration is to be performed with $\gamma = 0$ at $y_s = 0$. Thus we can use the values of γ calculated above in § 7.2.4 (i) (p. 112). In view of (7.73), (7.77), and (7.78) we have

$$\gamma = \beta(C_0 - C_s) = r, \quad y_s = \infty, \tag{7.79}$$

$$c = \gamma/r. \tag{7.80}$$

Here r is the natural logarithm of the ratio of the diffusion coefficients for the concentrations C_0 and C_s. Fig. 7.3 shows the values of r, i.e. the limiting values of γ at $y = \infty$, as a function of the auxiliary parameter g used for the integration of (7.59). Numerical values of c are compiled in Table 7.3.

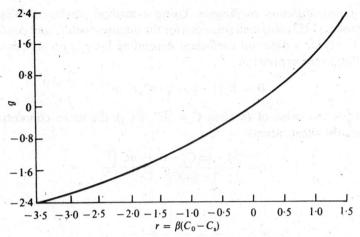

FIG. 7.3. Auxiliary parameter g as a function of $r = \beta(C_0 - C_s)$.

In order to obtain the concentration distribution for diffusion from the surface to the interior of a sample (or vice versa) one therefore proceeds as follows:

(1) read the value of the parameter g for the value of $r = \beta(C_0 - C_s)$ from Fig. 7.3;

(2) Plot c for this value of g as a function of y_s using Table 7.3.

The concentration C at any distance x and any time t follows, since from (7.77)

$$C = C_s + (C_0 - C_s)c, \qquad (7.81)$$

and y_s is given by (7.76)

Instead of using Fig. 7.3 one may calculate the values of g with the aid of the empirical interpolation formula

$$g = \frac{1 \cdot 128r}{1 - 0 \cdot 177r}. \qquad (7.82)$$

Using (7.76) and (7.81), the flux across the surface $x = 0$ becomes

$$-D_s(\partial C/\partial x)_{x=0} = -\tfrac{1}{2}(D_s/t)^{\frac{1}{2}}(C_0 - C_s)(dc/dy_s)_{y_s=0} \qquad (7.83)$$

and from (7.61), (7.80), and (7.82) it follows that

$$\left(\frac{dc}{dy_s}\right) = \frac{g}{r} = \frac{1 \cdot 128}{1 - 0 \cdot 177r}, \qquad y_s = 0. \qquad (7.84)$$

If β is positive (i.e. the diffusion coefficient increases with concentration) and if the substance diffuses from the surface into the interior of a sample with zero initial concentration, i.e. $C_0 - C_s < 0$, the value of g is negative.

(iii) *Linear diffusion coefficients*. Using a method similar to Wagner's, R. H. Stokes (1952) obtained solutions for the infinite medium and conditions (7.50), (7.51) for a diffusion coefficient depending linearly on concentration according to the expression

$$D = D_a\{1 + \tfrac{1}{2}a(C_1 + C_2) - aC\}. \tag{7.85}$$

Here D_a is the value of D when $C = \tfrac{1}{2}(C_1 + C_2)$, the mean concentration. By using the substitutions

$$v = \left\{\frac{1 + \tfrac{1}{2}a(C_1 + C_2) - aC}{1 + \tfrac{1}{2}a(C_1 - C_2)}\right\}^2 \tag{7.86}$$

and

$$w = \frac{x}{2\{1 + \tfrac{1}{2}a(C_1 - C_2)\}^{\frac{1}{2}}(D_a t)^{\frac{1}{2}}}, \tag{7.87}$$

the diffusion equation (7.49) reduces to

$$\frac{d^2 v}{dw^2} = -\frac{2w}{v^{\frac{1}{2}}}\frac{dv}{dw}, \tag{7.88}$$

with the boundary conditions

$$v = 1, \qquad w = +\infty, \tag{7.89}$$

$$v = \left\{\frac{1 - \tfrac{1}{2}a(C_1 - C_2)}{1 + \tfrac{1}{2}a(C_1 - C_2)}\right\}^2, \qquad w = -\infty. \tag{7.90}$$

For convenience, we denote by b^2 the value of v at $w = -\infty$. From (7.85) we see that b is the ratio of the diffusion coefficient at C_1 to that at C_2. Stokes obtained numerical solutions of (7.88) by starting at $w = 3$, $v = 1$ (the value $w = 3$ is suggested by the known solution for a constant diffusion coefficient) and a chosen *small* value of dv/dw, say less than 0·01. The integration proceeds in the direction of decreasing w till a value of v is reached which is constant to within the accuracy of working. This value of v is b^2, and is reached at some negative value of w between -2 and -5. Different values of the initial gradient dv/dw lead to different values of b^2 which are known only when the solutions are completed. This family of solutions gives, for different bs, values of v and dv/dw at closely spaced intervals in w (the usual interval used by Stokes was 0·1 in w). From these solutions, values of concentration and concentration gradient are readily obtained. They are conveniently expressed in terms of a new independent variable y_a defined by

$$y_a = \frac{x}{2(D_a t)^{\frac{1}{2}}} = w\sqrt{\left(\frac{2}{1+b}\right)}, \tag{7.91}$$

so that we have

$$\frac{dC}{dy_a} = -\frac{(C_1-C_2)\{\frac{1}{2}(1+b)\}^{\frac{1}{2}}}{2(1-b)} \frac{1}{v^{\frac{1}{2}}} \frac{dv}{dw},$$ (7.92)

$$C = C_2 + (C_1 - C_2)(1-v^{\frac{1}{2}})/(1-b).$$ (7.93)

Tables 7.4 and 7.5 show $-(dC/dy_a)/(C_1-C_2)$ and $(C-C_2)/(C_1-C_2)$ as functions of y_a for different values of b. All the solutions shown correspond to $b < 1$, i.e to diffusion coefficients which decrease as the concentration increases. The corresponding solutions for $b > 1$ may be obtained by reversing the sign of y_a; thus, for example, the solution for $b = 0.1407$ becomes the solution for $b = 1/0.1407 = 7.106$ when the sign of y_a is changed. Solutions for intermediate values of b can be obtained by interpolation.

(iv) *Diffusion coefficient directly proportional to concentration.* The problem treated by Wagner (1950) of a diffusion coefficient given by

$$D = D_0 C/C_0,$$ (7.94)

has some features of particular interest. Considering first the semi-infinite medium in which

$$C = C_0, \qquad x = 0, \qquad t > 0,$$ (7.95)

$$C = 0, \qquad x > 0, \qquad t = 0,$$ (7.96)

and introducing the variables

$$y = \tfrac{1}{2}x/(D_0 t)^{\frac{1}{2}},$$ (7.97)

$$c = C/C_0,$$ (7.98)

the diffusion equation

$$\frac{\partial C}{\partial t} = \frac{\partial}{\partial x}\left(D_0 \frac{C}{C_0}\frac{\partial C}{\partial x}\right),$$ (7.99)

becomes

$$c\frac{d^2 c}{dy^2} + \left(\frac{dc}{dy}\right)^2 + 2y\frac{dc}{dy} = 0.$$ (7.100)

Further simplification is achieved by substituting

$$v = c^2,$$ (7.101)

when (7.100) becomes

$$\frac{d^2 v}{dy^2} = -\frac{2y}{v^{\frac{1}{2}}}\frac{dv}{dy},$$ (7.102)

with boundary conditions

$$v = 1, \qquad y = 0, \tag{7.103}$$

$$v = 0, \qquad y = \infty. \tag{7.104}$$

Using numerical methods of integration and disregarding temporarily the second condition (7.104), a family of solutions of (7.102) can be obtained for arbitrarily chosen values of $(dv/dy)_{y=0} = v'_0$. Clearly if the condition (7.104) at $y = \infty$ is to be satisfied, v'_0 must be negative. Then from (7.102) the second derivative of v with respect to y is positive. If $|v'_0| \ll 1$, the curve of v against y tends to a constant value at large y as shown in Fig. 7.4, curve A.

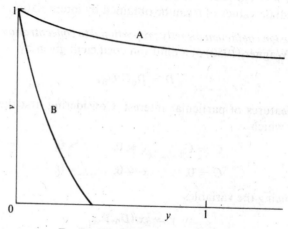

FIG. 7.4. Schematic solutions of (7.102).

A solution of this kind does not satisfy (7.104), however. On the other hand, if $|v'_0| \gg 1$, curves which approach the axis $v = 0$ with finite slopes as shown schematically by curve B in Fig. 7.6 are obtained. Tentatively we assume that a special value of $|v'_0|$ yields a solution for which v vanishes at a certain abscissa $y = y_0$ and satisfies the additional condition

$$\frac{dv}{dy} = 0, \qquad v = 0, \qquad y = y_0. \tag{7.105}$$

This is the condition that the rate of transport of diffusing substance shall be zero when $C = 0$, since by eqns (7.97) (7.98), and (7.101) we have for the rate of transport

$$D_0 \frac{C}{C_0} \frac{\partial C}{\partial x} = \frac{D_0^{\frac{1}{2}} C_0}{4t^{\frac{1}{2}}} \frac{dv}{dy}. \tag{7.106}$$

A function satisfying (7.102), (7.103), and (7.105) can then be combined with the trivial solution $v = 0$, $y \geqslant y_0$, so that a solution for the whole range $y \geqslant 0$ results.

With the tentative assumption that condition (7.105) is satisfied at a finite value $y = y_0$, the factor y on the right side of (7.102) can be replaced approximately by y_0 if $v \ll 1$. Thus

$$\frac{d^2v}{dy^2} = -\frac{2y_0}{v^{\frac{1}{2}}}\frac{dv}{dy}, \tag{7.107}$$

if $v \ll 1$, with the boundary conditions

$$v = 0, \qquad y = y_0, \tag{7.108}$$

$$dv/dy = 0, \qquad y = y_0. \tag{7.109}$$

Integration yields

$$v = \{2y_0(y_0 - y)\}^2, \qquad v \ll 1. \tag{7.110}$$

Consequently, in the vicinity of $y = y_0$, where $v \ll 1$ and hence $c \ll 1$ and $C \ll C_0$, the function $v(y)$ is represented approximately by a parabola and the concentration ratio $c = v^{\frac{1}{2}}$ by a straight line.

To obtain a solution of (7.102) for the whole range $0 < y < y_0$, numerical integration is required. Since the boundary conditions (7.103) and (7.105) refer to different values of y it is useful to make the substitutions

$$\zeta = v/y_0^4, \tag{7.111}$$

$$\xi = 1 - (y/y_0), \tag{7.112}$$

whereupon eqns (7.102), (7.108), and (7.109) become

$$\frac{d^2\zeta}{d\xi^2} = \frac{2}{\zeta^{\frac{1}{2}}}(1-\xi)\frac{d\zeta}{d\xi}, \tag{7.113}$$

$$\zeta = 0, \qquad \xi = 0, \tag{7.114}$$

$$\frac{d\zeta}{d\xi} = 0, \qquad \xi = 0. \tag{7.115}$$

In order to integrate (7.113) from $\xi = 0$, the second derivative $d^2\zeta/d\xi^2$ at $\xi = 0$ must be known. This cannot be obtained directly from (7.113) because for $\xi = 0$ the right side is indeterminate. By substituting (7.111) and (7.112) in (7.110) and differentiating twice, we obtain

$$d^2\zeta/d\xi^2 = 8, \qquad \xi = 0. \tag{7.116}$$

On carrying out the numerical integration of (7.113) starting with conditions (7.114), (7.115), (7.116) at $\xi = 0$, we obtain $\zeta = 2\cdot34$ when $\xi = 1$, i.e. when

$y = 0$. Then it follows from (7.111) and (7.103) that

$$y = (\zeta_{\xi=1})^{-\frac{1}{4}} = 0.81. \tag{7.117}$$

Transforming (7.112) and substituting (7.117) we have

$$y = y_0(1-\xi) = 0.81(1-\xi). \tag{7.118}$$

Finally it follows from (7.98), (7.101), (7.111), and (7.117) that

$$C/C_0 = c = v^{\frac{1}{4}} = \zeta^{\frac{1}{4}} y_0^2 = (\zeta/\zeta_{y=1})^{\frac{1}{4}} \tag{7.119}$$

A graph of C/C_0 against $\frac{1}{2}x/(D_0 t)^{\frac{1}{2}}$ is shown as curve I in Fig. 7.5.

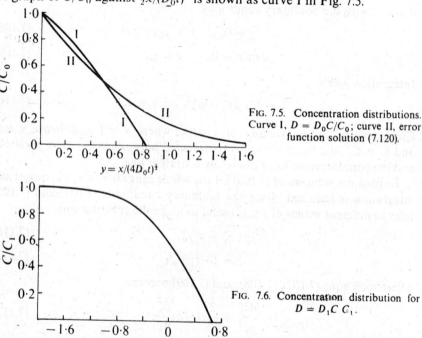

FIG. 7.5. Concentration distributions. Curve I, $D = D_0 C/C_0$; curve II, error function solution (7.120).

FIG. 7.6. Concentration distribution for $D = D_1 C C_1$.

In Fig. 7.5 curve II shows the corresponding graph for a constant diffusion coefficient, calculated from

$$C/C_0 = \mathrm{erfc}\left\{\frac{1}{2}x/(D_0 t)^{\frac{1}{2}}\right\}. \tag{7.120}$$

Curve II approaches the abscissa asymptotically, whereas curve I reaches the abscissa at the finite value $y = 0.81$ with a finite slope. Thus, in this special case, an advancing velocity of the diffusing component is strictly definable,[†] and from (7.97) and (7.117) it is given by

$$(\mathrm{d}x/\mathrm{d}t)_{c=0} = y_0 D_0^{\frac{1}{2}}/t^{\frac{1}{2}} = 2y_0^2 D_0/x = 1.31 D_0/x. \tag{7.121}$$

† Shampine (1973b) discusses solutions $c(\eta) = (1-\alpha\eta)^{1/\beta}$, $0 \leqslant \eta \leqslant 1/\alpha$, $c(\eta) = 0$, $\eta > 1/\alpha$, produced by $D(c) = (\beta c^\beta/2\alpha^2)\{1 - c^\beta/(1+\beta)\}$, where $\eta = x/t^{\frac{1}{2}}$ and α, β are positive parameters.

Furthermore, (7.113) may be integrated up to $\xi = +\infty$, i.e. $x = -\infty$, when the solution approaches a limiting value ζ_1, where $\zeta_1 = 6\cdot82$. We then have the solution of the problem in an infinite medium for the initial conditions

$$C = C_1, \qquad x < 0, \tag{7.122}$$

$$C = C_2, \qquad x > 0, \tag{7.123}$$

where C_1 and C_2 are both constant, and the diffusion coefficient D is given by

$$D = D_1 C/C_1,$$

where D_1 is the diffusion coefficient when $C = C_1$. Then from equations (7.98), (7.101), (7.111), and (7.112) we find

$$C/C_1 = (\zeta/\zeta_{\xi = \infty})^{\frac{1}{2}} = (\zeta/\zeta_1)^{\frac{1}{2}} \tag{7.124}$$

This concentration ratio can be related to the non-dimensional variable $\frac{1}{2}x/(D_1 t)^{\frac{1}{2}}$, which through (7.97) is given by

$$\frac{x}{2(D_1 t)^{\frac{1}{2}}} = y\left(\frac{D_0}{D_1}\right)^{\frac{1}{2}} = y\left(\frac{\zeta_{\xi=1}}{\zeta_{\xi=\infty}}\right)^{\frac{1}{4}} = 0\cdot765y, \tag{7.125}$$

where y is calculated from ξ using (7.118). A graph of C/C_1 as a function of $\frac{1}{2}x/(D_1 t)^{\frac{1}{2}}$ is shown in Fig. 7.6. The concentration at $x = 0$ is $0\cdot59C_1$ instead of $0\cdot50C_1$ which is the value obtained when the diffusion coefficient is constant.

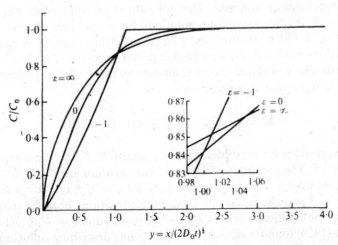

FIG. 7.7. Graph of concentration ratio C/C_0 versus dimensionless variable $y = x/(2D_0 t)^{\frac{1}{2}}$ for desorption problem with diffusion coefficient directly proportional to concentration. $D = D_0 C/C_0 (\varepsilon = \infty)$, constant ($\varepsilon = 0$), and directly proportional to concentration difference. $D = D_0(1 - C/C_0)(\varepsilon = -1)$.

Wilkins (1963) obtained a solution in the form of the very convenient power series

$$u = \xi - \frac{\xi^2}{4} + \frac{\xi^3}{72} + \frac{\xi^4}{576} + \frac{11\xi^5}{86\,400} - \frac{\xi^6}{115\,200} - \frac{1573\xi^7}{304\,819\,200}$$

$$- \frac{4877\xi^8}{4\,877\,107\,200} - \frac{4469\xi^9}{48\,771\,072\,000} + \frac{311\,543\xi^{10}}{43\,893\,964\,800\,000} + \dots \quad (7.126)$$

in which $\xi = 1 - y/y_0$ and $u = c/y_0^2$. It follows that $y_0 = 1\cdot1428$, $\xi = 1 - 0\cdot87506y$, and $c = 1\cdot3059u$.[†] The series solution clearly cannot be used to find the limiting value of c at $x = -\infty$ in the infinite medium. Wilkins, however, refers to values of u tabulated by Christian (1961) in relation to the Blasius problem in fluid flow which are in good agreement with his series in the range over which it is valid.

The previously-neglected desorption problem was also discussed by Wilkins (1963), who showed that the diffusion system could be transformed into the form

$$\frac{d^3f}{d\eta^3} + f\frac{d^2f}{d\eta^2} = 0,$$

$$\eta = 0, \quad f = df/d\eta = 0; \quad \eta = \infty, \quad df/d\eta = 2,$$

and that $c = \frac{1}{2}df/d\eta$ and $y = 2^{-\frac{1}{2}}f$. The quantities f, f', and f'' have been tabulated as functions of η by Howarth (1938), and were used by Wilkins to plot the desorption solution. This solution is reproduced in Fig. 7.7, together with the corresponding solutions for a constant D and for $D = D_0(1 - C/C_0)$. The inset shows on a larger scale the region of intersection of the solutions analysed also by a further power series (see § 9.3.2, p. 176).

Wilkins also calculated the total amount of material that enters the half-space $x > 0$ per unit area up to time t, i.e.

$$M = \int_0^\infty C\,dx = 2C_0(D_a t)^{\frac{1}{2}} \int_0^\infty c(y)\,dy.$$

His expressions for the dimensionless quantity $M/C_0(D_a t)^{\frac{1}{2}}$ are reproduced in Table 7.6. The conditions for the infinite medium are $t = 0$, $C = C_0$ for $x < 0$, $C = 0$ for $x > 0$; for sorption they are $C = C_0$, $x = 0$, $t \geqslant 0$; $C = 0$, $x \geqslant 0$, $t = 0$ and the usual corresponding conditions hold for desorption. The diffusion coefficient is $D = D_a(1 + \frac{1}{2}aC_0 - aC)$ and $\varepsilon = -aC_0/(1 + \frac{1}{2}aC_0)$.

Pattle (1959) found a class of exact solutions describing diffusion from a point source when D has the form

$$D = D_0(C/C_0)^b, \quad b > 0.$$

[†] Heaslet and Alksne (1961) give a series solution for $D = c^n$.

The equation

$$\frac{\partial C}{\partial t} = s^{1-n}\frac{\partial}{\partial s}\left(s^{n-1}D\frac{\partial C}{\partial s}\right)$$

has the solution

$$C = C_0\left(\frac{t}{t_0}\right)^{-n/(nb+2)}\left(1-\frac{s^2}{s_1^2}\right), \qquad t > 0, \quad s^2 \leqslant s_1^2,$$

$$C = 0, s^2 \geqslant s_1^2. \tag{7.127}$$

These equations represent an expanding diffusion zone with a definite edge, of radius s_1 given by

$$s_1 = s_0(t/t_0)^{1/(nb+2)}.$$

When a quantity Q is liberated at time $t = 0$ at a point in an infinite line, plane, or volume; the parameters s_0, t_0 are determined by

$$s_0^n = \frac{Q}{C_0\pi^{n/2}}\left\{\frac{\Gamma(b^{-1}+1+\tfrac{1}{2}n)}{\Gamma(b^{-1}+1)}\right\}, \tag{7.128a}$$

$$t_0 = \frac{s_0^2 b}{2D_0(nb+2)}. \tag{7.128b}$$

When $b > 1$, the concentration gradient at the boundary of the region is infinite; when $b = 1$, it is finite; and when $b < 1$, it is zero.

7.2.5. Analytical iterative methods

Kidder (1957) developed a perturbation method for desorption when $D(C)' = aC$ with a constant. Successive coefficients in a series solution are the analytical solutions of increasingly complicated ordinary differential equations. The method has been extended by Knight (1973a) to include power-law and exponential relationships between D and C for both sorption and desorption problems.

Parlange (1971a, b) described a method of successive approximations based on an inverted form of the diffusion equation in which the space coordinate x is taken as the dependent variable. This has been improved by Philip (1973b) and Knight and Philip (1973) using a formulation in terms of an integral equation which identifies Parlange's iterative scheme as equivalent to Picard's method of successive approximations (1893).

7.2.6. Exact formal solutions

By exact formal solutions we mean those that can be expressed without approximation in terms of the functions of mathematical analysis. Unfortunately, they are mostly of implicit or parametric form rather than explicit relationships between concentration and the space and time variables.

This means that they do not reveal general features, which is one of the attractions of analytical solutions, nor are they particularly simple to evaluate numerically. The solutions explored by Philip (1960) come nearest to overcoming these disadvantages (§ 7.2.6 (i)).

Fujita (1952, 1954) obtained solutions in parametric form for diffusion into a semi-infinite medium at the surface of which the concentration is held constant. He considered three expressions for $D(C)$ which are (1) $D_0/(1-\lambda C)$; (2) $D_0/(1-\lambda C)^2$; (3) $D_0/(1+2aC+bC^2)$. The last form clearly includes the first two. In all three, D_0, λ, a, b are constants. Some solutions evaluated by Fujita's methods are presented in Figs. 9.12 and 9.13. Figs. 9.20, 9.21, and 9.22 show an example of the behaviour associated with a diffusion coefficient which passes through a maximum value as C is increased.

Some of Philip's (1968) solutions in connection with swelling soils can be related to Fujita's solutions. Storm (1951) proposed a method which can provide an exact solution for diffusion into a semi-infinite medium, with a specified flux on the surface, when the diffusion coefficient has the same form as that of Fujita's case (2) above. Storm reduced the problem to the solution of a linear equation with linear boundary conditions but again his solution is in parametric form. Knight (1973b) extended Storm's method to more general problems including that of Fujita (1952) and gave explicit exact solutions in some cases including the problems of the instantaneous source and of redistribution in the finite region.

Lee (1971b) considered a time-lag, concentration-dependent diffusion coefficient of the form $D_0\{C(x, t-h)\}$.

(i) *A general class of exact solutions.* Philip (1960) showed that there exists an indefinitely large class of functional forms of $D(C)$ for which formal solutions of the diffusion equation may be found analytically. He starts from a postulated solution and derives the $D(C)$ relationship which produces it.

In this section we adopt the nomenclature of Philip's paper (1960) for ease of reference to the functions he develops and tabulates. He is concerned first with the problem

$$\frac{\partial \theta}{\partial t} = \frac{\partial}{\partial x}\left(D\frac{\partial \theta}{\partial x}\right) \qquad (7.129)$$

subject to the conditions

$$\theta = 0, \qquad x > 0, \qquad t = 0, \qquad (7.130)$$

$$\theta = 1, \qquad x = 0, \qquad t > 0. \qquad (7.131)$$

Substitution of the variable

$$\phi = x/t^{\frac{1}{2}} \qquad (7.132)$$

allows (7.129) to be written as

$$D = -\frac{1}{2}\frac{d\phi}{d\theta}\int_0^\theta \phi \, d\theta. \tag{7.133}$$

It follows that the solution of (7.129) subject to (7.130) and (7.131) exists so long as $D(\theta)$ is of the form

$$D = -\frac{1}{2}\frac{dF}{d\theta}\int_0^\theta F \, d\theta, \tag{7.134}$$

where F is any single-valued function of θ in the interval $0 < \theta \leqslant 1$, which satisfies certain conditions discussed below.

First the function F must satisfy the boundary conditions imposed on ϕ, and so

$$F(1) = 0. \tag{7.135}$$

Other conditions are imposed by the requirement that D exists. In fact, it is necessary that $\int_0^\theta F \, d\theta$ and $dF/d\theta$ exist throughout the range $0 \leqslant \theta \leqslant 1$. We note, however, that sometimes a finite number of discontinuities in D may be allowable or D may be permitted to be infinite at a finite number of points in the range $0 \leqslant \theta \leqslant 1$, in which cases $dF/d\theta$ may either not exist or may be infinite at the appropriate finite number of points in the θ-range.

Furthermore, we do not wish D to become negative. The flux of diffusant in the direction of x positive is $-D \, \partial\theta/\partial x = -t^{-\frac{1}{2}}D \, d\theta/d\phi$. For the conditions (7.130) and (7.131) we require $D \geqslant 0$ and the flux to be non-negative, i.e. $d\theta/d\phi \leqslant 0$. This calls for the further condition

$$dF/d\theta \leqslant 0, \qquad 0 \leqslant \theta \leqslant 1. \tag{7.136}$$

Philip (1960) gives a short table of solutions of (7.129), with their associated forms of D, obtained simply by selecting F functions satisfying the above conditions and evaluating (7.134). Unfortunately, all of them have the property that $\lim_{\theta \to 0} D(\theta)$ is either zero or infinite. Whilst diffusion coefficients which are zero at zero concentration are sometimes of interest it would seriously limit the importance and generality of Philip's method if all his solutions turned out to have this property. He examines the difficulty in the following way.

The well-known solution (see eqn (2.45)) is equivalent in the present nomenclature to the statement that for

$$F = 2D_0^{\frac{1}{2}} \text{ inverfc } \theta, D = D_0 = \text{constant.} \tag{7.137}$$

Here inverfc denotes the inverse of erfc defined and examined by Philip (1960), who therefore puts

$$F = 2D_0^{\frac{1}{2}} \text{ inverfc } \theta + f(\theta), D_0 \text{ finite, non-zero} \tag{7.138}$$

in (7.134) and obtains

$$D = D_0 + \tfrac{1}{2}D_0^{\frac{1}{2}}\left(\frac{2}{B}\int_0^\theta f\,\mathrm{d}\theta - B\frac{\mathrm{d}f}{\mathrm{d}\theta}\right) - \frac{1}{2}\frac{\mathrm{d}f}{\mathrm{d}\theta}\int_0^\theta f\,\mathrm{d}\theta, \qquad (7.139)$$

where

$$B(\theta) = 2\pi^{-\frac{1}{2}}\exp\left\{-(\mathrm{inverfc}\,\theta)^2\right\}. \qquad (7.140)$$

The properties of inverfc θ and $B(\theta)$ are treated in detail by Philip (1960), and the functions are tabulated. In obtaining (7.139) we used the relationships

$$\frac{\mathrm{d}}{\mathrm{d}\theta}(\mathrm{inverfc}\,\theta) = -\frac{1}{B}; \qquad \int_0^\theta \mathrm{inverfc}\,\theta\,\mathrm{d}\theta = \tfrac{1}{2}B. \qquad (7.141)$$

We are concerned with the conditions under which $D(0)$ is finite and non-zero. A sufficient condition is seen from (7.139) to be that both

$$\lim_{\theta\to 0}\left(\frac{1}{B}\int_0^\theta f\,\mathrm{d}\theta\right) \quad \text{and} \quad \lim_{\theta\to 0}\left(B\frac{\mathrm{d}f}{\mathrm{d}\theta}\right)$$

be finite (including zero). The zero condition ensures that $D(0) = D_0$. If both the above limits are zero it follows that the inverfc term in (7.138) accounts completely for the finite, non-zero value of $D(0)$ in (7.139). It is not difficult to show that the single condition

$$\lim_{\theta\to 0}\frac{f}{\mathrm{inverfc}\,\theta} = 0 \qquad (7.142)$$

ensures that both the limits referred to above are zero. By using the relation (Philip 1960)

$$\lim_{\theta\to 0}\frac{\mathrm{inverfc}\,\theta}{(-\ln\theta)^{\frac{1}{2}}} = 1, \qquad (7.143)$$

we find that (7.142) is equivalent to

$$\lim_{\theta\to 0}\frac{f}{(-\ln\theta)^{\frac{1}{2}}} = 0. \qquad (7.144)$$

Thus f can become infinite at $\theta = 0$ but must approach infinity more slowly than $(-\ln\theta)^{\frac{1}{2}}$.

We conclude that there exists a formal solution of (7.129) subject to (7.130) and (7.131) and for which $D(0) = D_0$ is finite and non-zero corresponding to every $D(\theta)$ of the form (7.139) for which $2D_0^{\frac{1}{2}}\,\mathrm{inverfc}\,\theta + f$ satisfies the appropriate conditions on F discussed above and satisfies (7.142).

A slight extension is that for

$$F = 2D_0^{\frac{1}{2}} \operatorname{inverfc} \theta \pm f, \quad (D_0 \text{ finite, non-zero})$$

$$D = D_0 \pm \frac{1}{2} D_0^{\frac{1}{2}} \left(\frac{2}{B} \int_0^{\theta} f \, d\theta - B \frac{df}{d\theta} \right) - \frac{1}{2} \frac{df}{d\theta} \int_0^{\theta} f \, d\theta,$$

but now we require the stronger condition

$$|df/d\theta| \leqslant 2D_0^{\frac{1}{2}}/B.$$

We finally note that the condition (7.135) implies

$$f(1) = 0.$$

There is little difficulty in generating functions f which satisfy the requirements. Philip (1960) presents a few typical solutions which are reproduced in Table 7.7. He also gives corresponding solutions for the infinite medium. Unfortunately, it still seems to be true (Philip 1969a) that the functional pairs most readily generated by this promising method are rather complicated in form and are not particularly well adapted to fitting experimental data on $D(\theta)$.

7.3. Goodman's integral method

Later in § 13.6.2 (p. 312) we obtain approximate analytical solutions by an 'integral method'. Goodman (1964) has applied his method to concentration-dependent and other non-linear systems. The technique is described in detail in § 13.5.2 with reference to problems in which a physically defined boundary surface moves through the medium. The application to concentration-dependent diffusion is so similar that no details are given in this chapter. It suffices to point out that an artificial boundary at what is called the 'penetration distance' $\delta(t)$, is introduced and thereafter takes the place of the physical boundary of § 13.5.2. Its properties are such that for $x > \delta(t)$ the medium is, to the accuracy of working, at its initial concentration. A full account is given by Goodman (1964), who also discusses several ways of improving the accuracy of a solution obtained by the integral method. One of these is the method of moments, which we now describe.

7.4. Methods of moments

If we substitute some approximate solution c_n into the one-dimensional diffusion equation we are left with a residual ε_n, say, such that,

$$\frac{\partial c_n}{\partial T} - \frac{\partial}{\partial X} \left\{ F(c_n) \frac{\partial c_n}{\partial X} \right\} = \varepsilon_n. \tag{7.145}$$

We seek a solution which makes ε_n small in some sense. In the version of the method of moments which we shall describe we multiply by a weighting factor X^n and average over the total range in X. On setting this average residual to zero we obtain

$$\int_0^1 \left[\frac{\partial c_n}{\partial T} - \frac{\partial}{\partial X} \left\{ F(c_n) \frac{\partial c_n}{\partial X} \right\} \right] X^n \, dX = 0, \qquad n = 0, 1, 2, \ldots. \qquad (7.146)$$

We choose c_n to satisfy the boundary conditions and to contain k unknown parameters. By using k different weighting factors we have enough equations of type (7.146) to determine the k parameters. Clearly, the case $n = 0$ is identical with the Goodman's integral method. Fujita (1951) inspired by the work of Yamada (1947) applied the method of moments to a sorption problem in a plane sheet of thickness $2l$. His equation is (7.145) with $\varepsilon_n = 0$ and the non-dimensional variables are defined as $C/C_0 = c, x/l = X, D_0 t/l^2 = T$, $D(C)/D(0) = F(c)$, where $D = D_0$ when $C = C_0$. The problem is specified by the conditions

$$c = 0, \qquad 0 < X < 1, \qquad T = 0, \qquad (7.147)$$

$$c = 1, \qquad X = 1 \qquad T > 0, \qquad (7.148)$$

$$\partial c/\partial X = 0, \qquad X = 0, \qquad T > 0. \qquad (7.149)$$

Fujita uses the zero and first moments only and the problem is reduced to that of finding $c(X, T)$ satisfying the two moment equations

$$\int_0^1 \left[\frac{\partial c}{\partial T} - \frac{\partial}{\partial X} \left\{ F(c) \frac{\partial c}{\partial X} \right\} \right] dX = 0, \qquad (7.150)$$

and

$$\int_0^1 \left[\frac{\partial c}{\partial T} - \frac{\partial}{\partial X} \left\{ F(c) \frac{\partial c}{\partial X} \right\} \right] X \, dX = 0, \qquad (7.151)$$

together with the conditions (7.147) to (7.149). We proceed by considering that in the early stages of diffusion the concentration–distance curve may be represented approximately by a curved portion near the surface of the sheet, followed by a horizontal part coinciding with the x-axis. Strictly, according to the diffusion equation, the concentration becomes finite, though it may be small, everywhere in the sheet at the instant diffusion commences. The region over which the concentration may be assumed zero depends, of course, on the accuracy of working. For a prescribed accuracy, we denote by x_0 the point at which the concentration becomes zero and clearly x_0 is a function of time. Thus $x_0 = l$ when $t = 0$ and $x_0 = 0$ when $t = t_1$, say. The time t_1 is that at which the concentration first becomes finite at the centre of the sheet to the prescribed accuracy of working. Denoting by X_0 the value of X

corresponding to x_0, i.e. $X_0 = x_0/l$, we assume that in the region $X_0 \leqslant X \leqslant 1$ the concentration is given by a cubic expression of the form

$$c(X, T) = B(T)\{X - X_0(T)\}^2 + E(T)\{X - X_0(T)\}^3, \qquad (7.152)$$

where $B(T)$, $E(T)$, and $X_0(T)$ are functions of T to be determined. We have furthermore

$$c = 0, \qquad 0 \leqslant X \leqslant X_0. \qquad (7.153)$$

The condition (7.149) is satisfied by (7.152). Also (7.147) may be written as

$$X_0 = 1, \qquad T = 0. \qquad (7.154)$$

Introducing (7.148) into (7.152) and putting

$$B(1 - X_0)^2 = U, \qquad E(1 - X_0)^3 = V, \qquad (7.155)$$

we obtain the relation between U and V,

$$U + V = 1. \qquad (7.156)$$

Inserting (7.152) and (7.153) into (7.150), integrating, and using (7.148) we obtain

$$\frac{d}{dT}\{(1 - X_0)(\tfrac{1}{3}U + \tfrac{1}{4}V)\} = \frac{(2U + 3V)}{1 - X_0}F(1), \qquad (7.157)$$

which, on eliminating V by eqn (7.156), becomes

$$\frac{d}{dT}\{(1 - X_0)(1 + \tfrac{1}{3}U)\} = \frac{12}{1 - X_0}(1 - \tfrac{1}{3}U)F(1). \qquad (7.158)$$

Similarly from the first-moment equation (7.151) we find

$$\frac{d}{dT}\{(1 - X_0)^2(1 + \tfrac{2}{3}U)\} = 20G(1), \qquad (7.159)$$

where

$$G(c) = \int_0^c F(c')\,dc'. \qquad (7.160)$$

Here $F(1)$ and $G(1)$ denote the values of F and G when $c = 1$. On integrating, (7.159) becomes

$$(U + \tfrac{3}{2})\xi^2 = \alpha T, \qquad (7.161)$$

where

$$\alpha = 30G(1), \qquad \xi = 1 - X_0, \qquad (7.162)$$

and the integration constant has been determined by (7.154). Inserting

(7.161) into (7.158) we find

$$\frac{d}{dT}\left(\tfrac{1}{2}\xi + \frac{\alpha T}{3\xi}\right) = \frac{12}{\xi}\left(\frac{3}{2} - \frac{\alpha T}{3\xi^2}\right)F(1). \tag{7.163}$$

The solution of (7.163) which satisfies the condition $\xi = 0$, $T = 0$, in accordance with (7.162) and (7.154), is

$$\xi = \sqrt{(T/\beta)}, \tag{7.164}$$

where β is a constant determined from the quadratic equation

$$24\alpha\beta^2 F(1) + \{\alpha - 108F(1)\}\beta + \tfrac{3}{2} = 0. \tag{7.165}$$

Clearly for (7.164) to have a physical meaning β must be positive. The question as to which of the two possible roots of (7.165) should be taken if both are positive is decided in Fujita's treatment by considering the special case of a constant diffusion coefficient. When the solution (7.170) below, derived by the method of moments, is evaluated for a constant diffusion coefficient it is found that better agreement with the formal solution is obtained by taking the larger of the two roots (actually $\beta = \tfrac{1}{12}$) of (7.165). When the diffusion coefficient is concentration-dependent we take that root of (7.165) which tends to the value $\tfrac{1}{12}$ as the range in the diffusion coefficient is decreased and we approach a constant coefficient. Denoting by T_1 the time at which $X_0 = 0$, i.e. at time T_1 the advancing front of the diffusion reaches the centre of the sheet, we have $\xi = 1$, $T = T_1$ and hence from (7.164)

$$\beta = T_1. \tag{7.166}$$

Inserting (7.164) into (7.161) yields

$$U = \alpha\beta - \tfrac{3}{2}, \tag{7.167}$$

and then

$$V = -\alpha\beta + \tfrac{5}{2}. \tag{7.168}$$

Thus both U and V are found to be constants. The expression for X_0 is

$$X_0 = 1 - \sqrt{(T/\beta)}. \tag{7.169}$$

Substituting these equations into (7.152) and remembering (7.155) we find

$$c(X, T) = (\alpha\beta - \tfrac{3}{2})\{X - 1 + \sqrt{(T/\beta)}\}^2(\beta/T)$$
$$- (\alpha\beta - \tfrac{5}{2})\{X - 1 + \sqrt{(T/\beta)}\}^3(\beta/T)^{\frac{3}{2}}, \quad 1 - \sqrt{(T/\beta)} \leqslant X \leqslant 1, \tag{7.170}$$

and

$$c(X, T) = 0, \quad 0 \leqslant X \leqslant 1 - \sqrt{(T/\beta)}. \tag{7.171}$$

From (7.170) we readily obtain an expression for the concentration distribution when the concentration just ceases to be zero at the centre of the sheet. It is

$$c(X, T) = (\alpha\beta - \tfrac{3}{2})X^2 - (\alpha\beta - \tfrac{5}{2})X^3, \qquad 0 \leqslant X \leqslant 1. \tag{7.172}$$

The next step is to derive an approximate solution which holds for later times when the concentration at the centre of the sheet has become appreciable, according to the accuracy of working. Such a solution must satisfy (7.149) and (7.148), and must also agree with (7.172) when $T = T_1$. We therefore assume a cubic equation again of the form

$$c(X, T) = A_1(T) + X^2 B_1(T) + X^3 E_1(T), \qquad 0 \leqslant X \leqslant 1, \tag{7.173}$$

where $A_1(T)$, $B_1(T)$, $E_1(T)$ are functions of T to be determined. They are of course different from the corresponding functions in (7.152). We see immediately that $A_1(T)$ in (7.173) is the concentration at the centre of the sheet. Proceeding as before, by inserting (7.148) in (7.173) we obtain

$$A_1 + B_1 + E_1 = 1. \tag{7.174}$$

The zero and first-moment equations in this case lead to

$$\frac{\mathrm{d}}{\mathrm{d}T}(A_1 + \tfrac{1}{3}B_1 + \tfrac{1}{4}E_1) = (2B_1 + 3E_1)F(1), \tag{7.175}$$

$$\frac{\mathrm{d}}{\mathrm{d}T}(\tfrac{1}{2}A_1 + \tfrac{1}{12}B_1 + \tfrac{1}{20}E_1) = G(1) - G(A_1). \tag{7.176}$$

Eliminating E_1 by using (7.174) and putting

$$\mathrm{d}A_1/\mathrm{d}T = \phi, \tag{7.177}$$

we have

$$\tfrac{3}{4}\phi + \tfrac{1}{12}\phi\frac{\mathrm{d}B_1}{\mathrm{d}A_1} = (3 - B_1 - 3A_1)F(1). \tag{7.178}$$

$$\tfrac{9}{20}\phi + \tfrac{1}{30}\phi\frac{\mathrm{d}B_1}{\mathrm{d}A_1} = G(1) - G(A_1). \tag{7.179}$$

We can eliminate B_1 from these two equations, remembering that $\mathrm{d}G/\mathrm{d}A_1 = F$, and obtain

$$-\frac{\mathrm{d}\phi}{\mathrm{d}A_1} = 28F(1) + \tfrac{20}{3}F(A_1) + \frac{80}{\phi}\{G(A_1) - G(1)\}F(1). \tag{7.180}$$

This can be simplified to the form

$$\psi\frac{\mathrm{d}\psi}{\mathrm{d}\phi} = \tfrac{49}{5}\{\psi r(f) - q(f)\}. \tag{7.181}$$

by introducing the following variables

$$f = 1 - A, \tag{7.182}$$

$$\psi = 7\phi/\{20F(1)\}, \tag{7.183}$$

$$r(f) = 1 + \frac{5}{21}\frac{F(A_1)}{F(1)}, \tag{7.184}$$

$$q(f) = \int_0^f \frac{F(A_1)}{F(1)}\,df' \tag{7.185}$$

In order that the solution (7.173) should reduce to (7.172) when $T = T_1$, over the whole range of X, we must have

$$A_1(T_1) = 0, \tag{7.186}$$

$$B_1(T_1) = \alpha\beta - \tfrac{3}{2}, \tag{7.187}$$

$$E_1(T_1) = -\alpha\beta + \tfrac{5}{2}. \tag{7.188}$$

Eliminating the term $\phi\,dB_1/dA_1$ from (7.178) and (7.179) we obtain, in terms of the new variables just introduced,

$$B_1 = \tfrac{15}{14}\psi(f) + 3f - \tfrac{5}{2}q(f). \tag{7.189}$$

From (7.182), (7.186), and (7.187) it follows that $B_1 = \alpha\beta - \tfrac{3}{2}$ when $f = 1$, and so from (7.189) we have

$$\psi(1) = \tfrac{14}{15}\{\alpha\beta - \tfrac{9}{2} + \tfrac{5}{2}q(1)\}. \tag{7.190}$$

For any given concentration-dependent diffusion coefficient the quantities $r(f)$ and $q(f)$ can be evaluated from (7.184) and (7.185), the relevant range of f being $0 \leqslant f \leqslant 1$ since $f = 1$ corresponds to $T = T$. and $f = 0$ to $T = \infty$. Thus (7.190) gives the initial condition to be satisfied by the solution of the differential equation (7.181). In general, numerical integration of (7.181) will be necessary.

When the solution of (7.181) is known, $\phi(A_1)$ follows from (7.183). Also by integrating (7.177) and using the condition $A_1 = 0$, $T = T_1$ we obtain

$$T = T_1 + \int_0^{A_1} \frac{dA_1'}{\phi(A_1')}, \tag{7.191}$$

which can be evaluated once $\phi(A_1)$ has been obtained from (7.183). Since from (7.173) A_1 represents the concentration at the centre of the sheet, from (7.191) we can calculate how this concentration varies with time. Once the relationship between A_1 and T, and hence between f and T is known, B_1 is easily determined as a function of T from (7.189) using (7.183). Finally, knowing $A_1(T)$ and $B_1(T)$, $E_1(T)$ is obtained from (7.174), and hence all functions of T on the right-hand side of (7.173) are known.

The total amount of diffusing substance M_t taken up by the sheet per unit area at time t is given by

$$M_t = 2 \int_0^l C(x, t)\, dx, \qquad (7.192)$$

which in terms of the non-dimensional variables becomes

$$M_t = 2lC_0 \int_0^1 c(X, T)\, dX. \qquad (7.193)$$

For early times, when the concentration at the centre of the sheet is effectively zero, i.e. $T < T_1$, c is given by (7.170) and (7.171), and so we have

$$M_t = 2lC_0 \left(\frac{1}{8\beta^{\frac{1}{2}}} + \frac{\alpha\beta^{\frac{1}{2}}}{12} \right) T^{\frac{1}{2}}, \qquad 0 < T < T_1. \qquad (7.194)$$

For later times we find

$$M_t = 2lC_0 \{1 - \tfrac{2}{3}B_1(T) - \tfrac{3}{4}E_1(T)\}. \qquad (7.195)$$

We note from (7.194) that, in the early stages, M_t is proportional to the square root of time, irrespectively of how the diffusion coefficient depends on concentration. We shall see later, in § 9.4 (p. 179), that this is a characteristic feature of concentration-dependent diffusion.

7.5. Approximation by orthogonal functions

Tsang (1960; 1961) developed a method of obtaining approximate analytic solutions which are simple in form and reasonably accurate for moderately concentration-dependent diffusion coefficients.

He discusses the equation

$$\frac{\partial \phi}{\partial t} = \frac{\partial}{\partial z} \left(\alpha \frac{\partial \phi}{\partial z} \right) \qquad (7.196)$$

subject to the conditions

$$\phi = 1, \qquad 0 \leqslant z \leqslant \pi, \qquad t = 0 \qquad (7.197)$$

$$\phi = 0, \qquad z = 0 \quad \text{and} \quad \pi, \qquad t \geqslant 0. \qquad (7.198)$$

These are conditions for desorption in a sheet of thickness l expressed in the non-dimensional variables,

$$z = \pi x/l, \qquad \phi = (C - C_0)/(C_1 - C_0), \qquad \alpha = D/D_0,$$

where C_0 is the surface concentration and C_1 the initial concentration.

Tsang starts from the expansion

$$\phi(z, t) = \sum_1^\infty f_n(t)(2/\pi)^{\frac{1}{2}} \sin nz. \qquad (7.199)$$

Substitution in (7.196) followed by the usual procedure for evaluating Fourier coefficients leads to an infinite set of first-order, ordinary differential equations for the functions $f_n(t)$, i.e.

$$df_n(t) dt = \sum_{n=1,3,5}^{\infty} \{\alpha_{kn}(t) + \beta_{kn}(t)\} f_n(t), \qquad (7.200)$$

where

$$\alpha_{kn} = (-2n^2/\pi) \int_0^\pi \alpha \sin kz \sin nz \, dz, \qquad (7.201)$$

$$\beta_{kn} = (2n/\pi) \int_0^\pi (\partial\alpha/\partial z) \sin kz \cos nz \, dx. \qquad (7.202)$$

The required initial conditions are

$$f_k(0) = (8/\pi k^2)^{\frac{1}{2}}, \qquad k = 1, 3, 5, \dots . \qquad (7.203)$$

Tsang demonstrates that provided α varies slowly with z a reasonable approximation is to neglect all the f_n terms in (7.200) for $n > k$. He further replaces all f_ks for $k \neq 1$ by $f_k(0) \exp(-k^2 t)$ implying that the harmonics die away rapidly.

As an example, he considers a linear concentration-dependent D for which

$$\alpha = 1 + \lambda\phi = 1 + (2/\pi)^{\frac{1}{2}}\lambda f_1(t) \sin z + \dots . \qquad (7.204)$$

Using only these two terms for α and the equation for $k = 1$ in (7.200) Tsang obtains the solution

$$\phi = \phi_0 - \left(\frac{64\lambda}{\pi}\right) \frac{e^{-t} - e^{-2t}}{3\pi^2 + 16\lambda - 16\lambda e^{-t}} \sin z, \qquad (7.205)$$

where ϕ_0 is the required solution of (7.196) with $\alpha = 1$, i.e.

$$\phi_0(z, t) = \sum_{n=1,3,5}^{\infty} \frac{4}{n\pi} \exp(-n^2 t) \sin nz. \qquad (7.206)$$

For the total amount of penetrant lost $M(t)$ we have

$$M_t = \int_0^\pi \phi \, dz = M_0(t) - \left(\frac{128\lambda}{\pi^2}\right) \frac{e^{-t} - e^{-2t}}{3\pi^2 + 16\lambda - 16\lambda e^{-t}}, \qquad (7.207)$$

where

$$M_0(t) = \frac{8}{\pi^2} \sum_{n=1,3,5}^{\infty} \frac{1}{n^2} \exp(-n^2 t). \qquad (7.208)$$

Values of $M(t)$ calculated from (7.207) for $\lambda = 2 \cdot 5, 4 \cdot 8, 10 \cdot 0$ by Tsang agree to within a few per cent with the corresponding solutions shown in Fig. 9.16 (p. 180).

8

NUMERICAL METHODS

8.1. Numerical solutions

In the preceding chapters a large number of mathematical solutions have been presented, most of them in the form of infinite series. Useful though these solutions are their application to practical problems can present difficulties. First, the numerical evaluation of the solutions is usually by no means trivial. Secondly, the analytical methods and solutions are, for the most part, restricted to simple geometries and to constant diffusion properties such as the diffusion coefficient. In other words, they apply strictly only to linear forms of the diffusion equations and the boundary conditions. This can be a severe limitation where the diffusion coefficient in polymer systems, for example, is often markedly concentration dependent.

The solution of mathematical equations which more closely model experimental and practical situations is possible by the methods of numerical analysis. In the days when the calculations had to be performed on a desk calculating machine use of these methods was a formidable task. The advent of the high speed digital computer revolutionized the situation. Recently, considerable developments have taken place in the subject of numerical analysis and in the construction of efficient computer programs to obtain numerical solutions. It is well beyond the scope of this book to attempt even a general introduction to the subject of numerical analysis. Most scientists and engineers, however, at sometime or other feel the need to solve problems for which they cannot find an analytical solution. An experiment is often conducted under conditions dictated by the analytical mathematical solution available even though the experimentalist would prefer an alternative arrangement. Those who have taken courses only in traditional mathematics often find it hard to appreciate that the simpler numerical methods offer, as it were, a new start not necessarily presented as an advanced course or as an extension of analytical methods. It is not necessary to establish a mathematical solution first, into which numbers are then inserted. These remarks have a special relevance to the subject of partial differential equations. This usually comes towards the end of an undergraduate course and is regarded as a branch of higher mathematics.

In order to try to overcome these mental barriers, and to give non-mathematicians an insight into the simpler numerical methods of solving the diffusion equation, we introduce the subject first from a physical point of view. After this, the more mathematical derivation aims to show how real problems can be tackled and to illustrate some of the pitfalls to be avoided. The reader should then be able to handle simpler problems himself: in more difficult cases he should at least know how computers obtain numerical solutions and what help the specialists in numerical analysis are able to offer. A few of the many books available for further reading are those by Fox (1962), Mitchell (1969), and Cohen (1973). Smith's (1965) book is a textbook written primarily for students with no previous knowledge of numerical methods whatsoever. The example in § 8.4.1 (p. 143) below is one of many to be found in his book.

8.2. Non-dimensional variables

Inspection of the analytical solutions in the preceding chapters reveals that the group of variables Dt/l^2 occurs frequently. The diffusion coefficient D is measured in $cm^2 s^{-1}$, say; it has the dimensions $(length)^2(time)^{-1}$. Thus if we write the group $Dt/l^2 = T$, we see that T is a dimensionless variable. There are a number of advantages in using non-dimensional variables in numerical work, though it is not essential to do so. Usually, a good deal of arithmetic is involved in large problems. A whole set of solutions with different physical parameters, e.g. for sheets of different thicknesses, can often be obtained from one basic solution in non-dimensional variables by simple scaling, with considerable economy of computer time. The fundamental parameters are often high-lighted and analogies with physically different systems become clearer, e.g. the distribution of diffusant and electric potential under similar conditions.

For diffusion in a plane sheet of thickness l, when the diffusion coefficient D is constant, convenient variables are

$$X = x/l, \qquad T = Dt/l^2, \qquad c = C/C_0, \tag{8.1}$$

where C_0 is some standard concentration such as the value at the surface of the sheet if it is constant. Then

$$\frac{\partial C}{\partial x} = \frac{\partial C}{\partial X}\frac{dX}{dx} = \frac{\partial C}{\partial X}\frac{1}{l}; \qquad \frac{\partial^2 C}{\partial x^2} = \frac{\partial^2 C}{\partial X^2}\frac{1}{l^2}$$

and

$$\frac{\partial C}{\partial t} = \frac{\partial C}{\partial T}\frac{dT}{dt} = \frac{\partial C}{\partial T}\frac{D}{l^2}.$$

Thus, the simple diffusion equation

$$\frac{\partial C}{\partial t} = D\frac{\partial^2 C}{\partial x^2} \tag{8.2}$$

becomes

$$\frac{\partial c}{\partial T} = \frac{\partial^2 c}{\partial X^2},\tag{8.3}$$

where we have substituted $C = cC_0$.

We note that in the non-dimensional variables the thickness becomes unity, e.g. the sheet lies between $X = 0$ and $X = 1$.

8.3. Physical derivation of a numerical solution

Consider a plane sheet in which a concentration gradient of diffusant exists initially. We wish to calculate how the distribution of diffusant changes with time. We divide the sheet into layers each of thickness h as in Fig. 8.1 and denote by C_0, C_1, C_2 the concentrations at three neighbouring interfaces. The dotted lines at R and S denote the mid sections of the two adjacent layers.

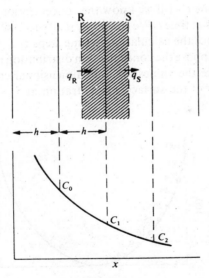

FIG. 8.1.

We recall Fick's first law of diffusion: rate of transfer of diffusant through unit area is proportional to gradient of concentration and the constant of proportionality is the diffusion coefficient D. Thus in a short time τ the amount of diffusant which has entered the shaded layer (Fig. 8.1) through unit area of the surface R is given approximately by $q_R = -D\tau(C_1 - C_0)/h$.

In the same time, the amount flowing out through the face at S is approximately $q_S = -D\tau(C_2 - C_1)/h$. The net amount of diffusant accumulated in the shaded element in time τ is

$$q_R - q_S = -\frac{D\tau}{h}(C_1 - C_0 - C_2 + C_1) = \frac{D\tau}{h}(C_0 - 2C_1 + C_2). \qquad (8.4)$$

If we now take C_1 to represent the average concentration in the narrow shaded element, the net gain of diffusant by the element can be written approximately as $(C_1' - C_1)h$, where C_1' is the concentration at the end of the interval τ. Equating this with (8.4) we have

$$C_1' - C_1 = \frac{D\tau}{h^2}(C_0 - 2C_1 + C_2). \qquad (8.5)$$

By choosing $D\tau/h^2 = \tfrac{1}{2}$ we find

$$C_1' = \tfrac{1}{2}(C_0 + C_2). \qquad (8.6)$$

This relationship enables us to calculate by simple arithmetic the concentration at a point at time $t + \tau$ if we know the concentrations C_0, C_2 at the two neighbouring points at time t. We can apply (8.6) successively at each point of the sheet and advance the calculation in time steps τ.

As an example, suppose the concentration distribution is initially parabolic as in Fig. 8.2, where the values are non-dimensional concentrations c expressed as fractions of the surface concentration at $x = 0$. The diagram is

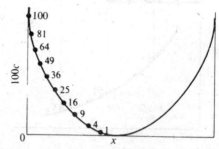

$100c$

100
81
64
49
36
25
16
9
4
0

x

FIG. 8.2. Initial parabolic concentration distribution.

symmetrical about the centre of the sheet and the sheet has been divided into 20 layers each of non-dimensional width $h = 0.05$. This means that $D\tau/(0.05)^2 = \tfrac{1}{2}$ and hence for the time step we have chosen $D\tau = 0.00125$. The time τ itself in seconds can easily be found for a known value of D. This does not enter into our basic calculation, however, for which we need only

(8.6). We first identify C_0, C_1, C_2 as the initial concentrations ($t = 0$) at the surface of the sheet and at two layers inside, $C_0 = 1$, $C_1 = 0.81$, $C_2 = 0.64$. Then by (8.6) we have

$$C_1' = \tfrac{1}{2}(1.00 + 0.64) = 0.82,$$

for the concentration one layer inside the sheet at $t = \tau$. We next apply (8.6) to the points at which the initial concentrations are 0.81, 0.64, and 0.49 respectively and calculate the new concentration at a depth $2h$ inside the sheet as 0.65. Having calculated a complete line of concentration values at $t = \tau$, we repeat the whole process to obtain a second line of values at $t = 2\tau$, and so on. A computer can carry out this repetitive arithmetic easily and quickly.

So far, we have used no calculus nor have we directly referred to the partial differential equation which expresses Fick's second law. Nevertheless, the relation (8.5) is one of the most commonly used formulae in the numerical solution of diffusion or heat-flow equations. But we have made certain approximations, and we have little idea of how accurate our solution is. Other difficulties, too, can arise in using this simple method, and so we now give a more mathematical derivation of (8.5) and of a slightly more complicated alternative.

8.4. Finite-difference solution: explicit method

We return to the non-dimensional equation (8.3) and take the sheet to occupy the space $0 \leqslant X \leqslant 1$. Let the range in X be divided into equal intervals δX and the time into intervals δT, so that the $X - T$ region is covered by a grid of rectangles, as in Fig. 8.3, of sides δX, δT. Let the coordinates of a representative grid point (X, T) be $(i\delta X, j\delta T)$, where i and j are integers. We denote the value of c at the point $(i\delta X, j\delta T)$ by $c_{i,j}$ with corresponding values at neighbouring points labelled as in Fig. 8.3. By using Taylor's series in the T direction but keeping X constant, we can write

$$c_{i,j+1} = c_{i,j} + \delta T\left(\frac{\partial c}{\partial T}\right)_{i,j} + \tfrac{1}{2}(\delta T)^2\left(\frac{\partial^2 c}{\partial T^2}\right)_{i,j} + \cdots, \tag{8.7}$$

from which it follows that

$$\left(\frac{\partial c}{\partial T}\right)_{i,j} = \frac{c_{i,j+1} - c_{i,j}}{\delta T} + O(\delta T), \tag{8.8}$$

where $O(\delta T)$ signifies that the leading term to have been neglected is of the order of δT when we have divided both sides of (8.7) to get (8.8).

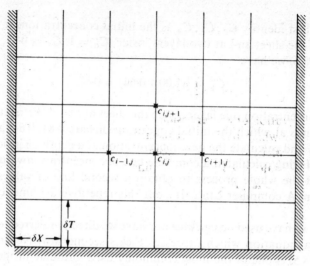

FIG. 8.3.

Similarly, by applying Taylor's series in the X direction, keeping T constant, we have

$$c_{i+1,j} = c_{i,j} + \delta X \left(\frac{\partial c}{\partial X}\right)_{i,j} + \tfrac{1}{2}(\delta X)^2 \left(\frac{\partial^2 c}{\partial X^2}\right)_{i,j} + \dots , \tag{8.9}$$

$$c_{i-1,j} = c_{i,j} - \delta X \left(\frac{\partial c}{\partial X}\right)_{i,j} + \tfrac{1}{2}(\delta X)^2 \left(\frac{\partial^2 c}{\partial X^2}\right)_{i,j} - \dots . \tag{8.10}$$

On adding we find

$$\left(\frac{\partial^2 c}{\partial X^2}\right)_{i,j} = \frac{c_{i+1,j} - 2c_{i,j} + c_{i-1,j}}{(\delta X)^2} + O(\delta X)^2. \tag{8.11}$$

By substituting (8.8) and (8.11) in (8.3) and neglecting the error terms we find after slight re-arrangement

$$c_{i,j+1} = c_{i,j} + r(c_{i-1,j} - 2c_{i,j} + c_{i+1,j}), \tag{8.12}$$

where $r = \delta T/(\delta X)^2$. If we choose $r = \tfrac{1}{2}$, we regain the equivalent of (8.6). With reference to Fig. 8.3, we can use (8.12) with a chosen value of r to calculate the values of c at all points along successive time rows of the grid provided we are given some initial starting values at $T = 0$, and some conditions on each of the boundaries $X = 0$, $X = 1$. A formula such as (8.12) which enables *one* unknown value to be expressed directly in terms of known values is called an 'explicit finite-difference formula'.

8.4.1. *Example*

Smith (1965) worked out a simple numerical example in which the initial distribution in non-dimensional form consists of two straight lines

$$c = 2X, \qquad 0 \leqslant X \leqslant \tfrac{1}{2}, \qquad T = 0,$$
$$c = 2(1 - X), \qquad \tfrac{1}{2} \leqslant X \leqslant 1, \qquad T = 0, \tag{8.13}$$

and the two faces of the sheet are maintained at zero concentration so that

$$c = 0, \qquad X = 0 \quad \text{and} \quad 1, \qquad T \geqslant 0. \tag{8.14}$$

The initial distribution (8.13) could arise if two separate sheets, each of thickness $\tfrac{1}{2}$, had separately been allowed to reach a steady state with one surface at unit concentration and the other at zero and then they had been placed firmly together. The problem is symmetric about $X = \tfrac{1}{2}$ and so we need the solution only for $0 \leqslant X \leqslant \tfrac{1}{2}$. Take $\delta X = \tfrac{1}{10}$ so that the initial and boundary values are as in the top row and first column of values in Table 8.1. Increasing values of T, i.e. of j, are shown moving downwards for convenience of calculation. The values of c tabulated in Table 8.1 correspond to $\delta X = \tfrac{1}{10}$, $\delta T = 1/1000$ so that $r = \tfrac{1}{10}$ (8.12).

The analytical solution satisfying (8.13) and (8.14) is

$$c = \frac{8}{\pi^2} \sum_{n=1}^{\infty} \frac{1}{n^2} \sin \tfrac{1}{2} n\pi \sin n\pi X \exp\left(-n^2 \pi^2 T\right). \tag{8.15}$$

In Table 8.2, the finite-difference solution is compared with values calculated from (8.15) at $X = 0.3$ and 0.5. The percentage error is the difference of the solutions expressed as a percentage of the analytical solution. The comparisons are seen to be reasonably good on the whole. The largest errors are found at $X = 0.5$ for small times. This is because of the discontinuity in the initial gradient $\partial c/\partial X$ which changes from $+2$ to -2 at $X = 0.5$. Table 8.2 shows, however, that the errors die away as T increases. This is the way a discontinuity usually effects the finite-difference solution of a diffusion equation.

We shall refer again in § 8.10.1 (p. 152) to the special treatment of discontinuities which is needed if we want accurate values of the function near the singularity.

We turn now to a serious limitation on the value of $r = \delta T/(\delta X)^2$ in (8.12). The reader will have no difficulty in repeating the calculations keeping $\delta X = \tfrac{1}{10}$ but increasing δT to $\tfrac{5}{1000}$ so that $r = \tfrac{1}{2}$. Smith (1965) tabulates the solution and finds the errors to be somewhat larger than for $r = \tfrac{1}{10}$ but still well within the accuracy of many experimental measurements. However, Figs. 8.4 and 8.5 taken from Smith (1965) demonstrate that the value of $r = \tfrac{1}{2}$ is critical. In Fig. 8.4, for $r = 0.48$, the finite difference solutions agree with the analytical reasonably well, but in Fig. 8.5 ($r = 0.52$) oscillations have developed. If a still larger value of r is taken, say $r = 1$, the finite-difference

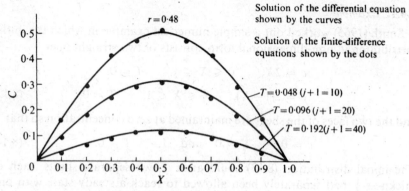

FIG. 8.4. After Smith 1965.

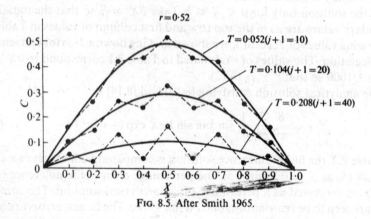

FIG. 8.5. After Smith 1965.

solution bears no resemblance to the analytical solution. It has become 'unstable' in the sense that the errors increase without limit. We shall discuss the subject of instability further in § 8.11 (p. 157). Here we note that a severe limitation is imposed on the value of r, and hence on δT, for a given δX. We are forced to take a large number of small time steps when using the simple explicit method. These considerations lead us to search for better formulae.

8.5. Crank–Nicolson implicit method

A method which is widely used was proposed by Crank and Nicolson (1947). They replaced $\partial^2 c/\partial X^2$ by the mean of its finite-difference representations on the jth and $(j+1)$th time rows and approximated eqn (8.3) by

$$\frac{c_{i,j+1}-c_{ij}}{\delta T} = \frac{1}{2}\left\{\frac{c_{i+1,j}-2c_{i,j}+c_{i-1,j}}{(\delta X)^2}+\frac{c_{i+1,j+1}-2c_{i,j+1}+c_{i-1,j+1}}{(\delta T)^2}\right\}, \quad (8.16)$$

where the total error is $O\{(\delta T)^2 + (\delta X)^2\}$. We can write (8.16) in the form

$$-rc_{i-1,j+1} + (2+2r)c_{i,j+1} - rc_{i+1,j+1} = rc_{i-1,j} + (2-2r)c_{i,j} + rc_{i+1,j} \quad (8.17)$$

where $r = \delta T / (\partial X)^2$.

The left side of (8.17) contains three unknowns on the time level $j+1$, and on the right side the three values of c are known on the jth time level. If there are N internal grid points along each time row, then for $j = 0$ and $i = 1$, $2, \ldots, N$, eqn (8.17) gives N simultaneous equations for N unknown values along the first time row expressed in terms of the known initial values and the boundary values at $i = 0$ and $N+1$. Similarly for $j = 1$, unknown values are expressed along the second time row in terms of those calculated along the first. This type of method, in which the solution of a set of simultaneous equations is called for at each time step, is described as an 'implicit method'. More work is involved at each time step but the Crank–Nicolson method has the strong advantage of remaining stable for all values of r. We can thus proceed with larger and hence fewer time steps, the limit being set by the accuracy required, having in mind that higher-order terms in the Taylor series have been neglected in deriving (8.17).

8.5.1. Example

Smith (1965) discusses the solution of the problem of § 8.4.1 (p. 143) by the Crank–Nicolson method, taking $\delta X = \frac{1}{10}$ and $\delta T = \frac{1}{100}$ so that $r = 1$. This is a convenient choice of r for which (8.17) reduces to

$$-c_{i-1,j+1} + 4c_{i,j+1} - c_{i+1,j+1} = c_{i-1,j} + c_{i+1,j}. \quad (8.18)$$

Taking $j = 0$, the values of $c_{i,0}$ ($i = 1, 2, \ldots, 9$) are given by the first row of Table 8.1. We remember the symmetry of the problem which makes $c_6 = c_4, c_7 = c_3$, etc. Since all the unknowns in the simultaneous equations are on the left side there need be no confusion if we write for any time step $c_{i,j+1} = c_i (i = 1, 2, \ldots, 9)$.

The set of equations for the first time step is

$$-0 + 4c_1 - c_2 = 0 + 0.4$$

$$-c_1 + 4c_2 - c_3 = 0.2 + 0.6$$

$$-c_2 + 4c_3 - c_4 = 0.4 + 0.8$$

$$-c_3 + 4c_4 - c_5 = 0.6 + 1.0$$

$$-2c_4 + 4c_5 = 0.8 + 0.8. \quad (8.19)$$

These equations are easily solved by a well-known method of elimination. The first equation is used to eliminate c_1 from the second equation; the new second equation is used to eliminate c_2 from the third equation, and so on. Finally, the new fourth equation can be used to eliminate c_4 from the last equation, giving one equation with only one unknown c_5. The other unknowns c_4, c_3, c_2, c_1 are then found by back-substitution. The solution is

$$c_1 = 0.1989, \qquad c_2 = 0.3956, \qquad c_3 = 0.5834,$$

$$c_4 = 0.7381, \qquad c_5 = 0.7691. \tag{8.20}$$

The simultaneous equations for the values of c along the second time row have left sides identical with those of (8.19) but with values from (8.20) substituted for $c_{i-1,j}$ and $c_{i+1,j}$ on the right sides. Part of this finite-difference solution is given in Table 8.3 and the comparison with the analytical solution at $T = 0.1$ is good. The implicit method for this example is about as accurate as the explicit method which uses ten times as many time steps for the same total time range. The remarks in § 8.4.1 (p. 143) about the discontinuity in the initial gradient at $X = 0.5$ apply also to the results of the implicit method.

More details of the elimination method of solving the equations and of alternative iterative methods are to be found in Smith (1965) and in standard works on numerical analysis.

8.6. Other boundary conditions

8.6.1. Surface concentration a given function of time

The way in which the concentration at the surface of the sheet varies with time may be given in the form of an algebraic formula, a graph or a set of tabulated values. The data can be substituted directly at each time step into (8.12) or (8.17) and the methods proceed as described.

8.6.2. Derivative condition

We have seen that boundary conditions involving the gradient of concentration occur frequently. As an example, we consider the evaporation condition of § 3.3.1 (p. 35), which we write as

$$\partial c / \partial X = \alpha(c_s - c_e), \qquad X = 0, \tag{8.21}$$

where c_s is the surface concentration at time T and c_e is the equilibrium surface concentration corresponding to the vapour pressure in the atmosphere remote from the surface. The algebraic signs need careful attention (§ 1.4.2 (iii) (d), p. 9). The left side of (8.21) is the rate of transfer away from the surface i.e. in the direction of negative X, and for a condition of loss by evaporation this is a positive quantity. Also for evaporation $c_s > c_e$ and so the right side of (8.21) is also positive. Fig. 8.6 depicts the relevant grid points at $X = 0$ and $X = 1$ if a similar condition were to hold on both surfaces.

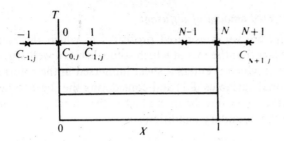

F̄ɪɢ. 8.6. After Smith 1965.

The simplest way to represent the condition (8.21) at $X = 0$ in finite differences is

$$(c_{i,j} - c_{0,j})/\delta X = \alpha(c_{0,j} - c_e), \qquad X = 0. \tag{8.22}$$

This gives one extra equation for $c_{0,j}$ at any time step, to be used in the explicit or implicit schemes instead of a given value of $c_{0,j}$.

A more accurate replacement of (8.21) is possible by introducing a 'fictitious' concentration $c_{-1,j}$ at the external grid point $(-\delta X, j\delta T)$ (Fig. 8.6). We imagine the sheet extended one layer. The condition (8.21) becomes

$$(c_{1,j} - c_{-1,j})/(2\delta X) = \alpha(c_{0,j} - c_e). \tag{8.23}$$

The value of $c_{-1,j}$ is unknown, but it can be eliminated by using the replacement of the partial differential equation. For the surface point this is

$$\frac{\partial c_{0,j}}{\partial T} = \frac{c_{-1,j} - 2c_{0,j} + c_{1,j}}{(\delta X)^2}. \tag{8.24}$$

Elimination of $c_{-1,j}$ from (8.23) and (8.24) gives

$$\frac{\partial c_{0,j}}{\partial T} = \frac{2}{\delta X} \left\{ \frac{c_{1,j} - c_{0,j}}{\delta X} - \alpha(c_{0,j} - c_e) \right\} \tag{8.25}$$

The explicit form of (8.25) is

$$c_{0,j+1} = c_{0,j} + 2r\{c_{1,j} - c_{0,j} - \alpha\delta X(c_{0,j} - c_e)\}, \tag{8.26}$$

and the Crank–Nicolson formula is

$$c_{0,j+1} = c_{0,j} + r\{(c_{1,j+1} + c_{1,j}) - (c_{0,j+1} + c_{0,j}) - \alpha\delta X(c_{0,j+1} + c_{0,j} - 2c_e)\}, \tag{8.27}$$

which can easily be rearranged to become the first equation of the set (8.17).

The special case of an impermeable surface corresponds to $\alpha = 0$ in the above equations.

8.6.3. *A restricted amount of diffusant*

In §4.3.5 (p. 56) we considered problems in which the condition (4.36) expresses the fact that the rate at which solute leaves a well-stirred solution is equal to that at which it enters a sheet immersed in the solution. Using the non-dimensional variables (8.1) and considering the sheet to be of thickness l and the surface $X = 0$ to be in contact with a stirred solution of extent a, we can write the boundary conditions as

$$a\, \partial c_{0,j}/\partial T = l\, \partial c_{0,j}/\partial X, \qquad X = 0, \qquad T > 0. \tag{8.28}$$

In this problem, C_0 in (8.1) is conveniently taken to be the initial uniform concentration in the solution. The simplest finite-difference replacement of (8.28) is

$$a\frac{\partial c_{0j}}{\partial T} = \frac{l}{\delta X}(c_{1,j} - c_{0,j}). \tag{8.29}$$

Alternatively, by introducing a fictitious concentration, as in (8.23), we find

$$\frac{\partial c_{0,j}}{\partial T}\left(1 + \frac{2a}{l\,\delta X}\right) = \frac{2}{(\delta X)^2}(c_{1,j} - c_{0,j}). \tag{8.30}$$

The explicit and implicit forms of (8.29) and (8.30) follow at once by approximate replacement of $\partial c_{0,j}/\partial T$.

8.7. Finite-difference formulae for the cylinder and sphere

In terms of the variables

$$R = r/a, \qquad T = Dt/a^2, \tag{8.31}$$

the equation for radial diffusion in a cylinder of radius a becomes

$$\frac{\partial c}{\partial T} = \frac{1}{R}\frac{\partial}{\partial R}\left(R\frac{\partial c}{\partial R}\right) = \frac{\partial^2 c}{\partial R^2} + \frac{1}{R}\frac{\partial c}{\partial R}. \tag{8.32}$$

The finite-difference approximations corresponding to (8.11), omitting the error term, are

$$\frac{1}{R}\frac{\partial}{\partial R}\left(R\frac{\partial c}{\partial R}\right) = \frac{1}{2i(\delta R)^2}\{(2i+1)c_{i+1,j} - 4ic_{i,j} + (2i-1)c_{i-1,j}\}, \quad i \neq 0, \tag{8.33}$$

$$= \frac{4}{(\delta R)^2}(c_{1,j} - c_{0,j}), \quad i = 0. \tag{8.34}$$

The nomenclature is as for the plane sheet, namely that $c_{i,j}$ is the concentration at the point $(i\delta R, j\delta T)$, the range $0 \leqslant R \leqslant 1$ having been divided into equal intervals δR. Explicit and implicit formulae follow by combining (8.33) and (8.34) with appropriate replacements for $\partial c/\partial T$.

An alternative method of dealing with a hollow cylinder of internal radius b, suggested by Eyres *et al.* (1946), is to use

$$X_1 = \ln(R/b) \tag{8.35}$$

as independent variable. Then (8.32) becomes

$$\frac{\partial c}{\partial T} = \frac{\exp(-2X_1)}{b^2} \frac{\partial^2 c}{\partial X^2}, \tag{8.36}$$

and we have diffusion in a plane sheet with a variable diffusion coefficient.

The non-dimensional equation for radial diffusion in a sphere of radius a is

$$\frac{\partial c}{\partial T} = \frac{1}{R^2} \frac{\partial}{\partial R}\left(R^2 \frac{\partial c}{\partial R}\right) = \frac{\partial^2 c}{\partial R^2} + \frac{2}{R} \frac{\partial c}{\partial R}, \tag{8.37}$$

and the finite-difference approximations are

$$\frac{1}{R^2} \frac{\partial}{\partial R}\left(R^2 \frac{\partial c}{\partial R}\right) = \frac{1}{i(\delta R)^2}\{(i+1)c_{i+1,j} - 2ic_{i,j} + (i-1)c_{i-1,j}\}, \qquad i \neq 0, \tag{8.38}$$

$$= \frac{6}{(\delta R)^2}(c_{1,j} - c_{0,j}), \qquad i = 0. \tag{8.39}$$

The treatment of the various boundary conditions for the cylinder and sphere is precisely analogous to that of § 8.6 (p. 146).

8.8. Composite media

Conditions at the interface between two different media can be dealt with by a slight extension of the methods of § 8.6.2 (p. 146). Considering the plane case first, let $x = x_s$ be the interface and let the suffices a, b refer to the left side $(x < x_s)$ and right side $(x > x_s)$ of the interface respectively. The space intervals δx_a and δx_b may be different if desired. Let $F(c_s, t)$ denote the flux across the interface. The conditions to be satisfied are

$$D_a \, \partial c_a/\partial x = D_b \, \partial c_b/\partial x = F(c_s, t), \qquad x = x_s, \tag{8.40}$$

which expresses the fact that diffusant enters one medium at the same rate as it leaves the other, together with some relation between the concentrations on the two sides of the interface. We shall consider the simplest case of

$$c_a = c_b = c_s \tag{8.41}$$

and the modification made necessary by a more general relationship such as

$$c_a = Qc_b + R$$

will be obvious.

As in § 8.6.2 (p. 146) the simplest procedure is to replace (8.40) by

$$\frac{D_a}{\delta x_a}(c_{s,j} - c_{s-1,j}) = \frac{D_b}{\delta x_b}(c_{s+1,j} - c_{s,j}), \qquad (8.42)$$

which relates c_s to the concentrations at the two neighbouring points on any time row.

As in § 8.6.2 (p. 146) a more accurate replacement is possible by imagining the medium to the left of the boundary to be extended one step δx_a to the right of x_s and eliminating the fictitious concentration there by using (8.40). We obtain

$$\frac{\partial c_{s,j}}{\partial t} = \frac{2D_a}{\delta x_a}\left(\frac{c_{s-1,j} - c_{s,j}}{\delta x_a} - \frac{F}{D_a}\right). \qquad (8.43)$$

Similarly, by extending the medium to the right of the interface one step δx_b to the left, we find

$$\frac{\partial c_{s,j}}{\partial t} = \frac{2D_b}{\delta x_b}\left(\frac{c_{s+1,j} - c_{s,j}}{\delta x_b} + \frac{F}{D_b}\right). \qquad (8.44)$$

By eliminating F from (8.43) and (8.44) we obtain

$$\tfrac{1}{2}(\delta x_a + \delta x_b)\frac{\partial c_{s,j}}{\partial t} = \frac{D_b}{\delta x_b}(c_{s+1,j} - c_{s,j}) - \frac{D_a}{\delta x_a}(c_{s,j} - c_{s-1,j}). \qquad (8.45)$$

This equation can be written in non-dimensional variables, and explicit or implicit finite difference formula derived by appropriate substitution for $\partial c_s/\partial t$. Corresponding formulae for composite cylinders and spheres are derived in the same way.

8.9. Two- and three-dimensional diffusion

There is no difficulty in principle in extending the methods described in §§ 8.4 and 8.5 (pp. 141 and 144) to two and three dimensions. In practice, however, the considerable increase in computational labour called for has stimulated attempts to find more efficient methods. These are still the subject of active research and we can only indicate briefly their general nature. A typical, simple problem is to obtain solutions of

$$\frac{\partial c}{\partial T} = D\left(\frac{\partial^2 c}{\partial x^2} + \frac{\partial^2 c}{\partial y^2}\right) \qquad (8.46)$$

over the rectangular region $0 \leqslant x \leqslant a, 0 \leqslant y \leqslant b$, subject to the conditions that c is known over the whole region at $t = 0$ and is prescribed on the boundary for $t \geqslant 0$. We cover the region with a rectangular space grid at each time step and denote coordinates by

$$x = i\delta x, \qquad y = j\delta y, \qquad t = n\delta t, \qquad (8.47)$$

where i, j, n are positive integers. The values of c at the grid points are denoted by

$$c(i\delta x, j\delta y, n\delta t) = c_{i,j,n}. \qquad (8.48)$$

The explicit finite-difference scheme of § 8.4 (p. 141) becomes

$$\frac{c_{i,j,n+1} - c_{i,j,n}}{\delta t} = \frac{D}{(\delta x)^2}(c_{i-1,j,n} - 2c_{i,j,n} + c_{i+1,j,n})$$

$$+ \frac{D}{(\delta y)^2}(c_{i,j-1,n} - 2c_{i,j,n} + c_{i,j+1,n}). \qquad (8.49)$$

This is computationally straightforward but the stability restriction is

$$D\left\{\frac{1}{(\delta x)^2} + \frac{1}{(\delta y)^2}\right\}\delta t \leqslant \tfrac{1}{2}, \qquad (8.50)$$

which is more severe than $r \leqslant \tfrac{1}{2}$ as in § 8.4. In three dimensions the restriction is even more severe. Extremely small steps δt must be used.

The Crank–Nicolson method becomes

$$\frac{c_{i,j,n+1} - c_{i,j,n}}{\delta t} = \tfrac{1}{2}D\left\{\left(\frac{\partial^2 c}{\partial x^2} + \frac{\partial^2 c}{\partial y^2}\right)_{i,j,n} + \left(\frac{\partial^2 c}{\partial x^2} + \frac{\partial^2 c}{\partial y^2}\right)_{i,j,n+1}\right\}, \qquad (8.51)$$

and requires the solution of $(M-1)(N-1)$ simultaneous algebraic equations for each time step δt, where $N\delta x = a$, $M\delta y = b$. Because five unknown values of c in general appear in each equation they are less amenable to solution by direct elimination methods and are usually solved iteratively.

One of the earliest and still probably the most widely-used method which offers considerably improved efficiency is that of Peaceman and Rachford (1955). The essential feature of Peaceman and Rachford's method and of several later variants is to replace only one second-order derivative, say $\partial^2 c/\partial x^2$, by an implicit difference approximation, leaving the other derivative $\partial^2 c/\partial y^2$ to be treated explicitly. By applying the resulting difference equation to each of the $(N-1)$ points along a grid line parallel to the x-direction we obtain $(N-1)$ simultaneous equations, each one containing only three unknown values of c, e.g. for the point (i, j, n), $c_{i,j,n+1}$, $c_{i-1,j,n+1}$, $c_{i+1,j,n+1}$ are unknowns but $c_{i,j+1,n+1}$, $c_{i,j-1,n+1}$ do not appear. We have a set of $(N-1)$ equations to solve $(M-1)$ times for each of the $(M-1)$ grid lines parallel to the x-direction. This is much easier than the solution of the $(N-1)(M-1)$ simultaneous equations needed in an implicit method. The solution is advanced from the $(n+1)$th to the $(n+2)$th step by interchanging the treatment of the second-order derivatives, i.e. $\partial^2 c/\partial x^2$ is treated explicitly and $\partial^2 c/\partial y^2$ implicitly. This and similar methods are generally referred to as alternating direction implicit methods (ADI methods) and are usually stable for all ratios $\delta t/(\delta x)^2$ and $\delta t/(\delta y)^2$. Mitchell (1969) discusses several methods including three dimensional variants and gives guidance on which to choose for different types of problem. Even greater efficiency of calculation is promised by using the 'hopscotch methods' developed by Gourlay (1970, 1971). For other geometries and curved boundaries special finite-difference formulae are needed. (Smith 1965, Mitchell 1969, Fox 1962).

8.10. Singularities: local solutions

8.10.1. *One dimension: discontinuity at origin*

It is possible to start a finite-difference solution directly from the prescribed initial conditions which give a row of values of c at $T = 0$. In some cases, e.g. when the surface concentration is a continuously varying function of time which is known, starting this way will probably be quite satisfactory. In other cases the conditions are such that a singularity exists at $T = 0$. For example, if the conditions are

$$c = 0, \quad 0 < X < 1, \quad T = 0, \tag{8.52}$$

$$c = 1, \quad X = 0, \quad T \geqslant 0, \tag{8.53}$$

the limiting value of c is unity as T tends to zero for $X = 0$, whereas c tends to zero as X approaches zero for $T = 0$. In other words, the concentration is discontinuous at $(0, 0)$. As we neglected higher terms in the Taylor series when deriving the finite-difference replacements for derivatives, the difference solution is likely to be a poor one near a discontinuity where some derivatives will be infinitely large. Although this is confirmed in practice, it is a fortunate property of the diffusion equation that a difference solution quickly approaches the analytical solution and its accuracy is probably acceptable after a few time steps.

We can secure accuracy in the neighbourhood of the singularity either by developing an analytical or a series solution applicable for small times, or by transforming the variables so that the singularity is removed.

It is often not difficult to find a suitable small-time solution because a medium of any shape behaves as if it were semi-infinite in the early stages of diffusion when the measurable penetration of diffusant is small. Thus the solutions of Chapters 3 or 7 can be used.

For the conditions (8.52) and (8.53) the singularity at $(0, 0)$ can be removed by using the variables

$$\xi = X/T^{\frac{1}{2}}, \quad \tau = T^{\frac{1}{2}}. \tag{8.54}$$

The effect is to expand the origin $X = 0$, $T = 0$ into the positive half of the

ξ axis and to remove the whole of the positive half of the X-axis to $\xi = \infty$. The discontinuity in c at $X = 0$, $T = 0$ is transformed into a smooth change along the positive ξ-axis. Finite-difference solutions can proceed in terms of the variables ξ, τ, without difficulty, though the diffusion equation is slightly more complicated after the transformation. Thus, the simple diffusion equation (8.3) becomes

$$2\frac{\partial^2 c}{\partial \xi^2} = \tau \frac{\partial c}{\partial \tau} - \xi \frac{\partial c}{\partial \xi}, \tag{8.55}$$

and (8.52) and (8.53) are now

$$c = 0, \qquad \xi = \infty, \qquad \tau = 0, \tag{8.56}$$

$$c = 1, \qquad \xi = 0, \qquad \tau \geqslant 0. \tag{8.57}$$

The numerical process starts with the solution of the ordinary differential equation for $c(\xi)$ obtained by putting $\tau = 0$ in (8.55). In this example, an analytic solution exists: in other cases, e.g. for a variable diffusion coefficient, the methods of Chapter 7 can be used. There will come a time τ when the boundary condition on $X = 1$ will need to be taken into account. We can continue in the transformed plane or return to the original variables (X, T) according to which is more convenient.

Discontinuities in derivatives rather than in the function c itself may also need special treatment, though usually the trouble is less marked. Discontinuities may occur also in the initial condition but not on the boundary as in the example of §8.4.1 (p. 143).

We have said that the accuracy of finite-difference solutions improves as T increases and usually they have been found to approach the corresponding analytical solutions where these are known. For certain initial and derivative boundary conditions, however, it can happen that a small difference persists for all time T between the numerical and analytical solutions. This rather specialist topic is discussed by Mitchell (1969).

8.10.2. *Two dimensions: boundary discontinuities*

Difficulties can arise at a corner on the boundary of the region within which diffusion is taking place. Re-entrant corners, where, the boundary changes direction through an angle exceeding π as at P in Fig. 8.7 are particularly troublesome. Some or all of the derivatives of the concentration may become infinitely large at the corner.

(i) *Series solutions.* Motz (1946) described how an analytical solution which incorporates the singularity can be welded onto the usual finite-difference solutions of steady-state problems away from the corner. Fox and Sankar (1969) generalized this method to include other boundary conditions. Bell

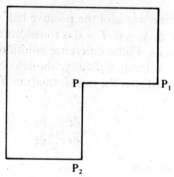

FIG. 8.7. Re-entrant corner.

and Crank (1973) combined the same idea to solve transient diffusion problems in two-dimensions. They express the diffusion equation (8.46) in polar coordinates r, θ centred at P (Fig. 8.8) in the form

$$D\left(\frac{\partial^2 C}{\partial r^2} + \frac{1}{r}\frac{\partial C}{\partial r} + \frac{1}{r^2}\frac{\partial^2 C}{\partial \theta^2}\right) = \frac{\partial C}{\partial t} \tag{8.58}$$

As an example the boundary conditions in the neighbourhood of P are taken to be

$$\partial C/\partial \theta = 0 \quad \text{on} \quad \theta = 0, \theta_0. \tag{8.59}$$

Consider a separation of variables solution

$$C = \exp\left(-\alpha^2 Dt\right)R(r)\psi(\theta) + W(r, \theta),$$

where $W(r, \theta)$ is a solution of Laplace's equation representing the steady-state form of the singularity. We obtain for R and ψ

$$\psi'' = -\omega^2\psi, \tag{8.60}$$

$$R'' + \frac{1}{r}R' + R\left(\alpha^2 - \frac{\omega^2}{r^2}\right) = 0. \tag{8.61}$$

Eqn (8.60) has a solution

$$\psi = a\cos\omega\theta + b\sin\omega\theta$$

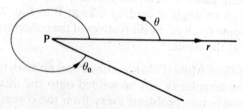

FIG. 8.8. Corner in polar form.

and (8.61) is Bessel's equation for which

$$R = \beta J_\omega(\alpha r), \qquad \omega \geqslant 0,$$

which is finite at P.

Thus a solution of (8.58) is

$$C(r, \theta, t) = \exp(-\alpha^2 Dt)J_\omega(\alpha r)\{A \cos \omega\theta + B \sin \omega\theta\} + W(r, \theta), \qquad (8.62)$$

in which the boundary conditions (8.59) require

$$B = 0, \qquad \omega = k\pi/\theta_0, \qquad k = 0, 1, 2, \dots,$$

and following Motz (1946)

$$W(r, \theta) = \sum_{k=0}^{\infty} c_k r^{k\lambda} \cos k\lambda\theta$$

with $\lambda = \pi/\theta_0$.

As an illustration we consider the re-entrant corner at P in Fig. 8.7 for which $\theta_0 = 3\pi/2$, i.e. $\lambda = 2/3$ and the normal derivatives of C are zero on PP_1 and PP_2. The general solution obtained by summing solutions of type (8.62) for this case is

$$C(r, \theta, t) = \sum_{j=0}^{\infty} \sum_{k=0}^{\infty} A_{kj} \exp(-\alpha_j^2 Dt)J_{2k/3}(\alpha_j r) \cos \frac{2k\theta}{3}$$

$$+ \sum_{k=0}^{\infty} c_k r^{2k/3} \cos \frac{2k\theta}{3}. \qquad (8.63)$$

We note that although C itself is finite at P, all derivatives with respect to r contain singular terms at P. Hence the need to avoid using finite-differences near P. Instead we use an approximation for small r based on (8.63) which we can express as a single power series in r by expanding the Bessel functions and collecting like terms in r. Thus

$$J_{2k/3}(z) = \sum_{m=0}^{\infty} \frac{(-1)^m (\tfrac{1}{2}z)^{\frac{2}{3}k + 2m}}{m!\,\Gamma(\tfrac{2}{3}k + m + 1)},$$

and (8.63) can be written

$$C = a_0(t) + a_1(t) \cos \tfrac{2}{3}\theta \, r^{\frac{2}{3}} + a_2(t) \cos \tfrac{4}{3}\theta \, r^{\frac{4}{3}}$$

$$+ r^2\{a_2(t) \cos 2\theta - b_0(t)\} + O(r^{\frac{8}{3}}). \qquad (8.64)$$

The as are functions of t and represent the coefficients of the leading term in each Bessel function expansion together with the corresponding coefficient of the steady-state solution.

The series (8.64) is used, at time t, to obtain function values C at points near the corner in terms of those further away. The two sets of points are referred to as 'near' and 'far' points. One far point is needed to determine each unknown coefficient in the series. In the simple case of a three-term approximating series, for example, we see that three far points are needed. We can select three near points, both sets being displayed in Fig. 8.9. Eqn (8.64) becomes

$$C = a_0 + a_1 \cos \frac{2\theta}{3} r^{\frac{2}{3}} + a_2 \cos \frac{4\theta}{3} r^{\frac{4}{3}}. \tag{8.65}$$

In principle, we use values of C and of r, θ at F_1, F_2, F_3 in Fig. 8.9 to determine a_0, a_1, a_2 from (8.65). We then evaluate C at N_1, N_2, N_3 using appropriate values of r, θ and those of a_0, a_1, a_2 just determined. These values of C are substituted in the explicit finite-difference-formula when required. Elsewhere, the simple difference scheme is used. Practical details of the method and some refinements together with an assessment of accuracy are given by Bell and Crank (1973).

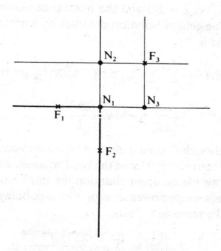

FIG. 8.9. Near and far points.

(ii) *Heat-balance equation.* A simpler treatment of a corner uses the method of § 8.3 (p. 139) and the two-dimensional equivalent of Fig. 8.1. The domain is divided into elementary areas, which are mainly rectangular except where they are made to fit the shape of the boundary. Bell (1973) has treated heat flow in a chamfered billet by considering elements as shown in Fig. 8.10. In diffusion terms, the corner concentration C_P at P is assumed to be the mean concentration for the corner element, for which a mass-balance equation is used. The rate at which diffusant enters the corner element is approximated by $-D(C_1 - C_P)\Delta y/\Delta x$ over the face RS and by $-\frac{1}{2}D(C_2 - C_P)\Delta x/\Delta y$ over QR. The flux over PQ and PS is specified by the

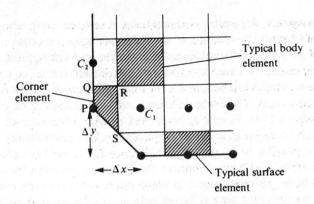

FIG. 8.10. Chamfered corner.

boundary conditions. The net flux into the element through all faces is equated to $V_C \, \partial C_P / \partial t$, where V_C is the volume of the corner element. The resulting equation is used in conjunction with the corresponding ones for all other elements and which are identical with the usual finite-difference equations. This method seems to have been mentioned for the first time by Eyres, Hartree, Ingham, Jackson, Sarjant, and Wagstaff (1946).

8.11. Compatibility, convergence, stability

We have assumed that finite-difference methods provide reasonable approximations to the true solution of the partial differential equation. The investigation of this assumption is an important part of the work of professional numerical analysts. We shall outline the nature of the problems in a general way, leaving the reader to consult standard references for the details. It is broadly true that the investigation of linear systems is reasonably satisfactory but that relatively little progress has been made with general treatments of non-linear systems. In these cases, for the most part, the analysis is based on the assumption of localized linearity.

We are concerned with three related properties:

(a) *Compatibility.* In deriving the finite-difference equation we neglected higher-order terms in the Taylor series. These constitute a truncation error. We require that the truncation error should tend to zero as δX and δT, for example, approach zero. If this is not so, the difference scheme is said to be incompatible or inconsistent with the partial differential equation. In this case, the difference solution is not likely to approach the solution sought. It is not usually difficult to establish the conditions necessary for the compatibility of a difference scheme.

(b) *Convergence*. Assuming compatability is ensured there next arises the question of whether the solution of the difference equations converges to the solution of the partial differential equation as the grid size approaches zero.

Let U represent the *exact* solution of a partial differential equation, with x and t independent variables, and u the *exact* solution of the approximating difference equations. The finite-difference solution is said to be convergent when u tends to U as δx and δt both tend to zero. Convergence is usually more difficult to investigate than compatability. Once established, however, it should be possible to decrease the difference $U - u$, usually called the discretization error, and hence improve the difference solution by decreasing δx or δt or both, perhaps subject to some relationship between them. When two solutions obtained with different grid sizes agree to within the desired accuracy it is usually assumed that the process of solution is complete. Fox and Mayers (1968) have drawn attention to pathological cases in which the assumption may be dangerous.

(c) *Stability*. If it were possible to carry out calculations to an infinite number of decimal places and if the initial and boundary data were specified exactly, the numerical calculations would produce the exact solution u of the *difference* equations. In practice, of course, each calculation is carried out to a finite number of figures and 'round off' errors are introduced. The solution actually computed is not u but N, say, which we can call the 'numerical solution'. The subject of stability is concerned with the possible build up of errors in the calculation which would cause N to be significantly different from u.

A set of finite-difference equations is said to be stable when the cumulative effect of all rounding errors is negligible. If the magnitudes of the errors introduced at the various grid points are each less than δ, then the finite-difference equations are usually considered stable if the maximum value of $u - N$ tends to zero as δ tends to zero, and does not increase exponentially with the number of time rows in the computation.

Usually it is not possible to determine the magnitude of $u - U$ at a given mesh point for an arbitrary distribution of errors. Standard methods of investigating stability mainly study the growth of an isolated error or a single row of errors. On the whole stability estimates tend to be conservative and numerical solutions are usually more accurate than estimated.

More extensive treatments of these matters are given, e.g. by Smith (1965), Fox (1962), and Mitchell (1969).

8.12. Steady-state problems

Steady-state solutions have been obtained in earlier chapters for one-dimensional diffusion in a plane sheet, cylinder, or sphere. In two and three dimensions for these shapes, and particularly for a constant diffusion coefficient, analytical solutions can sometimes be obtained as in § 2.5 (p. 24). Even these are usually cumbersome for computation and are not available at all for more complicated shapes or non-linear problems. The numerical methods, which have been used extensively, are mostly referred to as finite-difference methods or finite-element methods.

8.12.1. *Finite-difference methods*

We are here referring to problems for which $\partial C/\partial t = 0$ everywhere. The finite-difference methods use the same replacements for the space derivatives in the diffusion equation and for the boundary conditions as in § 8.9 (p. 150). Large sets of simultaneous algebraic equations are solved by direct or iterative methods including the ADI methods. The practical details and discussions of accuracy and convergence are discussed e.g. by Smith (1965), Fox (1962), and Mitchell (1969). Singularities may occur both on the boundary or within it and these need special consideration. We have referred in § 8.10.2 (p. 153) to the Motz method (1946). This and other methods are discussed by Woods (1953), Fox and Sankar (1969), Whiteman (1970), Whiteman and Papamichael (1972), and Symm (1966).

8.12.2. *Finite-element methods*

These methods have been widely used for several years by engineers interested in stress problems and in steady-state heat flow. More recently, mathematicians have attempted to put the methods on a rigorous mathematical basis. The finite-element methods can be described as adaptations of the calculus of variations suited to numerical evaluation by a computer. So far, they have been applied to transient diffusion problems only in a few isolated instances (see e.g. Bruck and Zyvoloski (1973)). None of the results presented in this book have been obtained by using finite elements. The subject is in a very active state of development, however, and may well make a major contribution to the mathematics of diffusion in the next few years.

Meanwhile, the interested reader will find available simple introductions to both finite-difference and finite-element methods in the book by Myers (1971), a more general account by Zienkiewicz (1967, 1971), and the proceedings of a conference on the mathematics of finite elements and applications edited by Whiteman (1973).

9

SOME CALCULATED RESULTS FOR VARIABLE DIFFUSION COEFFICIENTS

9.1. Steady state

IN the steady state the concentration distribution through a plane sheet, in which diffusion is assumed to be one-dimensional, is given by the solution of the equation

$$\frac{d}{dx}\left(D\frac{dC}{dx}\right) = 0, \tag{9.1}$$

where D is the diffusion coefficient, not necessarily constant. The corresponding equations for the hollow cylinder and sphere are obvious.

9.1.1. D a function of concentration

If D is a given function of concentration, i.e.

$$D = D_0\{1+f(C)\},$$

the general solution of (9.1) can be written as

$$D_0 \int \{1+f(C)\} \, dC = Ax+B, \tag{9.2}$$

where A and B are constants to be determined by the boundary conditions. The corresponding solution for the hollow cylindrical tube is

$$D_0 \int \{1+f(C)\} \, dC = A \ln r + B, \tag{9.3}$$

and for the hollow spherical shell,

$$D_0 \int \{1+f(C)\} \, dC = -\frac{A}{r} + B. \tag{9.4}$$

When the boundary conditions for the plane sheet are

$$C = C_1, \qquad x = 0, \tag{9.5}$$

$$C = C_2, \qquad x = l, \tag{9.6}$$

eqn (9.2) becomes, using Barrer's (1946) nomenclature,

$$\frac{C_1 + F(C_1) - C - F(C)}{C_1 + F(C_1) - C_2 - F(C_2)} = \frac{x}{l}, \tag{9.7}$$

where

$$F(C) = \int_0^C f(C')\,dC'. \tag{9.8}$$

Similarly, for the cylindrical tube or spherical shell with boundary conditions

$$C = C_1, \qquad r = r_1, \tag{9.9}$$

$$C = C_2, \qquad r = r_2, \tag{9.10}$$

we find for the cylinder

$$\frac{C_1 + F(C_1) - C - F(C)}{C_1 + F(C_1) - C_2 - F(C_2)} = \frac{\ln r_1 - \ln r}{\ln r_1 - \ln r_2}, \tag{9.11}$$

and for the sphere

$$\frac{C_1 + F(C_1) - C - F(C)}{C_1 + F(C_1) - C_2 - F(C_2)} = \frac{r_2}{r_1 - r_2} \frac{r_1 - r}{r}. \tag{9.12}$$

For any given relationship between D and C, the integrals $F(C)$ are readily evaluated either analytically, graphically, or numerically, and the concentration distribution follows immediately from the above equations. Some typical examples calculated by Barrer (1946) are reproduced in Figs. 9.1

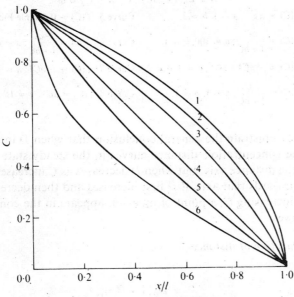

Fig. 9.1. Typical steady-state concentration distributions across a membrane when
$D = D_0\{1 + f(C)\}$. $C_1 = 1$, $C_2 = 0$.

Curve 1: $f(C) = aC$; $a = 100$. Curve 4: $f(C) = 0$ (simple Fick law obeyed).
Curve 2: $f(C) = aC$; $a = 10$. Curve 5: $f(C) = -aC$; $a = 0.5$.
Curve 3: $f(C) = aC$; $a = 2$. Curve 6: $f(C) = -aC$; $a = 1.0$.

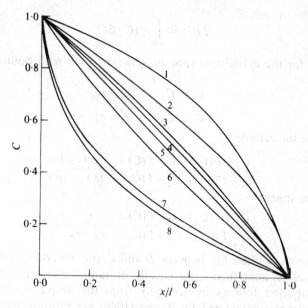

FIG. 9.2. Typical steady-state concentration distributions across a membrane when
$$D = D_0\{1 + f(C)\}. \quad C_1 = 1, C_2 = 0.$$

Curve 1: $f(C) = a\,e^{bC}$; $a = 1, b = 3$.   Curve 5: $f(C) = 0$ (simple Fick law obeyed).

Curve 2: $f(C) = \dfrac{aC}{1+bC}$; $a = 100, b = 1$.   Curve 6: $f(C) = \dfrac{-aC}{1+bC}$; $a = 0.9, b = 1$.

Curve 3: $f(C) = -aC^{\frac{1}{2}} + bC^2$; $a = 1, b = 2$.   Curve 7: $f(C) = -aC^{\frac{1}{2}}$; $a = 1$.

Curve 4: $f(C) = \dfrac{aC}{1+bC}$; $a = 1, b = 1$.   Curve 8: $f(C) = aC^{\frac{1}{2}}$; $a = 1$.

and 9.2. They illustrate the general conclusion that when D increases as C increases the concentration–distance curves in the steady state are convex away from the distance axis; but when D decreases as C increases the curves are convex towards that axis. If D first increases and then decreases or vice versa, with increasing C, a point of inflexion appears in the concentration–distance curves.

9.1.2. *D a function of distance*

If we have

$$D = D_0\{1 + f(x)\}, \tag{9.13}$$

or

$$D = D_0\{1 + f(r)\}, \tag{9.14}$$

for the sheet or cylinder and sphere, the general solutions (9.2), (9.3), and

(9.4) are to be replaced by

$$D_0 C = A \int \frac{dx}{1 + f(x)} + B, \tag{9.15}$$

$$D_0 C = A \int \frac{dr}{r\{1 + f(r)\}} + B, \tag{9.16}$$

or

$$D_0 C = A \int \frac{dr}{r^2\{1 + f(r)\}} + B, \tag{9.17}$$

respectively. Denoting by I the integral on the right-hand side of each of these equations, taking $x = 0$ or $r = 0$ as the lower limit and I_1, I_2 the values of I at the two boundary surfaces, we find

$$\frac{C_1 - C}{C_1 - C_2} = \frac{I_1 - I}{I_1 - I_2}. \tag{9.18}$$

Concentration distributions follow immediately for given $f(x)$ or $f(r)$. Barrer (1946) shows typical curves for $f(x) = ax$ and $f(x) = bx + ax^2$. They are reproduced in Fig. 9.3. When D is an increasing function of x the curves are convex towards the axis of x and when D is a decreasing function of x they are convex away from that axis.

9.1.3. Rate of flow

We saw in § 4.2.2 (p. 46) that the rate of flow F through unit area of a plane membrane of thickness l, when the concentrations at the two faces are C_1, C_2 is given by

$$F = (1/l) \int_{C_2}^{C_1} D \, dC, \tag{9.19}$$

where D is a function of concentration C. The corresponding argument for a cylindrical shell of inner radius r_1 and outer radius r_2 is as follows. Let F denote the rate of flow per unit-length of cylinder. Then

$$F = -2\pi r D \, dC/dr, \tag{9.20}$$

which on integration becomes

$$\int_{r_1}^{r_2} \frac{F}{2\pi r} \, dr = \int_{C_2}^{C_1} D \, dC. \tag{9.21}$$

But in the steady state F is independent of r and hence we find

$$F = \frac{2\pi}{\ln (r_2/r_1)} \int_{C_2}^{C_1} D \, dC. \tag{9.22}$$

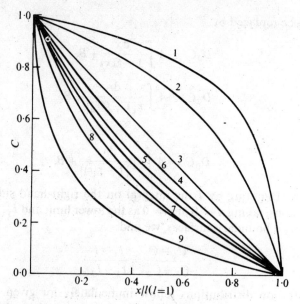

FIG. 9.3. Typical steady-state concentration distributions when $D = D_0\{1 + f(x)\}$. $C_1 = 1$, $C_2 = 0$.

Curve 1: $f(x) = -ax$; $a = 0.99$.
Curve 2: $f(x) = -ax$; $a = 0.90$.
Curve 3: $f(x) = 0$ (simple Fick law obeyed).
Curve 4: $f(x) = ax$; $a = 1.0$.
Curve 5: $f(x) = ax$; $a = 2.0$.

Curve 6: $f(x) = bx + ax^2$; $a = 1, b = 2$.
Curve 7: $f(x) = bx + ax^2$; $a = 2.25, b = 3$.
Curve 8: $f(x) = ax$; $a = 9$.
Curve 9: $f(x) = ax$; $a = 99$.

The corresponding result for a spherical shell is

$$F = 4\pi \frac{r_1 r_2}{r_2 - r_1} \int_{C_2}^{C_1} D \, dC, \tag{9.23}$$

where F now refers to the total flow through the shell. When D is a function of x or r given by (9.13) or (9.14) it is easy to see from (9.18) that

$$F = -D \frac{dC}{dx} = -D_0 \frac{C_1 - C_2}{I_1 - I_2}, \quad \text{for the plane sheet,} \tag{9.24}$$

$$F = -2\pi D_0 \frac{C_1 - C_2}{I_1 - I_2}, \quad \text{for the cylindrical tube,} \tag{9.25}$$

and

$$F = -4\pi D_0 \frac{C_1 - C_2}{I_1 - I_2}, \quad \text{for the spherical shell.} \tag{9.26}$$

9.1.4. *Asymmetrical diffusion through membranes*

Hartley [1948] has indicated that in any membrane for which the diffusion coefficient depends both on concentration and on distance through the membrane, different rates of penetration in the forward and backward direction are to be expected in general. Writing $D = f(x, C)$ we have for the rate of flow

$$F = -f(x, C)\, dC/dx. \tag{9.27}$$

If the function $f(x, C)$ is expressible as the product of two separate functions $f_1(x)$ and $f_2(C)$, (9.27) becomes

$$F \int_0^l \frac{dx}{f_1(x)} = \int_{C_2}^{C_1} f_2(C)\, dC, \tag{9.28}$$

where the boundaries of the sheet are at $x = 0$, $x = l$. In this case, since

$$\int_0^l \frac{dx}{f_1(x)} = \int_0^l \frac{dx}{f_1(l-x)}, \tag{9.29}$$

the permeability of the membrane will be symmetrical. The function $f(x, C)$ will not usually be separable in this way, however, except in the simple cases in which $f_1(x)$ or $f_2(C)$ is constant, and so it is unlikely that the membrane will be symmetrical unless D is constant or a function of either x or C alone.

Sternberg and Rogers (1968) set up a mathematical model of a membrane with a solubility coefficient which varied asymmetrically with position. The model gives a quantitative description of the directional transport process. Peterlin and Williams (1971) discuss four different theoretical models. The subject of selective permeation through asymmetric membranes is reviewed and discussed by Rogers and Sternberg (1971).

9.1.5. *Diffusion of one substance through a second substance which is confined between membranes*

A point of considerable interest concerning steady-state diffusion was raised by Hartley and Crank (1949). Let a substance B be confined between membranes which are impermeable to B. Also let the concentrations of A in contact with the membranes be maintained by supply and removal of A through the membranes from and to reservoirs of vapour or of solutions of A in B or in any other substance which cannot penetrate the membranes. It is convenient here to use the ideas and nomenclature of § 10.3 (p. 205). Then in the steady state, sections at fixed distances from the membranes are fixed with respect to amount of component B, and the rates of transfer of A across all such sections must be equal so that

$$-D_A^B\, \partial C_A^B/\partial \xi_B = \text{constant}. \tag{9.30}$$

We will assume that the partial volumes are constant. Eqns (10.21), (10.26), (10.28) are

$$d\xi_B = V_B^0 C_B^V dx; \qquad dC_A^V/dC_A^B = (V_B C_B^V)^2; \qquad D_A^B = D^V (V_B C_B^V)^2, \qquad (9.31)$$

and on substituting these relationships in (9.30) we have

$$\frac{-D^V}{V_B C_B^V} \frac{\partial C_A^V}{\partial x} = \text{constant.} \qquad (9.32)$$

On using

$$V_A \frac{\partial C_A^V}{\partial x} + V_B \frac{\partial C_B^V}{\partial x} = 0, \qquad (9.33)$$

(9.32) becomes

$$\frac{D^V}{V_A C_B^V} \frac{\partial C_B^V}{\partial x} = \text{constant.} \qquad (9.34)$$

If, instead, we maintain the steady state by substituting membranes permeable only to B and supplying and removing B in the reverse direction we find, by interchanging A and B in (9.32) and reversal of signs to allow for the reversal of direction,

$$\frac{D^V}{V_A C_A^V} \frac{\partial C_B^V}{\partial x} = \text{constant.} \qquad (9.35)$$

It is evident that eqns (9.34) and (9.35) require different concentration–distance functions. The steady state is therefore different, even between the same concentration limits, according to which component is constrained and which is free to diffuse.

This difference is particularly evident in dilute solutions. We shall assume here, in the interests of simplicity, that D^V is constant. The concentration of A being very low, $V_B C_B^V$ can be assumed to be unity. Eqn (9.32) now becomes

$$D^V \frac{\partial C_A^V}{\partial x} = \text{the constant rate of transfer of } A. \qquad (9.36)$$

This is valid for A diffusing and B restrained.

Substituting from (9.33) we may modify (9.35) to

$$\frac{D^V}{V_B C_A^V} \frac{\partial C_A^V}{\partial x} = \frac{D^V}{V_B} (\ln C_A^V) = \text{constant rate of transfer of } B. \qquad (9.37)$$

This is valid for A restrained and B diffusing between the same concentration limits.

It will be seen that when the dilute component is diffusing its concentration gradient under these simple conditions is linear. When the same limits are maintained by diffusing the 'solvent' through membranes impermeable to the 'solute', the gradient of the logarithm of concentration of the latter is linear and hence the concentration itself varies exponentially with distance.

When the total fall of concentration ΔC_A^V of the dilute constituent is small compared with its mean concentration, (9.36) becomes

$$\text{rate of transfer of } A = \frac{D^V}{l} \Delta C_A^V, \qquad (9.38)$$

and (9.37) becomes

$$\text{equivalent rate of transfer of } B = \frac{D^V}{l} \frac{\Delta C_A^V}{C_A^V V_B}, \qquad (9.39)$$

where l is the distance between the membranes. The rate of diffusion of water in such a system down a given, small, mean vapour pressure gradient is therefore not expected to be constant but to be inversely proportional to the mean concentration of the solute. It will further be proportional to the diffusion rate of the solute given by (9.38). Thus if water diffuses from 99·1 per cent to 99·0 per cent relative humidity through a layer containing hydrogen chloride, it will do so about twice as rapidly as when the layer contains sodium chloride. If the diffusion occurs from 90·1 per cent to 90·0 per cent relative humidity, each rate will fall to one-tenth of its former value.

These conclusions refer to the case where the membranes are separated by a fixed distance. If, as is more likely to be true in practice, the membranes confine a given amount of component A so that the volume between them will vary inversely as the mean concentration of A, the distance l in (9.39) will be more nearly inversely proportional to C_A than constant. In this case the rate of transfer of B will be, to a first approximation, dependent on ΔC_A only and not on the mean value of C_A. This result may be obtained directly from equations in ξ in the treatment given of the swelling membrane in § 10.6.5. (p. 239). Diffusion of solvent through a constant amount of solute per unit area thus behaves more simply than diffusion through a constant thickness. This conclusion is not at once obvious and may have some important applications in physiological processes.

A second conclusion of interest may be drawn when we consider what happens in such a steady-state system if the membranes are suddenly rendered completely impermeable. Sections fixed with respect to the membranes are now fixed with respect to volume of solution. The change of concentration with time will therefore from now on be governed by

$$\frac{\partial C_A^V}{\partial t} = \frac{\partial}{\partial x} \left(D^V \frac{\partial C_A^V}{\partial x} \right), \qquad (9.40)$$

but, at the instant of change of membranes, (9.32) still holds if A has been the diffusing component, whence

$$\frac{\partial}{\partial x}\left(D^V \frac{\partial C^V}{\partial x}\right) = -\text{constant} \times V_B \frac{\partial C_B^V}{\partial x} = \text{constant} \times V_A \frac{\partial C_A^V}{\partial x} \qquad (9.41)$$

Combining (9.40) and (9.41) we obtain

$$\left(\frac{dx}{dt}\right)_C = -\text{constant} \times V_A, \qquad (9.42)$$

since

$$\left(\frac{\partial C_A^V}{\partial t}\right) dt + \left(\frac{\partial C_A^V}{\partial x}\right) dx = dC_A^V = 0, \qquad (9.43)$$

for C_A^V constant. Now the constant in (9.30) and (9.42) is the rate of transfer of A at the steady state in standard units of amount. Multiplied by V_A it represents the volume rate of transfer of A, or, since we are always considering transfer across unit area, the linear velocity with which A appeared to pass through the system.

We thus find that, at the instant when A ceases to flow through, the whole concentration–distance distribution commences to move backwards at the velocity with which substance A previously passed through the membrane. With increasing time, the distribution will flatten out, of course, from the low concentration upwards as the substance A, diffusing down the gradient now, accumulates.

It is evident, therefore, that in the steady state during the passage of substance A there was superimposed on the true diffusion process a real flow of the whole system.

9.2. Non-steady-state conditions

We present here, in graphical form, a collection of solutions of the equation for diffusion in one dimension. They have been evaluated by one or other of the methods described in Chapters 7 and 8, and refer to various types of concentration-dependent diffusion coefficients and three simple boundary conditions. Solutions for the infinite medium refer to the initial condition

$$C = C_1, \qquad x < 0, \qquad (9.44)$$

$$C = C_2, \qquad x > 0. \qquad (9.45)$$

For sorption in a semi-infinite medium the conditions are

$$C = C_0, \qquad x = 0, \qquad t > 0, \qquad (9.46)$$

$$C = 0, \qquad x > 0, \qquad t = 0, \qquad (9.47)$$

and for desorption

$$C = 0, \qquad x = 0, \qquad t > 0, \qquad (9.48)$$

$$C = C_0, \qquad x > 0, \qquad t = 0. \qquad (9.49)$$

One example of each type of diffusion coefficient considered is shown in Figs. 9.4(a) and 9.4(b).

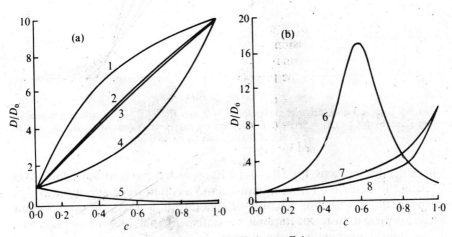

FIG. 9.4. Some typical diffusion coefficients.

1. $D/D_0 = 1 + 10(1 - e^{-2 \cdot 303c})$.
2. $D/D_0 = 1 + 50 \ln (1 + 0.5136c)$.
3. $D/D_0 = 1 + 9c$.
4. $D/D_0 = e^{2 \cdot 303c}$.

5. $D/D_0 = e^{-2 \cdot 303c}$.
6. $D/D_0 = 1/(1 - 3.292c + 2.877c^2)$.
7. $D/D_0 = 1/(1 - 0.6838c)^2$.
8. $D/D_0 = 1/(1 - 0.9c)$.

9.3. Concentration–distance curves

Shampine (1973a) used a Runge–Kutta computer library program to obtain numerical solutions of the Boltzmann diffusion equation (7.10) subject to $C = 1, \eta = 0$; $C = 0, \eta = \infty$. His 'shooting method' is analagous to that of Lee, § 7.2.2 (iii) (p. 108). Tabulated values of $C(\eta)$ for each of the eight diffusion coefficients listed below Fig. 9.4 may be obtained from L. F. Shampine. In his paper, he also examines the existence and uniqueness properties.

The shapes of the concentration–distance curves are characteristic of the diffusion coefficient and of the boundary conditions. They do not conform to quite such a simple classification as do the corresponding curves in the steady state. Thus, as we saw earlier in § 9.1.1. (p. 162) Barrer (1946) was able to write, with regard to the results of his steady-state calculations, that whenever D increases as C increases, concentration–distance curves are convex away

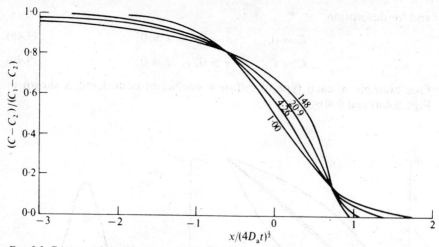

FIG. 9.5. Concentration–distance curves for $D = D_a \exp \beta\{C - \frac{1}{2}(C_1 + C_2)\}$, where β is positive and given by $\beta(C_2 - C_1) = \ln(D_2/D_1)$. Numbers on curves are values of D_1/D_2. D_1 and D_2 are values of D at C_1 and C_2 respectively.

from the distance axis. Figs. 9.8 and 9.10 show that this is true also for curves relating to desorption from a semi-infinite medium when D is a linearly or exponentially increasing function of concentration. Figs. 9.7 and 9.9 show that for these diffusion coefficients the statement holds also for sorption by a semi-infinite medium over the greater part of the concentration range, but that there is an important difference in behaviour in the region of low con-

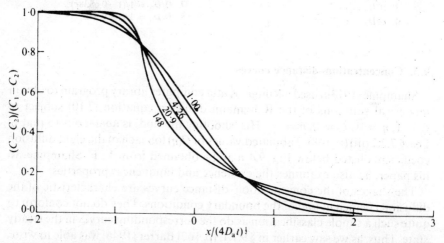

FIG. 9.6. Concentration–distance curves. $D = D_a \exp \beta\{C - \frac{1}{2}(C_1 + C_2)\}$, where β is negative and given by $(C_2 - C_1) = \ln(D_2/D_1)$. Numbers on curves are values of D_2/D_1. D_1 and D_2 are values of D at C_1 and C_2 respectively.

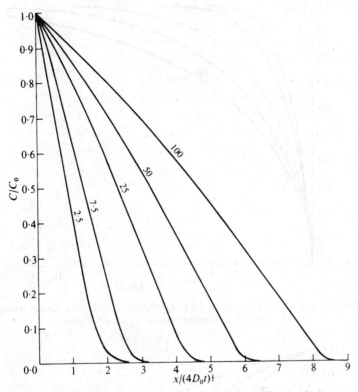

FIG. 9.7. Concentration–distance curves for linear diffusion coefficients during sorption for $D = D_0(1 + aC/C_0)$. Numbers on curves are values of a.

centration. This difference is a direct consequence of the boundary condition. In the steady state, the condition is that the concentration shall have some fixed value, possibly zero, at the face of the membrane through which the diffusing substance emerges. When diffusion occurs into a semi-infinite medium, however, the condition that the concentration shall approach zero at infinity means that the gradient of concentration tends to zero at the limit of penetration into the medium. This produces a point of inflexion in any concentration–distance curve which is convex away from the distance axis at high concentrations. The curves of Fig. 9.11 relate to a diffusion coefficient which increases as C increases but does so at a steadily decreasing rate (see Fig. 9.4(a)). We see that in this case the concentration–distance curves may be convex downwards. The curves of Figs. 9.12 and 9.13 conform with these statements.

The curves shown in Figs. 9.5 and 9.6 for infinite media follow the same general pattern as for the semi-infinite media, except that the boundary conditions ensure that the concentration gradient becomes zero at each end.

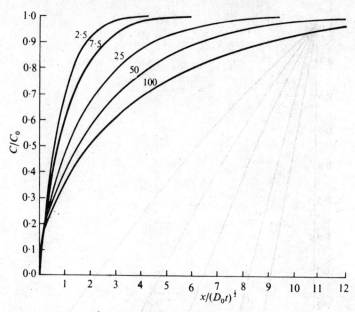

FIG. 9.8. Concentration–distance curves for linear diffusion coefficients during desorption for $D = D_0(1 + aC/C_0)$. Numbers on curves are values of a.

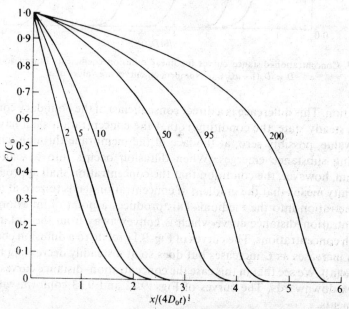

FIG. 9.9. Concentration–distance curves for exponential diffusion coefficients during sorption for $D = D_0 \exp(kC/C_0)$. Numbers on curves are values of e^k, being the ratio of D at $C = C_0$ to D at $C = 0$.

9.3.1. *Correspondence between sorption and desorption*

If $(C_0 - C)$ is written for C, the diffusion equation in one dimension is unchanged but the conditions (9.46) and (9.47) become respectively (9.48) and (9.49). This means that the solution for sorption when D is a given function of C is also the solution for desorption when D is the same function of $(C_0 - C)$ and vice versa. For example, the sorption curve for $D = D_0 \exp(kC)$ is also the desorption curve for

$$D = D_0 e^{k(C_0 - C)} = (D_0 e^{kC_0}) e^{-kC}$$

This allows general statements corresponding to those of §9.3 to be made for diffusion coefficients which decrease as concentration increases.

Helfferich (1963) has drawn attention to the general nature of this correspondence principle. One system is specified by

$$\partial C_i / \partial t = \operatorname{div}(D_i \operatorname{grad} C_i),$$

$$D_i = f_1(c_i), \tag{9.50}$$

$$C_i = f_2(x), \qquad x > 0, \qquad t = 0,$$

$$C_i = f_3(t), \qquad x = 0, \qquad t \geqslant 0,$$

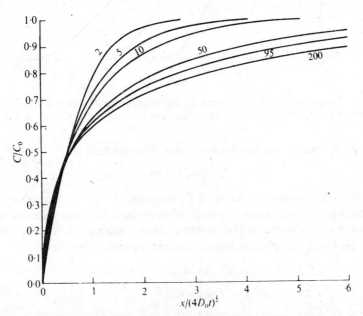

FIG. 9.10. Concentration–distance curves for exponential diffusion coefficients during desorption for $D = D_0 \exp(kC/C_0)$. Numbers on curves are values of e^k, being the ratio of D at $C = C_0$ to D at $C = 0$.

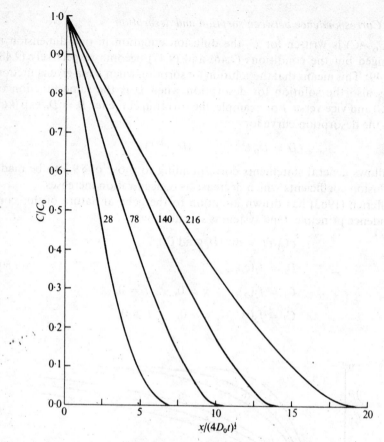

FIG. 9.11. Concentration–distance curves during sorption for $D = D_0\{1 + 50\ln(1 + kC/C_0)\}$. Numbers on curves are values of $1 + 50\ln(1 + k)$, being the ratio of D at $C = C_0$ to D at $C = 0$.

where f_1, f_2, and f_3 are known functions. The solution is

$$C_i = F(x, t).$$

The solution remains the same if C_i is replaced by $C_0 - C_j$, where C_j is variable and C_0 is a constant equal to or greater than the maximum concentration C_i occurring in the problem. Thus, once calculated, the solution $F(x, t)$ provides the solution also to the corresponding problem

$$\partial C_j/\partial t = \text{div}\,(D_j\,\text{grad}\,C_j),$$

$$D_j = g_1(C_j) = f_1(C_0 - C_j),$$

$$C_j = g_2(x) = C_0 - f_2(x), \qquad x > 0, \qquad t = 0,$$

$$C_j = g_3(t) = C_0 - f_3(t), \qquad x = 0, \qquad t \geqslant 0,$$

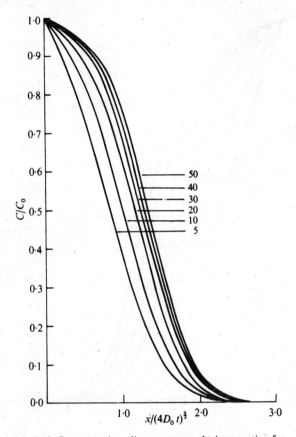

FIG. 9.12. Concentration–distance curves during sorption for

$$D = \frac{D_0}{(1 - \alpha C/C_0)}.$$

Numbers on curves are values of $1/(1-\alpha)$, being ratio of D at $C = C_0$ to D at $C = 0$.

and the solution is

$$C_j = G(x, t) = C_0 - F(x, t).$$

Helfferich lists some typical corresponding solutions for sorption and desorption. The correspondence principle applies in two and three dimensions and in anisotropic media provided that each diffusion coefficient along the principal axes obey a correspondence relation.

The diffusion coefficient may depend on time and position, in addition to concentration and the absolute value of the concentration gradient, provided the dependence is the same in the corresponding functions f_1 and g_1. Temperature- and stress-dependent diffusion systems can also correspond.

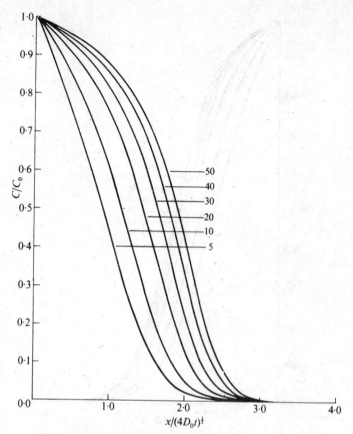

FIG. 9.13. Concentration–distance curves during sorption for

$$D = \frac{D_0}{(1 - \alpha C/C_0)^2}.\dagger$$

Numbers on curves are values of $1/(1-\alpha)^2$, being ratio of D at $C = C_0$ to D at $C = 0$.

9.3.2. Common points of intersection

Stokes (1952) drew attention to the fact that if a large-scale graph is prepared from the data of Table 7.3 (p. 384) an interesting property emerges. Two concentration-gradient curves are shown in Fig. 9.14, one for a constant diffusion coefficient and the other for a diffusion coefficient varying linearly with concentration, the value at the higher concentration being 0·1406 of that at the lower concentration. The other curves are omitted to avoid confusion of the diagram but on his large-scale plot Stokes found that whatever the

† Garg and Ruthven (1972) calculated concentration, and sorption and desorption curves in zeolite spherical crystals for this D and that of Fig. 9.12.

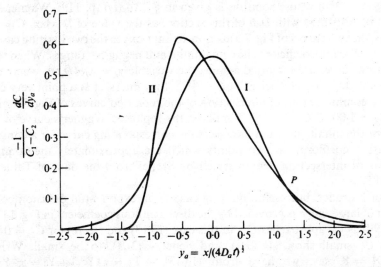

FIG. 9.14. Comparison of concentration-gradient curves. Curve I. For a constant diffusion coefficient. Curve II. For a diffusion coefficient varying linearly with concentration, the value at the higher concentration being 0·1406 of that at the lower concentration.

value of b, all his curves appeared to pass through the point P (Fig. 9.14) of coordinates

$$y_a = 1·205, \qquad \frac{dC/dy_a}{C_1 - C_2} = -0·133.$$

The nomenclature here is that of § 7.2.4. (iii) (p. 118). If a similar large-scale graph of $(C - C_2)/(C_1 - C_2)$ is prepared from the data of Table 7.4 (p. 385), there appear to be two common points with coordinates

$$y_a = +0·66, \qquad (C - C_2)/(C_1 - C_2) = 0·176;$$

$$y_a = -0·66, \qquad (C - C_2)/(C_1 - C_2) = 1 - 0·176 = 0·824.$$

If the sorption curves of Fig. 9.7 for linear D are replotted against the variable y_a they too pass through a 'common point' of coordinates $y_a = 1·00$, $C/C_0 = 0·157$. For the replotted desorption curves of Fig. 9.8 the common point is $y_a = 1·09$, $C/C_0 = 0·872$. Similar common points are found in other cases. For example, when the curves of Fig. 9.9 for exponential diffusion coefficients are replotted against $x/(4\bar{D}t)^{\frac{1}{2}}$, where $\bar{D}$ is the integrated mean value $D_0(e^k - 1)/k$, they all pass approximately through the point for which $\bar{C}/C_0 = 0·25$ and $x/(4\bar{D}t)^{\frac{1}{2}} = 0·85$.

We can obtain a partial understanding of the significance of these apparently common points by looking at the limiting case of a linear diffusion coefficient of very large range. Thus, a very large $D = D_0(1 + aC/C_0)$ is approximated by $D = D_0 aC/C_0$, except for very small C, and this is the form examined by

Wagner (1950) whose solution is given in § 7.2.4 (iv) (p. 119). Wagner's D_0 is to be identified with $D_0 a$ of this section, as the value of D when $C = C_0$. Thus, the two curves of Fig. 7.5 are approximations to the two limiting cases of linear diffusion coefficients having infinite and negligible ranges. When these limiting curves are replotted against the variable $y_a = x/(4D_a t)^{\frac{1}{2}}$, where now D_a is the diffusion coefficient at $C = \frac{1}{2}C_0$, they intersect at a point very close to the common point of intersection of the replotted curves of Fig. 9.7 given by $y_a = 1.00$, $C/C_0 = 0.157$. Similarly, the replotted Wagner's curve of Fig. 7.6 for the infinite medium intersects the corresponding curve for a constant diffusion coefficient in two points which are approximately the common points of intersection demonstrated by Stokes from the data of Table 7.4 (p. 385).

A much cruder, but useful, limiting case of an exponential diffusion coefficient of large range is provided by the discontinuous coefficient in Fig. 13.7(a) (p. 294) as C_X approaches C_1. In this case (13.13) and (13.18) for $C_2 = 0$ and $C_X - C_1$ small show $g(k/2D_1^{\frac{1}{2}})$, and hence $k/(2D_1^{\frac{1}{2}})$, to be small. Writing $k/2D_1^{\frac{1}{2}} = Y$, say, we have from (13.13), $g(Y) = \pi^{\frac{1}{2}} Y^2 (1 + Y) = \pi^{\frac{1}{2}} Y^2 = \pi^{\frac{1}{2}} k^2/(4D_1)$ approximately. Thus, using (13.11) and (13.18), we have

$$g = \frac{\pi^{\frac{1}{2}} X^2}{4D_1 t} = \frac{C_1 - C_X}{C_X} = 1 - \frac{C_X}{C_1},$$

since C_X is approaching C_1. But for the discontinuous diffusion coefficient we have

$$\bar{D} = \frac{1}{C_1} \int_0^{C_1} D \, dC = D_1 \left(1 - \frac{C_X}{C_1}\right)$$

and so finally

$$\frac{X}{(4\bar{D}t)^{\frac{1}{2}}} = \frac{1}{\pi^{\frac{1}{2}}} = 0.75 \text{ approximately.}$$

Now X is the position of the discontinuities in the concentration curves of Fig. 13.7, and replotting them against $x/(4\bar{D}t)^{\frac{1}{2}}$ confirms that as $C_X \to C_1$, $X/(4\bar{D}t)^{\frac{1}{2}}$ approximates to 0.75. The intersection with the concentration curve for a constant D will be at about $X/(4\bar{D}t)^{\frac{1}{2}} = 0.75$, $C/C_1 = 0.3$, which is not far from the 'common point' of the replotted curves of Fig. 9.9.

Wilkins (1963), however, carried out a more accurate and systematic study of the concentration curves associated with linear diffusion coefficients of the form $D = D_0(1 + aC/C_0)$. He obtained series solutions for $a = -1$ as well as $a = \infty$ (see § 7.2.4, (iv), p. 124) and for small values of a. Table 9.1 summarizes Wilkins' findings on the positions of the various intersection points and shows that they are quite insensitive to a but not strictly independent of a.

The application of these points to the measurement of concentration dependence as suggested by Stokes (1952) and Crank and Henry (1949b) is still possible provided the 'spread' demonstrated by Wilkins is within the errors of observation in a practical experiment.

9.4. Sorption– and desorption–time curves

On the basis of the sorption and desorption curves shown in Figs. 9.15, 9.16, and 9.17 and of corresponding curves published elsewhere (Crank and Henry 1949a) we may draw the following general conclusions for a system in which the diffusion coefficient increases as concentration is increased but does not depend on any other variable.

(i) In the early stages, when diffusion takes place essentially in a semi-infinite medium, the amount sorbed or desorbed is directly proportional to the square root of time. This is true, incidentally, whatever the relationship between the diffusion coefficient and concentration, and follows directly from the fact that for the prescribed boundary conditions concentration depends on the single variable $x/t^{\frac{1}{2}}$ (§ 7.2.1, p. 105). When D increases with concentration increasing, the linear behaviour may extend well beyond 50 per cent of the final equilibrium uptake in the case of sorption. The same is true for desorption when D decreases with concentration increasing.

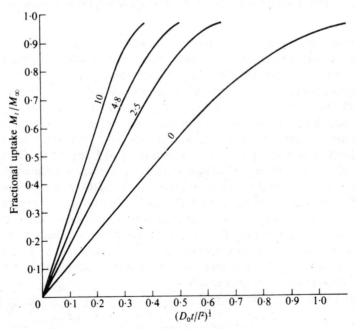

FIG. 9.15. Sorption curves for $D = D_0(1 + aC/C_0)$. Numbers on curves are values of a.

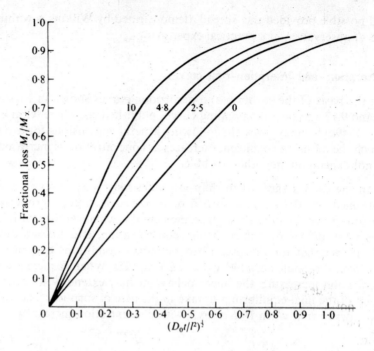

FIG. 9.16. Desorption curves for $D = D_0(1 + aC/C_0)$. Numbers on curves are values of a.

(ii) When they cease to be linear, the sorption and desorption curves plotted against (time)$^{\frac{1}{2}}$ each become concave towards the (time)$^{\frac{1}{2}}$ axis, and steadily approach the final equilibrium value. This is true for all the calculated results obtained for the initial and boundary conditions (9.46), (9.47), (9.48), (9.49) and it is reasonable to conclude that it is a quite general result though no satisfactory general proof has yet been produced. It is an important property because sorption curves have been observed experimentally (Crank and Park 1951; Mandelkern and Long 1951) which show points of inflexion when plotted against (time)$^{\frac{1}{2}}$, i.e. there is a region in which the curve is convex to the (time)$^{\frac{1}{2}}$ axis. This has been taken as evidence either that the boundary condition (9.46) does not describe the experimental conditions, or that the diffusion coefficient is a function of some variable other than concentrations. Prager (1951) has given a general proof that sorption and desorption curves when plotted against *time* can never become convex to the time axis, but this is a less stringent restriction than the one just put forward. His proof is as follows.

We wish to show that the integral

$$M_t = \int_0^\infty C(x, t)\,\mathrm{d}x, \qquad (9.51)$$

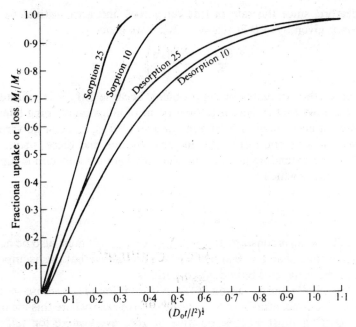

FIG. 9.17. Sorption and desorption curves for exponential diffusion coefficient $D = D_0 \exp(kC/C_0)$. Numbers on curves are values of e^k being ratio of D at $C = C_0$ to D at $C = 0$.

obtained as a solution of

$$\frac{\partial C}{\partial t} = \frac{\partial}{\partial x}\left(D\frac{\partial C}{\partial x}\right), \tag{9.52}$$

with the conditions

$$C = C_0 = \text{constant}, \quad x = 0, \quad x = l, \quad t > 0, \tag{9.53}$$

$$C = 0, \quad 0 < x < l, \quad t = 0, \tag{9.54}$$

cannot yield an inflexion point when plotted against t. The diffusion coefficient D is a function of concentration C only and is always positive. On writing

$$s(C) = \int_0^C D(C')\,dC', \tag{9.55}$$

(9.52) becomes

$$D(s)\frac{\partial^2 s}{\partial x^2} = \frac{\partial s}{\partial t}, \tag{9.56}$$

with conditions

$$s = S_0, \quad x = 0, \quad x = l, \quad t > 0, \tag{9.57}$$

$$s = 0, \quad 0 < x < l, \quad t = 0. \tag{9.58}$$

Furthermore, since the rate of transfer across unit area normal to the x-direction is given by $-D\,\partial C/\partial x = -\partial s/\partial x$, we have

$$\frac{\partial M_t}{\partial t} = -2\left(\frac{\partial s}{\partial x}\right)_{x=0} \tag{9.59}$$

allowing for the symmetry of the problem. Because dM_t/dt is infinite when $t = 0$, it follows that if there is to be a point of inflexion at some later time, dM_t/dt must go through a minimum, and so $(\partial s/\partial x)_{x=0}$ must go through a maximum as a function of t. This in turn requires that there is some finite length of time, extending just beyond that at which the inflexion is supposed to occur, during which

$$\frac{d}{dt}\left(\frac{\partial s}{\partial x}\right)_{x=0} < 0. \tag{9.60}$$

Also, since $s = S_0$ is constant at $x = 0$ and $x = l$, $\partial s/\partial t$ must also be negative during that time near $x = 0$ and $x = l$. We now show that this is impossible for the given initial and boundary conditions.

For very small t, the condition $\partial C/\partial t \geqslant 0$ holds for all x, because we cannot have negative concentrations. Since D is positive everywhere this means that $\partial s/\partial t = D\,\partial C/\partial t$ must also be positive or zero everywhere for sufficiently small t. But $\partial C/\partial t$, and hence $\partial s/\partial t$, is continuous everywhere and $\partial s/\partial t$ must therefore first be zero if it is to become negative. Let $x = X$ be the first point at which $\partial s/\partial t = 0$ excepting $x = 0$ or $x = l$ where this is always true. Then $\partial s/\partial t$ will be positive or zero on both sides of X, i.e. it will show a minimum at X when plotted against x so that

$$\left\{\frac{\partial^2}{\partial x^2}\left(\frac{\partial s}{\partial t}\right)\right\}_{x=X} \geqslant 0,$$

or

$$\left\{\frac{\partial}{\partial t}\left(\frac{\partial^2 s}{\partial x^2}\right)\right\}_{x=X} \geqslant 0. \tag{9.61}$$

But it follows from (9.56) that since $\partial s/\partial t = 0$ at $x = X$, so also $\partial^2 s/\partial x^2 = 0$ there. Eqn (9.61) then indicates that in the next instant $(\partial^2 s/\partial x^2)_{x=X}$ is positive or zero, and so using (9.56) again and remembering that $D > 0$ we see that $(\partial s/\partial t)_{x=X}$ becomes positive or zero. Thus $\partial s/\partial t$ can never become negative and so there can never be a point of inflexion in the plot of M_t against t, i.e. in the uptake curve plotted against time.

(iii) If D increases as concentration increases, the shape of the sorption curve when plotted against *time* is not very sensitive to the form of the diffusion coefficient. It is often not significantly different from the corresponding curve for a constant diffusion coefficient. This is, of course, because the sorption curves are parabolic, i.e. linear against $(\text{time})^{\frac{1}{2}}$, over most of their length. The desorption curves when plotted against time are much more sensitive to the form of the diffusion coefficient if this increases as concentration increases. If the diffusion coefficient *decreases* as concentration increases, then the *desorption* curve will approximate to that for a constant diffusion coefficient.

(iv) When D increases with concentration increasing throughout the relevant range of concentration, desorption is always slower than sorption, and conversely if D decreases with concentration increasing. This is illustrated by Fig. 9.17. In particular, the last stages of desorption are much slower than those of sorption if D increases with concentration increasing, and vice versa.

Crank and Henry (1949a) examined the sorption and desorption curves associated with a diffusion coefficient which passes through a maximum value at some intermediate concentration. Three such diffusion coefficients are shown in Fig. 9.18, where $c = C/C_0$ as usual. They correspond to the algebraic relationships:

$$D/D_0 = 1 + 14 \cdot 8c(1 - c), \tag{9.62}$$

$$D/D_0 = 1 + 100c^2 \exp(-10c^2), \tag{9.63}$$

$$D/D_0 = 1 + 100(1 - c)^2 \exp\{-10(1 - c)^2\}. \tag{9.64}$$

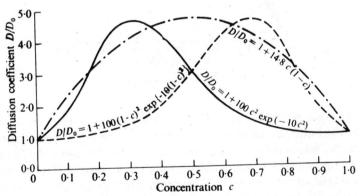

FIG. 9.18. Diffusion coefficient–concentration relationships.

All three satisfy the conditions $D/D_0 = 1$ when $c = 0$ and $c = 1$; and the maximum value of D/D_0 is approximately $4 \cdot 7$ in each case. Eqn (9.62) is a symmetrical form in which this maximum occurs at $c = 0 \cdot 5$, while the maximum values for (9.63) and (9.64) are at $c = 0 \cdot 3$ and $0 \cdot 7$ respectively, since (9.64) follows immediately from (9.63) by writing $(1 - c)$ for c. The sorption and desorption curves for these diffusion coefficients are presented in Fig. 9.19.

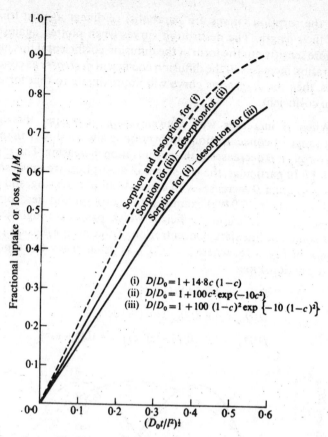

FIG. 9.19. Sorption and desorption curves.

Since (9.62) is symmetrical in c and $(1-c)$ it follows from what was said above that for the boundary conditions considered the sorption– and desorption–time curves are coincident, and this is, of course, true for any symmetrical relationship between D and c. The results show that for D given by (9.63) with a maximum at $c = 0.3$ the desorption curve lies wholly above that for sorption, while the opposite is true for the relation (9.64) where the maximum value of D occurs at $c = 0.7$.

The relative behaviour of the sorption– and desorption–time curves is affected by the relative values of the diffusion coefficient at $c = 0$ and $c = 1$, and also by the position of the maximum in the D–c curve if one occurs. This suggests that, in some cases, the desorption curve may be above the sorption curve in the early stages of diffusion but later may cross it so that the final stages of desorption are again slower than those of sorption. This is likely to occur when there is a maximum in the diffusion coefficient–

concentration relationship and when the value of D at $c = 1$ is greater than at $c = 0$.

In order to study this behaviour a diffusion coefficient–concentration relationship of the form

$$D/D_0 = 1 + \alpha \operatorname{erf}(\beta c) + \gamma c \qquad (9.65)$$

was used, where α, β, γ are constants for any one curve. There is no significance in the precise form of (9.65) except that it leads to diffusion coefficient–concentration curves of the desired form and is convenient to handle numerically. From (9.65) we have

$$\frac{d(D/D_0)}{dc} = \frac{2\alpha\beta}{\pi^{\frac{1}{2}}} \exp(-\beta^2 c^2) + \gamma = 0, \qquad (9.66)$$

when D has a maximum value. All curves given by (9.65) pass through the point $D/D_0 = 1$, $c = 0$. The three further conditions that D/D_0 shall have a prescribed value at $c = 1$ and a given maximum value at a prescribed value of c, can be satisfied by suitable choice of the parameters α, β, γ, the desired values being readily determined by use of (9.65) and (9.66). A typical curve of this family, actually the one given by

$$D/D_0 = 1 + 4 \cdot 62 \operatorname{erf}(6 \cdot 10c) - 3 \cdot 12c, \qquad (9.67)$$

which satisfies the conditions

$$\left. \begin{array}{llll} D/D_0 = 1, & c = 0; & D/D_0 = 2 \cdot 5, & c = 1; \\ d(D/D_0)/dc = 0, & D/D_0 = 4 \cdot 7, & c = 0 \cdot 25, \end{array} \right\} \qquad (9.68)$$

is shown in Fig. 9.21 together with the sorption– and desorption–time curves. The curves intersect at

$$(D_0 t/l^2)^{\frac{1}{2}} = 0 \cdot 29.$$

For values of $(D_0 t/l^2)^{\frac{1}{2}}$ less than this, desorption proceeds more rapidly than sorption but after the intersection the desorption curve lies below that for sorption. This is to be contrasted with the sorption and desorption curves shown in Fig. 9.22 which do not intersect and which are for a diffusion coefficient of the form

$$D/D_0 = 1 + 29 \cdot 86 \operatorname{erf}(0 \cdot 98c) - 23 \cdot 40c. \qquad (19.69)$$

This diffusion coefficient is also shown in Fig. 9.22 where it is seen to differ from that defined by (9.67) in that it has a maximum value at $c = 0 \cdot 6$ instead of $c = 0 \cdot 25$. Fig. 9.20 shows further curves for a diffusion coefficient of the same general shape having a maximum value at $c = 0 \cdot 125$. It is clear from inspection of these curves that, keeping the end points of the diffusion coefficient–concentration curves fixed at $D/D_0 = 1$ and $D/D_0 = 2 \cdot 5$ and the

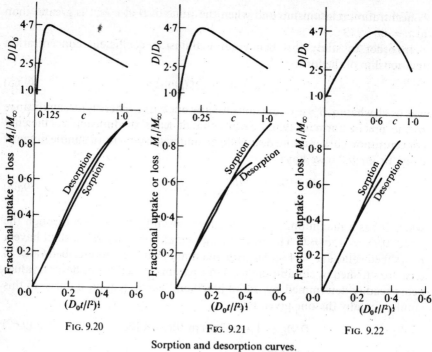

FIG. 9.20   FIG. 9.21   FIG. 9.22

Sorption and desorption curves.

FIG. 9.20. $D/D_0 = 1 + 4.06 \, \text{erf} \, (14.50c) - 2.56c$.

FIG. 9.21. $D/D_0 = 1 + 4.62 \, \text{erf} \, (6.10c) - 3.12c$.

FIG. 9.22. $D/D_0 = 1 + 29.86 \, \text{erf} \, (0.98c) - 23.40c$.

maximum value of $D/D_0 = 4.7$, as the position of the maximum values moves back from $c = 1$ to $c = 0$ there is first a range of positions of the maximum for which the whole process of desorption is slower than sorption. Continuing to move the position of the maximum towards $c = 0$ there is evidently a further range of positions for which the desorption and sorption curves intersect, the point of intersection occurring at successively larger values of $D_0 t/l^2$ as the position of the maximum moves towards $c = 0$. There is some intermediate position of the maximum for which the desorption curve crosses the sorption curve at the origin only and this is the limiting case between sorption and desorption curves which intersect for $t > 0$ and those which do not.

For this set of curves having $D/D_0 = 2.5$, $c = 1$, it is found that the limiting position of the maximum for which intersection of the sorption and desorption curves occurs at $D_0 t/l^2 = 0$ is about $c = 0.26$. When the maximum occurs at higher values of c than this, desorption is slower than sorption right from $D_0 t/l^2 = 0$, but when the maximum lies in the range $0 < c < 0.26$ the desorption curve is first above the sorption curve but crosses it later so that the final stages of desorption are again slower than those of sorption.

Still confining attention to the general form of variable diffusion coefficient expressed by (9.65), it is to be expected that the critical position of the maximum for which the sorption and desorption curves have equal gradients in the neighbourhood of $t = 0$ when plotted against $(D_0 t/l^2)^{\frac{1}{2}}$ will vary as the value of D at $c = 1$ is caused to vary, taking the maximum value of D to remain constant and the condition $D/D_0 = 1$, $c = 0$, to be satisfied in all cases. This expectation is confirmed by the results presented graphically in Fig. 9.23 where the critical position of the maximum is plotted as a function of the value of D/D_0 at $c = 1.0$. The critical position is seen to move towards $c = 0$ as the value of D at $c = 1$ is increased.

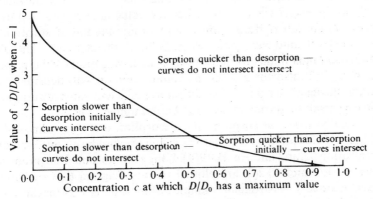

FIG. 9.23. Critical position of the maximum of D/D_0 as a function of D/D_0 at $c = 1.0$.

The points at each end of the curve of Fig. 9.23 were arrived at by general reasoning and the intermediate points by calculation. Thus considering first a D–c curve for which the value of D/D_0 at $c = 1$ is very close to 4.7, which is the value of the maximum D/D_0 for this family of curves, it is clear that, in general, for a D–c curve of this type the diffusion coefficient is effectively increasing over the whole range of concentration, and sorption will be quicker than desorption throughout. This is true for all positions of the maximum except when it is so near to $c = 0$ that the diffusion coefficient is effectively constant over the whole range of concentration, in which case the sorption– and desorption–time curves coincide throughout and $c = 0$ is therefore the limiting position of the maximum when $D/D_0 = 4.7$ at $c = 1$.

By an extension of this argument the critical position of the maximum can be determined when $D/D_0 = 0$ at $c = 1$. It was seen above that the sorption–time curve obtained when D is a certain function of c is the same as the desorption–time curve when D is the same function of $(1 - c)$ and vice versa. It follows immediately that if the initial rates of sorption and desorption are equal when D is a certain function of c, they will also be equal when D is the same function of $(1-c)$, i.e. if the critical position of the maximum value of D is $c = c_m$ in the first case, it will be $c = 1 - c_m$ in the second case. Now if D is a function of c such that D is very small when $c = 1$, the corresponding function of $(1-c)$ is such that the value of D at $1-c = 1$ is relatively very close to the maximum value of D. This function approximates to the type just considered above and the critical position of the maximum value of D is at $(1-c) = 0$ and therefore at $c = 1$ for the original D–c curve.

For any D–c curve of the general family under discussion, the relative behaviour of the sorption– and desorption–time curves can be deduced from Fig. 9.23 if the value of D/D_0 at $c = 1$ and the position of the maximum are known. Four regions are to be distinguished, in two of which the sorption– and desorption–time curves intersect at some time, $t > 0$, and in the other two they do not intersect when $t > 0$. When the value of D/D_0 at $c = 1$ exceeds unity, the initial rate of desorption is more rapid than that of sorption if the curves intersect, and vice versa when $D/D_0 < 1.0$ at $c = 1$. The critical position of the maximum no doubt depends on the magnitude of the maximum value of D, and also for a given maximum value it will depend to some extent on the detailed shape of the diffusion coefficient–concentration curve. These aspects of the problem have not yet been investigated.

The relative rates of sorption and desorption are also examined later (§ 13.2, p. 387) for diffusion coefficients which are discontinuous functions of concentration (Crank and Henry 1951). If the sorption and desorption curves of Figs. 13.1(d) and 13.2(d) are plotted on one diagram they are found to intersect markedly. A particular example is shown in Fig. 9.24 for the diffusion coefficient shown in the inset. These curves are for a diffusion coefficient which is infinite over an intermediate range of concentration and it is easy to see the condition necessary in this case for intersection. Referring again to Figs. 13.1 and 13.2 and using that nomenclature, we see that the height of the initial vertical part of the sorption curve of Fig. 13.1(d) is lC_X, and that of the corresponding part of the desorption curve (Fig. 13.2(d)) is $l(C_1 - C_Y)$. Hence, provided D is greater at high than at low concentrations, intersection occurs if $C_1 - C_Y$ exceeds C_X, so that

$$C_Y < \tfrac{1}{2}(C_1 - a), \tag{9.70}$$

where a is the concentration range over which D is infinite. Thus the limiting case is when this range is symmetrically situated with respect to the whole concentration range 0 to C_1. If the infinite region is mainly in the upper half of the concentration range, desorption is slower than sorption throughout, but if it is mostly in the lower half the sorption and desorption curves intersect. Clearly, the point of intersection occurs earlier the smaller D_2 is compared with D_1 in Fig. 13.1(d).

Fig. 9.25 shows sorption and desorption curves for the diffusion coefficient shown in the inset. These curves were calculated by the numerical methods

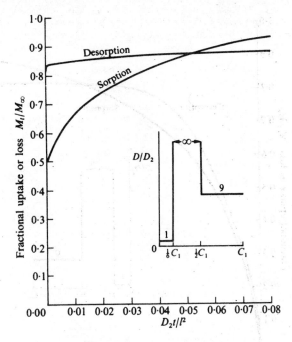

FIG. 9.24. Sorption and desorption curves.

described in Chapters 7 and 8. Although the value of D in the middle concentration range of the inset of Fig. 9.25 is still relatively high, the intersection is much less marked in Fig. 9.25 than in Fig. 9.24 and becomes even less pronounced if the largest value of D is decreased further.

9.4.1. *Effect of the initial concentration on the rate of sorption*

It might be expected that the initial rate of sorption by a sheet having a finite, uniform concentration of diffusing substance in it initially would be greater than the corresponding rate for zero initial concentration if the operative diffusion coefficient is small at low concentrations. Such an effect has been observed experimentally, e.g. by King (1945) for the uptake of methyl alcohol by wool. The effect should be most marked when the diffusion coefficient is zero at low concentrations as in Fig. 13.7(a) (p. 294). A convenient measure of initial rate of sorption is the initial gradient of the sorption curve when plotted against (time)$^{\frac{1}{2}}$ as in Fig. 13.7, i.e.

$$\frac{d}{dT^{\frac{1}{2}}}\left(\frac{M_t}{lC_1}\right), \quad \text{where } T = D_1 t/l^2. \tag{9.71}$$

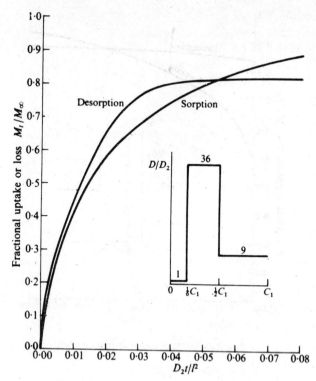

FIG. 9.25. Sorption and desorption curves.

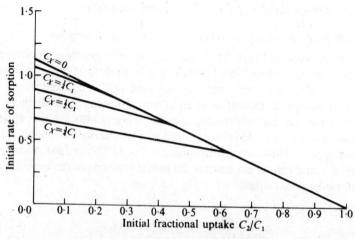

FIG. 9.26. Effect of initial uptake on initial rate of sorption.

The nomenclature is that of Chapter 13 and Figs. 13.6 and 13.7. The initial rate so calculated is shown in Fig. 9.26 as a function of C_2, the initial uniform concentration in the sheet (Crank and Henry 1951). The different curves refer to different values of C_X, the concentration at which D changes discontinuously. In all cases, the initial rate of sorption decreases as the initial regain is increased. This, then, is a further characteristic feature of a purely concentration-dependent system, and in systems which show qualitatively different behaviour some factor other than concentration-dependence must be sought as the cause.

9.5. Diffusion-controlled evaporation

Some evaporative processes are diffusion controlled in the sense that the rate of evaporation depends largely on the rate at which solvent or mixture is supplied to the evaporating surface by internal diffusion. An important consideration is the effect of the proportion of solvent vapour, or the relative humidity, in the atmosphere into which the evaporation takes place. Instances have been reported (Crank 1950), when evaporation takes place through an organic membrane or a polymer film, in which the rate observed in practice is increased by increasing the relative humidity at the evaporating surface. This behaviour has been attributed to the effect of a diffusion coefficient which is low at low solvent concentrations, the argument being that by maintaining some vapour in the outside atmosphere the concentration range over which diffusion within the sheet is difficult is removed. It has been shown theoretically (Crank 1950), by evaluating solutions of the diffusion equation for appropriate diffusion coefficients and boundary conditions, that such behaviour is not to be expected in a purely concentration-dependent system. Both the steady-state evaporation through a membrane and the loss of vapour from a sheet containing solvent have been examined.

9.5.1. Steady-state evaporation through a membrane

We saw in § 4.2.2 (p. 46) that if we have a membrane of thickness l separating a region of high from one of low vapour pressure, then the rate of evaporation through the membrane in the steady state is F, where

$$F = (1/l) \int_{C_1}^{C_0} D \, dC. \tag{9.72}$$

Here C_0 and C_1 are the concentrations just within the surfaces of the membranes on the high- and low-pressure sides respectively. The argument leading to (9.72) is true whether D is constant or not. In particular it is true when D is a function of concentration, and provided D is never negative in the range from C_0 to C_1, the integral in (9.72) must always increase or remain constant as the range of concentration is increased. Thus, as Park has pointed out (see

Crank 1950) if the high concentration remains fixed, the rate of evaporation F can never increase as a result of increasing the lower concentration C_1, i.e. the rate of evaporation is greatest into an atmosphere completely free of solvent vapour. Where the experimental facts genuinely differ from this, some alternative explanation must be found.

Nooney (1973) gives examples of hypothetical diffusion coefficients that yield anomalous behaviour. Their essential property is that they become zero or infinite on some surface within or on the entry surface of the membrane. Thus $D(x, C) = 2x^{\frac{3}{2}}(1 - C)^{-2}$ with $C(0) = 1$, $C(1) = 1 - h$ admits of a solution $C(x) = 1 - hx^{\frac{1}{2}}$, $F = 1/h$, so that $D = 2x^{\frac{1}{2}}h^{-2}$ and $D(0, 1) = 0$ while $dC(0)/dx = \infty$. In his second example, $D(x, C) = \frac{1}{2}x^{\frac{3}{2}}(1 - C)^{-\frac{3}{2}}$. Then $C(x) = 1 - hx^2$ and $F = h^{-4}$, so that $D(x, C) = h^{-\frac{3}{2}}x^{-1}$ and $D(0, 1) = \infty$ while $dC(0)/dx = 0$. In both examples, F increases as $C(1)$ increases.

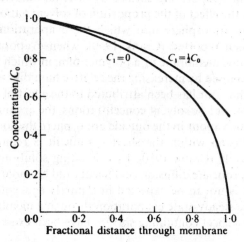

FIG. 9.27. Concentration distributions in steady state.

Further insight into the effect of changing the vapour pressure on the low-pressure side when D depends only on concentration is given by Fig. 9.27. Calculated distributions of concentration through the membrane are shown for a case in which D is an exponential function of concentration such that D increases by 50-fold from $C = 0$ to $C = C_0$. The concentration, C_0, on the high-pressure side is the same for each curve, but in one case $C_1 = 0$ and in the other $C_1 = \frac{1}{2}C_0$. The curves show how the concentration distribution adjusts itself so that the rate of flow is greater when $C_1 = 0$, even though the region of low diffusion coefficient is then included. Clearly, a low diffusion coefficient is compensated by a high concentration gradient.

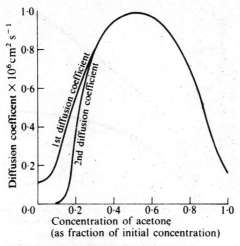

FIG. 9.28. Diffusion coefficients.

9.5.2. *Evaporation from films and filaments*

It is not possible to express the rate of loss of vapour from a film containing solvent in terms of a simple expression such as (9.72) and so it is less easy to examine the effect of introducing vapour into the outside atmosphere. However, some illustrative examples have been worked numerically (Crank 1950) for the first diffusion coefficient shown in Fig. 9.28. This is an experimental curve relating to the diffusion of acetone in cellulose acetate. The concentration distributions through a sheet initially containing solvent at a uniform concentration have been calculated for this diffusion coefficient and two different boundary conditions:

(i) The surface of the sheet is assumed to reach equilibrium with the outside atmosphere instantaneously when evaporation commences, i.e. if the atmosphere is free of vapour the concentration at the surface falls immediately to zero; if the vapour pressure in the atmosphere is p, the surface of the sheet immediately reaches the concentration which is in equilibrium with p.

(ii) A condition expressing the rate of evaporation from the surface is assumed. This is taken to be

$$D_s \, \partial C/\partial x = \alpha(C_s - C_0), \tag{9.73}$$

where C_s is the actual concentration in the surface of the sheet at any time and C_0 the concentration which would be in equilibrium with the vapour pressure remote from the surface. The diffusion coefficient D_s is the value corresponding to the concentration C_s.

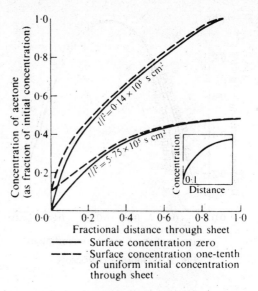

FIG. 9.29. Concentration distributions at various times.

Fig. 9.29 shows the calculated variation of concentration with distance through the sheet at two different times for the condition (i), that of vigorous surface evaporation. For each time, curves are shown for the cases in which (a) the surface concentration falls to zero instantaneously, (b) it falls to one-tenth of its initial value instantaneously. We see that there is little difference between the two curves at either time. In particular the total solvent content of the sheet at any time, represented by the area under the appropriate curve of Fig. 9.29, is much the same whether the surface concentration falls to zero or to one-tenth of its initial value, i.e. whether the concentration over which D is small is removed or not. In so far as there is a slight difference, the rate of loss of solvent by the sheet is greater in the vapour-free atmosphere. As in the steady state, the rate of evaporation is not increased by introducing vapour into the atmosphere.

The reason for this is clear from these calculations as it was in the steady state. When the diffusion coefficient has a small value at the surface the concentration gradient is correspondingly large so that the rate of loss of vapour, given by $D\, \partial C/\partial x$ at the surface, does not alter appreciably. This compensating effect is most obvious when the diffusion coefficient is zero over a range of low concentrations as in the second diffusion coefficient shown in Fig. 9.28. This is a hypothetical coefficient chosen to exaggerate the effect. The concentration–distance curve for this coefficient is sketched in the inset of Fig. 9.29 from general reasoning. The surface gradient is infinite over the concentration range for which D is zero (in this case for concentrations less than 0·10) so that the product can have a non-zero value. A finite gradient cannot develop for concentrations less than 0·10, since this would imply that solvent had been removed from a region of zero diffusion coefficient under the action of a finite gradient, which is not possible. For such a diffusion coefficient the rate of evaporation is precisely the same for all surface concentrations between zero and 0·1.

The same general conclusion holds for the surface condition (9.73), i.e. the rate of loss from the sheet is always increased by decreasing the external vapour pressure. Hansen (1968) tabulated concentration distributions and their means for the drying of a lacquer film by solvent evaporation. The concentration-dependent D is approximated by two exponential curves joined together. Various solvents and drying conditions are covered.

9.6. Effect of a surface skin

Many films and fibres show evidence of a surface skin having properties different from those of the underlying layer or core. In this section we examine the effect of such a skin on diffusion behaviour. The results presented were obtained as part of an attempt to understand some of the peculiar features which are sometimes observed when solvents diffuse into and out of polymer substances (Crank and Park 1951).

We shall restrict ourselves to cases in which the skin and the core are each homogeneous and the boundary between the two is sharply defined. In the skin the diffusion coefficient is assumed to be either constant or a discontinuous function of concentration, and beneath the skin to be everywhere infinite at all concentrations.

(i) The simplest case is one in which D is constant in the skin, and the medium is semi-infinite. The solution has been given in Chapter 3, eqn (3.57) (p 41), which becomes

$$\frac{M_t}{2lC_0} = \left(\frac{Dt}{l^2}\right)^{\frac{1}{2}}\left\{\frac{1}{\pi^{\frac{1}{2}}} + 2\sum_{1}^{\infty} \text{ierfc}\,\frac{nl}{(Dt)^{\frac{1}{2}}}\right\}, \tag{9.74}$$

when $D_2 = \infty$. Here M_t is the total amount of diffusing substance which enters the composite medium in time t if the surface is maintained at a constant concentration C_0. When $M_t/2lC_0$ is plotted as a function of $(Dt/l^2)^{\frac{1}{2}}$ the curve is linear for small times and its gradient later increases steadily as t increases. When plotted as a function of Dt/l^2 the curve is parabolic at first and then becomes linear, the gradient of the linear part being determined by the constant rate of flow across the outer surface in the final steady state. Both the curves shown in Fig. 9.30 are noticeably different from the corresponding ones for a semi-infinite homogeneous sheet when M_t is proportional to $t^{\frac{1}{2}}$ for all times.

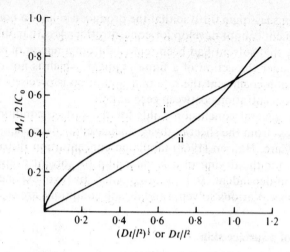

FIG. 9.30. Sorption curves for a composite semi-infinite sheet. D in skin is constant; D below skin is infinite. (i) $M_t/2lC_0$ against Dt/l^2, (ii) $M_t/2lC_0$ against $(Dt/l^2)^{\frac{1}{2}}$.

(ii) We consider next a finite sheet having on each surface a skin in which D is a discontinuous function of concentration C of the type shown in Fig. 9.31(a). The general shape of the sorption-time curve for this case is easily deduced from the concentration–distance curve for the same diffusion coefficient and a homogeneous sheet. Solutions are given in Chapter 13, Figs. 13.6 and 13.7; the sorption–time curve is parabolic for small times and

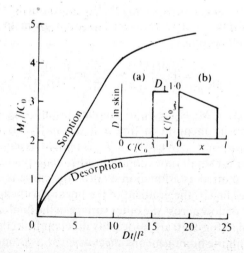

FIG. 9.31. Sorption and desorption curves for sheet with skin.

the concentration–distance curve is characterized by a sharp front which
advances according to the square root of time (Fig. 9.31(b)). It is clear that
until this sharp front has penetrated to the inner boundary of the skin the
sorption–time curve is the same as for a homogeneous sheet having the
properties of the skin throughout. After this the concentration on the outer
surface of the sheet remains at C_0 and that on the inner surface of the skin at
C_X till the uniform concentration throughout the region of infinite diffusion
coefficient has reached the value C_X. During this time there is approximately
a steady-state flow through the skin so that the rate of sorption by the whole
sheet is constant and the sorption–time curve is linear. When the whole sheet
beneath the skin has reached a concentration C_X there is a gradual decrease
in rate of sorption as the final equilibrium concentration C_0 is approached
throughout the sheet.

The details of the calculation for this case are as follows. For sorption
we require solutions of the usual equation

$$\frac{\partial C}{\partial t} = D\frac{\partial^2 C}{\partial x^2}, \tag{9.75}$$

where in the region $0 < x < l$, D is defined by

$$D = 0, \qquad 0 < C < C_X, \tag{9.76}$$

$$D = D_1, \qquad C_X \leqslant C < C_0, \tag{9.77}$$

and in the region $l < x < l+a$, D is infinite. The solutions are subject to the
conditions

$$C = C_0, \qquad x = 0, \qquad t > 0, \tag{9.78}$$

$$C = 0, \qquad 0 < x < l+a, \qquad t = 0. \tag{9.79}$$

For small times the concentration–distance curve falls discontinuously
from $C = C_X$ to $C = 0$, and until this vertical part of the curve reaches
$x = l$ at, say, time $t = t_0$, the medium is effectively semi-infinite and the
solution is

$$C = C_0 + A\,\mathrm{erf}\,\frac{x}{2\sqrt{(D_1 t)}}, \tag{9.80}$$

where

$$A\,\mathrm{erf}\,k/(2D_1^{\frac{1}{2}}) = C_X - C_0, \tag{9.81}$$

and the constant k is given by

$$\frac{C_X - C_0}{\mathrm{erf}\,(k/2D_1^{\frac{1}{2}})} + \frac{\pi^{\frac{1}{2}}k\,e^{k^2/4D_1}}{2D_1^{\frac{1}{2}}} C_X = 0. \tag{9.82}$$

After $t = t_0$ the concentration at $x = l$ remains at $C = C_X$ till an amount

aC_X has crossed $x = l$. During this time the usual solution, §4.3.1, eqn (4.16), (p. 47), for a sheet whose surface concentrations are fixed and in which the initial concentration distribution is $f(x)$, given by substituting $t = t_0$ in (9.80), applies, i.e.

$$C = C_0 + (C_X - C_0)\frac{x}{l} + \frac{2}{\pi}\sum_{n=1}^{\infty}\frac{C_X \cos n\pi - C_0}{n}\sin\frac{n\pi x}{l}\exp\{-D_1 n^2 \pi^2(t-t_0)/l^2\}$$

$$+ \frac{2}{l}\sum_{n=1}^{\infty}\sin\frac{n\pi x}{l}\exp\{-D_1 n^2 \pi^2(t-t_0)/l^2\}\int_0^l f(x)\sin\frac{n\pi x}{l}\,dx. \tag{9.83}$$

From (9.83) it is easily deduced that the amount M_0 crossing unit area at $x = 0$ from time $t = t_0$ onwards is given by

$$-\frac{lM_0}{D_1} = (C_X - C_0)(t - t_0) - 2\sum_{n=1}^{\infty}\frac{l^2}{D_1 n^2}[\exp\{-D_1 n^2 \pi^2(t-t_0)/l^2\} - 1]$$

$$\times \left\{C_X \cos n\pi - C_0 + \frac{n\pi}{l}\int_0^l f(x)\sin\frac{n\pi x}{l}\,dx\right\}. \tag{9.84}$$

The amount M_l crossing unit area at $x = l$ in the same time is given by

$$-\frac{l}{D_1}M_l = (C_X - C_0)(t - t_0)$$

$$- 2\sum_{n=1}^{\infty}(-1)^n\frac{l^2}{D_1 n^2}[\exp\{-D_1 n^2 \pi^2(t-t_0)/l^2\} - 1]$$

$$\times \left\{C_X \cos n\pi - C_0 + \frac{n\pi}{l}\int_0^l f(x)\sin\frac{n\pi x}{l}\,dx\right\}. \tag{9.85}$$

Eqn (9.84) expresses the total amount absorbed by the composite sheet from time t_0 onwards and these solutions apply till $M_l = aC_X$ as calculated from (9.85). For much of this time there will be approximately a steady-state flow across the region $0 < x < l$.

After the concentration throughout the region $l < x < l+a$ has reached C_X, the boundary condition on $x = l$ is

$$-D_1\, \partial C/\partial x = a\, \partial C/\partial t, \tag{9.86}$$

and the solution is conveniently continued by numerical methods described in Chapter 8 using for (9.86) the finite-difference form

$$\frac{\partial C}{\partial x} = \frac{C_{s-2} - 4C_{s-1} + 3C_s}{2\delta x}, \tag{9.87}$$

where the range $0 < x < l$ is divided into equal steps δx, and C_{s+p} is the value of C at $x = l + p\,\delta x$.

For desorption we require solutions of eqn (9.75) for the same diffusion coefficient but subject to the conditions

$$C = 0, \qquad x = 0, \qquad t > 0, \tag{9.88}$$

$$C = C_0, \qquad 0 < x < l+a, \qquad t = 0. \tag{9.89}$$

In this case the concentration–distance curve rises discontinuously at $x = 0$ from $C = 0$ to $C = C_X$, and in the early stages, when the concentration at $x = l$ is not appreciably different from C_0 to the required accuracy, the solution is simply that for a semi-infinite sheet having a surface concentration C_X, i.e.

$$C - C_X = (C_0 - C_X)\,\mathrm{erf}\,\frac{x}{2(D_1 t)^{\frac{1}{2}}}. \tag{9.90}$$

At later times the condition (9.86) again applies and the solution can be continued numerically as for sorption. A typical sorption–time curve is shown in Fig. 9.31 for $C_X = \frac{2}{3}C_0$ and a skin which forms one-fifth of the half-thickness of the sheet. The corresponding curve for desorption is also shown in Fig. 9.31.

The investigations into the cause of intersecting sorption- and desorption-time curves discussed in § 9.4 (iv) (p. 183) suggest that a diffusion coefficient in the skin of the form shown in Fig. 9.32(a) might be interesting. The concentration–distance curve for this diffusion coefficient has a sharp front as in Fig. 9.32(b), and the required mathematical solutions up to the time at which this front reaches the inner surface of the skin are presented in § 13.2.4.

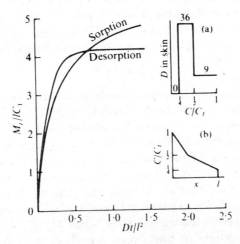

FIG. 9.32. Sorption and desorption curves for sheet with skin.

(p. 296), together with a numerical procedure for extending them to later times (Chapter 8). Desorption can be treated by the same equations and methods. Fig. 9.32 shows calculated sorption– and desorption–time curves for the diffusion coefficient of Fig. 9.32(a) and a skin one-fifth of the half-thickness of the sheet as before. Here we see that the curves intersect but the linear part has almost disappeared from the sorption curve. We can now see that this must always be so, for, as we have already noted, intersection of sorption and desorption curves only occurs when the region of high diffusion coefficient is in the lower half of the concentration range, which means that the diffusion coefficient is small or zero only at very low concentrations. But it is clear from what we have said and from Fig. 9.32(b) that for relatively thin skins the amount absorbed by the sheet when the sorption–time curve ceases to be linear is roughly proportional to the height of the sharp front in the concentration–distance curve, i.e. to the concentration range over which D is small or zero. It follows that, in general, when the sorption–time curve calculated for the model we have chosen has a long linear portion, desorption will be everywhere slower than sorption; conversely if the sorption and desorption curves intersect markedly no appreciable linear part will be observed in the sorption curve.

9.6.1. *Composite cylinder*

The effect of a skin on the uptake of dye by a cylindrical fibre has been studied (Crank and Godson 1947). The solutions obtained relate to a circular cylinder of infinite length and radius a immersed in a solution. Dye molecules diffuse into the cylinder and are deposited in capillaries of the cylinder. The concentration of the dye in the solution is always uniform, while the cylinder is initially free of dye. The cylinder has a core of radius b, in which the diffusion coefficient is D_b and for $b < r < a$ the diffusion coefficient is D_a. In any element of the cylinder the amount of dye S deposited in the capillaries and immobilized is related to C, the amount free to diffuse, by the relationship

$$S = RC^{\frac{1}{2}}, \tag{9.91}$$

where R is a constant which, in this example, is chosen so that in the final equilibrium state 90 per cent of the dye initially in the solution has entered the fibre. When the ratio of the volume of solution to that of the fibre is $25:1$, $R = 70.8$. If we denote by M_t the amount of dye in the cylinder after time t, and M_∞ the corresponding amount after infinite time, Fig. 9.33 shows M_t/M_∞ as a function of $\log (D_b t/a^2)$ for $b = \frac{9}{10}a$. The several curves correspond to different values of D_b/D_a. The effect of changing the ratio D_b/D_a is to displace the sorption curve parallel to the time axis, and this is accompanied by a slight change in the shape of the curves. Thus, when $M_t/M_\infty = 0.5$, in Fig. 9.33, the abscissae of points on the curves for $D_b/D_a = 2$, 10, and 30 differ from the corresponding abscissa for $D_a = D_b$ by the amounts 0.30,

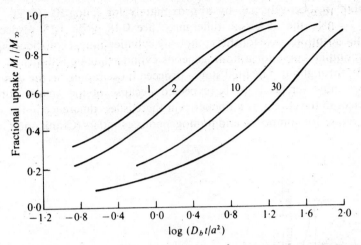

FIG. 9.33. Sorption curves for a composite cylinder $b = \frac{9}{10}a$. Numbers on curves are values of D_b/D_a.

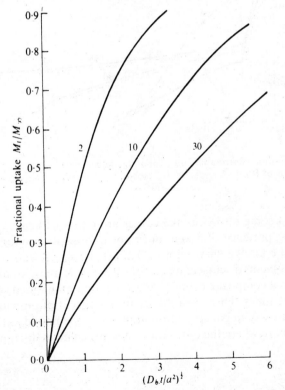

FIG. 9.34. Sorption curves for a composite cylinder, $b = \frac{9}{10}a$. Numbers on curves are values of D_b/D_a.

1·00, 1·45 respectively, i.e. by approximately $\log 2$, $\log 10$, $\log 30$. When $M_t/M_\infty = 0.8$, the abscissae differences are 0·18, 0·78, 1·18 respectively. Thus the sorption curve for this composite cylinder almost coincides with the corresponding curve for a homogeneous cylinder having a diffusion coefficient D_a throughout. The final stages proceed more rapidly in the composite cylinder, however, as is to be expected because of the greater diffusion coefficient in the core. For a thicker skin there is less difference in the shapes of the curves for composite and homogeneous cylinders (Crank and Godson 1947).

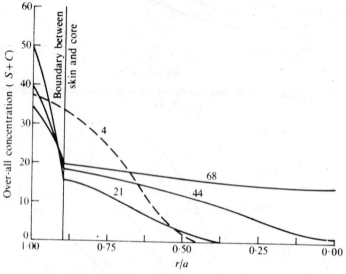

FIG. 9.35. Concentration distribution in a composite cylinder, $b = \frac{9}{10}a$; $D_b/D_a = 30$. Numbers on curves are values of $D_b t/a^2$. ———————— solution for homogeneous cylinder of diffusion coefficient D_b.

In Fig. 9.34 some of the uptake curves are replotted against $(D_b t/a^2)^{\frac{1}{2}}$ to show that the presence of a skin does not necessarily produce a point of inflexion in the uptake curve plotted against the square root of time, as it does in the problems discussed in § 9.6. Fig. 9.35 shows the over-all concentration $(S+C)$ in a composite cylinder as a function of the radial coordinate at three different times. There is a rapid fall in concentration within the skin and a discontinuity in concentration gradient at the boundary between the skin and the core where the diffusion coefficient changes discontinuously.

THE DEFINITION AND MEASUREMENT
OF DIFFUSION COEFFICIENTS

10.1. Definitions

QUANTITATIVE measurements of the rate at which a diffusion process occurs are usually expressed in terms of a diffusion coefficient. Before describing some methods of measurement, we shall examine the definition of the diffusion coefficients a little more carefully than in § 1.2 (p. 2). Confining attention to one dimension only, the diffusion coefficient is defined as the rate of transfer of the diffusing substance across unit area of a section, divided by the space gradient of concentration at the section. Thus, if the rate of transfer is F, and C the concentration of diffusing substance, and if x denotes the space coordinate, then

$$F = -D\, \partial C/\partial x, \tag{10.1}$$

and (10.1) is a definition of the diffusion coefficient D. In using this definition in practice, it is necessary to specify carefully the section used and the units in which F, C, and x are measured. Only the simplest system of practical importance is considered, which is a two-component system, since it is not possible to set and observe a concentration gradient of a single substance in itself without introducing complicating features such as pressure gradients, etc. The diffusion of isotopes is best regarded as a special case of a two-component system.

10.2. A frame of reference when the total volume of the system remains constant

Consider the inter-diffusion of two liquids A and B in a closed vessel and assume that there is no overall change of volume of the two liquids on mixing. Two diffusion coefficients, D_A^V, D_B^V, one for each liquid, may be defined by the relationships

$$F_A = -D_A^V\, \partial C_A/\partial x, \tag{10.2}$$

$$F_B = -D_B^V\, \partial C_B/\partial x. \tag{10.3}$$

C_A and C_B are the concentrations of A and B respectively, each expressed in the usual way in any convenient unit of amount (e.g. gram or, in the case of simple molecular substances, gram mole) per unit over-all volume. F_A and F_B are the rates of transfer of A and B measured in the same units of amount per unit time, across a section which is defined by the condition that the total volume on either side of it remains constant as diffusion proceeds. In the particular case under consideration it is therefore fixed with respect to the containing vessel. The origin from which x is measured is such that the x-coordinate of the section is constant; x is measured in normal units of length, e.g. centimetres and the same unit of length is used in measuring the volume which appears in the definition of concentration. If the unit of time adopted is the second it follows that the units of D_A^V and D_B^V are each $cm^2 s^{-1}$. These somewhat obvious statements are made here in full because it will be seen later in § 10.3 (p. 205) that other scales of length and alternative ways of measuring concentration are more suitable in some circumstances.

Let V_A and V_B denote the constant volumes of the unit amounts used in defining the concentrations of A and B. Thus if C_A is expressed in grams per unit volume, V_A is the volume of 1 g of A. In dilute solutions, where the volume changes in the range of concentration concerned can be considered negligible, V_A and V_B will be the partial specific or molar volumes. That of the solute may be very different from the specific volume in the pure state. The volume transfer of A per unit time across unit area of the section defined is therefore

$$-D_B^V V_A \, \partial C_A/\partial x,$$

and that of B is

$$-D_B^V V_B \, \partial C_B/\partial x.$$

By definition of the section as one across which there is no net transfer of volume we have immediately

$$D_A^V V_A \frac{\partial C_A}{\partial x} + D_B^V V_B \frac{\partial C_B}{\partial x} = 0. \tag{10.4}$$

The volume of A per unit over-all volume of solution is $V_A C_A$ and of B is $V_B C_B$, so that, since only molecules of A and B are present, we have

$$V_A C_A + V_B C_B = 1, \tag{10.5}$$

which, following differentiation with respect to x, becomes

$$V_A \frac{\partial C_A}{\partial x} + V_B \frac{\partial C_B}{\partial x} = 0. \tag{10.6}$$

In order that (10.4) and (10.6) shall both be satisfied it follows that

$$D_A^V \equiv D_B^V, \tag{10.7}$$

or else that

$$V_A = 0 \quad \text{or} \quad V_B = 0 \tag{10.8}$$

If $V_A = 0$ and $V_B \neq 0$, it follows from (10.6) that

$$\partial C_B/\partial x = 0, \tag{10.9}$$

and further reference is made to this case in § 10.6.4 (p. 234). In either case the behaviour of a two-component system, satisfying the condition of zero volume change on mixing, may be described in terms of a single diffusion coefficient, which may vary with composition. It is convenient to refer to it as the mutual diffusion coefficient denoted henceforth by D^V. This coefficient is familiar in the interdiffusion of gases (Jeans 1940). Its physical significance is considered later in § 10.4 (p. 209).

10.3. Alternative frames of reference

The definition of the volume-fixed section used in § 10.2 (p. 203) above is unambiguous only as long as the total volume of the diffusion system remains constant. If there is an over-all change of volume of the two components on mixing, the side of the section on which the volume is to remain constant must be chosen arbitrarily, and the diffusion coefficient becomes equally arbitrary. In such a case some alternative frame of reference must be used in defining the section across which transfer of diffusing substance is to be measured. There are clearly several possibilities. Thus, for example, the total mass of the system will always be conserved even though volume is not, and a section can be defined consistently such that the mass of the system on either side of the section remains constant during diffusion.

Where a convention other than that of constant volume on either side is used in defining a section, the second-order differential equation describing diffusion may not take the standard form of (7.5). It is clearly convenient if it can be made to do so since the standard form has frequently been used as the starting-point in calculations of diffusion behaviour. This can always be arranged by departing from the orthodox linear scale, e.g. centimetres, for measurement of the spatial coordinate so far denoted by x, and by measuring concentration in a certain way. Let some modified scale of length be denoted by ξ, and consider two sections, fixed on the same convention, at ξ and $\xi + d\xi$. The rate of entry of A into the volume enclosed between these sections is F_A and that of departure is $F_A + (\partial F_A / \partial \xi) d\xi$. The rate of accumulation is therefore

$$-(\partial F_A / \partial \xi) d\xi,$$

and this is always true independently of how F_A and ξ are measured. It can only be equated to $(\partial C_A / \partial t) d\xi$, however, when C_A and ξ are measured in certain consistent units. Thus, if the sections are fixed with respect to total mass, then ξ must be measured so that equal increments of ξ always include equal increments of total mass, and C_A must be defined as the amount of A per unit total mass. Similarly, if the sections are fixed with respect to volume or mass of component B, equal increments of ξ must include equal increments of amount of B, and C_A must be expressed as the amount of A per unit amount of B. In general, for all values of ξ and t, the element of unit length in terms of ξ, and of unit cross-sectional area, is that which contains an amount of A equal to the unit used in defining the concentration C_A. When the quantities C_A and ξ satisfy this condition the usual relationship

$$\frac{\partial C_A}{\partial t}\,d\xi = -\frac{\partial F_A}{\partial \xi}\,d\xi \qquad (10.10)$$

follows at once, and by substituting for F_A from the relationship

$$F_A = -D\,\partial C_A/\partial \xi \qquad (10.11)$$

we derive the familiar form of the diffusion equation

$$\frac{\partial C_A}{\partial t} = \frac{\partial}{\partial \xi}\left(D\,\frac{\partial C_A}{\partial \xi}\right). \qquad (10.12)$$

It is convenient that ξ should have the dimension of length and D the usual dimensions of $(\text{length})^2(\text{time})^{-1}$. This can be arranged without interfering with the generality or simplicity of (10.12), by multiplying the mass of component B (or the total mass of A and B together if this is the reference system being used) by an arbitary constant specific volume. The volume represented by the product of a mass of B, for example, and this arbitrary specific volume will be referred to for convenience as the basic volume of that mass of B.

The concentration of A was defined above as the amount of A per unit amount of B. We now redefine the concentration of A as the amount per unit basic volume of B, and unit ξ to contain unit basic volume of B per unit area. A convenient arbitrary specific volume is that of the pure component B, so that the basic volume of a certain mass of B is the volume that mass of B would occupy in the pure state.

The same arbitrary specific volume is used for concentrations expressed in the original definition per unit mass of A and B together, i.e. the basic volume of a mass of A and B together is obtained by multiplying the mass by the same arbitrary specific volume. This is true also for deriving the basic volume of A alone, so that the basic volume has a simple physical significance only in the case of the basic volume of B. Nevertheless, the use of this particular basic volume has the convenience that all the concentrations measured in the different frames of reference tend to the same value in dilute solutions.

Of course, concentration is frequently expressed in a number of different ways and so the symbol C_A is retained, but the appropriate index V, B, or M as superscript is added, so that the concentration of A is written C_A^V, C_A^B, or C_A^M according as the amount of A is contained in unit volume of solution, or in unit basic volume of B or in unit basic volume of total mass. According as unit ξ contains, per unit area, unit basic volume of B or of A and B together, the symbol ξ_B or ξ_M is used. The diffusion coefficients D_A^V, D_A^B, D_A^M also carry an index to indicate the frame of reference to which they refer. The arbitrary specific volume may be denoted by V_B^0, and then ξ_B and ξ_M are defined formally by the respective relationships

$$d\xi_B = V_B^0 C_B^V \, dx, \qquad (10.13)$$

$$d\xi_M = V_B^0 (C_A^V + C_B^V) \, dx. \qquad (10.14)$$

10.3.1. *Sections fixed with respect to total mass and mass of one component*

It was found in § 10.2 (p. 203) that the behaviour of a two-component system satisfying the condition of zero volume change on mixing can be represented in terms of a single diffusion coefficient D^V. A similar result follows readily for a system in which volume changes occur, provided the diffusion coefficients are defined with respect to a mass-fixed section. Thus the equation defining such a section is

$$D_A^M \frac{\partial C_A^M}{\partial \xi_M} + D_B^M \frac{\partial C_B^M}{\partial \xi_M} = 0, \qquad (10.15)$$

and the definitions of C_A^M and C_B^M lead immediately to the equation

$$C_A^M + C_B^M = 1/V_B^0. \qquad (10.16)$$

On differentiating (10.16) with respect to ξ_M and comparing with (10.15) we find

$$D_A^M \equiv D_B^M. \qquad (10.17)$$

If a section fixed with respect to one component, say B, is used, then clearly $D_B^B = 0$ and only the coefficient D_A^B is needed to describe the diffusion behaviour. Thus the statement that the diffusion behaviour of a two-component system can be described in terms of a single diffusion coefficient, is valid whether there is a change of volume of the whole system or not, provided the appropriate frame of reference is used in defining the diffusion coefficient. Frames of reference could be so chosen that the two coefficients are not identical and neither is zero, but they would be related through some function of the partial volumes and would not be independent measures of two separate diffusion processes. The possibility of measuring the diffusion of the two molecular species independently is discussed in ξ 10.4 (p. 211).

10.3.2. *Relations between the diffusion coefficients D_A^V, D_A^M, D_A^B*

The rate of transfer of A through a B-fixed section is greater than that through a total-mass-fixed section by an amount given by the concentration of A per unit mass of B multiplied by the flux of B across the mass-fixed section. Thus the flux of A across a B-fixed section in the direction of ξ

increasing is

$$-D^M \frac{\partial C_A^M}{\partial \xi_M} + D^M \frac{C_A^M}{C_B^M} \frac{\partial C_B^M}{\partial \xi_M} = -\frac{D^M}{C_B^M V_B^0} \frac{\partial C_A^M}{\partial \xi_M}, \tag{10.18}$$

using (10.16).

But the rate of transfer across a B-fixed section is $-D_A^B \partial C_A^B/\partial \xi_B$ so that we have

$$D_A^B = \frac{D^M}{V_B^0 C_B^M} \cdot \frac{\partial C_A^M}{\partial \xi_M} \frac{\partial \xi_B}{\partial C_A^B}. \tag{10.19}$$

From the definitions of C_A^M and C_A^B it is easy to show that

$$\frac{dC_A^M}{dC_A^B} = (V_B^0 C_B^M)^2. \tag{10.20}$$

Also since

$$d\xi_M = V_B^0(C_A^V + C_B^V)\, dx, \qquad d\xi_B = V_B^0 C_B^V\, dx, \tag{10.21}$$

we have

$$\frac{\partial \xi_B}{\partial \xi_M} = V_B^0 C_B^M, \tag{10.22}$$

so that finally, by substituting (10.22) and (10.20) in (10.19), we find

$$D_A^B = D^M (V_B^0 C_B^M)^2, \tag{10.23}$$

since re-arrangement of the partial derivatives in (10.19) is permissible. For a system in which there is zero volume change on mixing, so that V_A and V_B are constant, the relation between D_A^B and D^V can be similarly established. Thus the flux of A across a B-fixed section in the direction of ξ increasing is

$$-D^V \frac{\partial C_A^V}{\partial x} + D^V \frac{C_A^V}{C_B^V} \frac{\partial C_V^B}{\partial x} = -\frac{D^V}{V_B C_B^V} \frac{\partial C_A^V}{\partial x}, \tag{10.24}$$

using (10.5) and (10.6). But the rate of transfer of A across a B-fixed section is $-D_A^B \partial C_A^B/\partial \xi_B$, so that we have

$$D_A^B = \frac{D^V}{V_B C_B^V} \frac{\partial C_A^V}{\partial x} \frac{\partial \xi_B}{\partial C_A^B}. \tag{10.25}$$

From the definition of C_A^V and C_A^B it follows that

$$\frac{dC_A^V}{dC_A^B} = (V_B C_B^V)^2, \tag{10.26}$$

and from the second of (10.21) we have immediately

$$\frac{\partial \xi_B}{\partial x} = V_B C_B^V, \tag{10.27}$$

so that on substituting (10.26) and (10.27) in (10.25) we find

$$D_A^B = D^V(V_B C_B^V)^2 = D^V(\text{volume fraction of } B)^2. \qquad (10.28)$$

Since (10.28) applies only when V_B is constant, and therefore $V_B = V_B^0$, comparison of (10.28) and (10.23) shows that when there is no over-all volume change accompanying diffusion

$$D^M = D^V(C_B^V/C_B^M)^2 = D^V \text{ (basic total volume/true total volume)}^2. \qquad (10.29)$$

10.4. Intrinsic diffusion coefficients

We saw in § 10.3.1 that any two-component system can be described by a single or mutual diffusion coefficient, which may be a function of composition but will be the same function for both components. In the simplest case, where the molecules of the components A and B are identical in mass and size, the rates of transfer of A and B due to random motion across a volume-fixed section may reasonably be expected to be equal and opposite. In general, however, differences of mass and size of A and B molecules result in the transfer of A by random motions being greater or less than that of B. Consequently, a hydrostatic pressure tends to be built up in the region of the solution which contributes least to the volume rate of transfer. This pressure is relieved by a compensating bulk-flow of A and B together, that is of the whole solution (Meyer 1899, Jeans 1940, Hartley 1946). This existence of bulk flow can be demonstrated in the case of gases, when diffusion occurs across a porous plate which offers considerable viscous resistance. In this case, an increased pressure is found to arise in that part of the vessel occupied initially by the slower diffusing component. It has been demonstrated in metal systems (Darken 1948) and in polymer solvent systems (Robinson 1946) by the insertion of marker particles.† In the latter case, the large polymer molecules will diffuse far more slowly, as a result of random motions, than the small solvent molecules. Thus the polymer movement measured by the mutual coefficient is almost entirely a bulk-flow.

The over-all rate of transfer, say of component A, across a volume-fixed section may be expressed as the combined effect of bulk-flow and true diffusion resulting from the random motion of non-uniformly distributed A molecules. From the point of view of interpreting diffusion coefficients in terms of molecular motions, the mutual diffusion coefficient D^V thus appears to be unnecessarily complicated by the presence of the bulk-flow. It is desirable to define new diffusion coefficients, $\mathscr{D}_A$ and $\mathscr{D}_B$, in terms of the rate of transfer of A and B, respectively, across a section fixed so that no bulk-flow occurs through it. Such a section may be impossible to determine in practice, except in special conditions mentioned below. It is fixed in a different way

---

† The marker movement is often referred to as the Kirkendall effect, since it was first observed in metals by Kirkendall, E. O. (1942).

from any of the other sections previously dealt with, and it must follow the bulk-flow although this flow is not normally directly observable. These new diffusion coefficients will be referred to as 'intrinsic diffusion coefficients'. When the partial volumes are constant they are related to the mutual diffusion coefficient in the following way.

On one side of a section fixed so that no bulk-flow occurs through it, there is a rate of accumulation of total volume of solution, which may be denoted by ϕ, where

$$\phi = V_A \mathcal{D}_A \frac{\partial C_A^V}{\partial x} + V_B \mathcal{D}_B \frac{\partial C_B^V}{\partial x}. \tag{10.30}$$

As thus defined, ϕ is actually the rate of increase of volume on the side of smaller x, and this must be equal to the rate of transfer of total volume by bulk-flow across a volume-fixed section. Such a bulk-flow involves a rate of transfer of A of ϕC_A^V, so that, equating two expressions for the net rate of transfer of A across the volume-fixed section, we find

$$D^V \frac{\partial C_A^V}{\partial x} = \mathcal{D}_A \frac{\partial C_A^V}{\partial x} - \phi C_A^V. \tag{10.31}$$

On substituting for ϕ from (10.30) and using (10.6) we have finally

$$D^V = V_A C_A^V (\mathcal{D}_B - \mathcal{D}_A) + \mathcal{D}_A. \tag{10.32}$$

If the molal volumes vary with composition, the coefficient D^V has no physical significance, but $\mathcal{D}_A, \mathcal{D}_B$ can still be defined in terms of the rates of transfer of A and B respectively across a section which moves so that there is no bulk-flow of A and B together, through it. It is convenient in this case to relate the intrinsic diffusion coefficients to D_A^B. Since the net rate of transfer of B through a B-fixed section is, by definition, zero, it follows that the contributions to the transfer of B resulting from the over-all bulk-flow and from the true diffusion of B relative to the bulk-flow, must be equal and opposite. The rate of transfer of B by true diffusion relative to the bulk-flow is

$$-\mathcal{D}_B \partial C_B^V / \partial x,$$

in the direction of x increasing and hence the volume transfer of the whole solution accompanying bulk-flow with respect to the B-fixed section is given by

$$\frac{\mathcal{D}_B}{C_B^V} \frac{\partial C_B^V}{\partial x},$$

in the direction of x increasing. This produces a rate of transfer of A through the B-fixed section of

$$\frac{C_A^V}{C_B^V} \mathcal{D}_B \frac{\partial C_B^V}{\partial x},$$

due to the bulk-flow. This is to be combined with the rate of transfer of A relative to the bulk-flow which is given by

$$-\mathscr{D}_A\,\partial C_A^V/\partial x$$

to give the net rate of transfer of A across a B-fixed section, which is simply

$$-D_A^B\,\partial C_A^B/\partial \xi_B.$$

Thus we have the equation

$$-D_A^B\frac{\partial C_A^B}{\partial \xi_B} = -\mathscr{D}_A\frac{\partial C_A^V}{\partial x}+\frac{C_A^V}{C_B^V}\mathscr{D}_B\frac{\partial C_B^V}{\partial x} \tag{10.33}$$

When the molal volumes are not constant, the relationship

$$V_A C_A + V_B C_B = 1$$

still holds, but the differentiated form (10.6) is to be replaced by

$$\left(V_A+C_A^V\frac{\mathrm{d}V_A}{\mathrm{d}C_A^V}\right)\frac{\partial C_A^V}{\partial x}+\left(V_B+C_B^V\frac{\mathrm{d}V_B}{\mathrm{d}C_B^V}\right)\frac{\partial C_B^V}{\partial x}=0. \tag{10.34}$$

Since

$$\frac{\mathrm{d}C_A^V}{\mathrm{d}C_A^B} = V_B^0 V_B(C_B^V)^2,$$

it follows immediately from (10.27), (10.33), and (10.34) that

$$D_A^B = V_B C_B^V (V_B^0 C_B^V)^2\left\{\mathscr{D}_A+\mathscr{D}_B\frac{C_A^V}{C_B^V}\frac{V_A+C_A^V\,\mathrm{d}V_A/\mathrm{d}C_A^V}{V_B+C_B^V\,\mathrm{d}V_B/\mathrm{d}C_B^V}\right\}. \tag{10.35}$$

This reduces to (10.32) when V_A and V_B are constant.

It is clear from (10.32) or (10.35) that the values of $\mathscr{D}_A$ and $\mathscr{D}_B$ cannot be deduced separately, unless some information other than D^V or D_A^B is available. One possibility is to use an observation of the bulk-flow, as suggested by Darken (1948) and by Hartley and Crank (1949). Sometimes, e.g. in solvent-polymer systems, the intrinsic diffusion for one component, e.g. the polymer, is so much smaller than for the other that it can be assumed to be zero. With the assumption that $\mathscr{D}_B = 0$ we have from (10.32)

$$\mathscr{D}_A = D^V/(1-V_A C_A^V) = D^V/(V_B C_B^V) = D^V/(\text{volume fraction of } B), \tag{10.36}$$

which allows the intrinsic diffusion coefficient of component A and its dependence on concentration to be deduced from observations of D^V. The ideas discussed above in §§ 10.1–10.4 have been the subjects of papers by Darken (1948), Hartley and Crank (1949), Kuusinen (1935), Lamm (1943) and Pattle, Smith, and Hill (1967).

10.5. 'Self'-diffusion coefficients

By using radioactively-labelled molecules it is possible to observe the rate of diffusion of one component in a two-component system of uniform chemical composition. Since what is involved is an interchange of labelled and unlabelled molecules which are otherwise identical there is no bulk-flow and the true mobility of the labelled molecules with respect to stationary solution is measured. Nevertheless, the diffusion coefficient so deduced will in general differ from the corresponding intrinsic diffusion coefficient for the same chemical composition. Johnson (1942) has found this to be so in metal systems. Seitz (1948), regarding the diffusion process as a jumping of molecules from one equilibrium position to another, accounts for the difference on the basis that when there is a gradient of chemical composition the frequency with which a molecule jumps to the right is not the same as that with which it jumps to the left. For the labelled molecules, however, the two frequencies are identical. Darken (1948) and Prager (1953) have related the diffusion coefficient measured by an experiment using radioactive molecules to the intrinsic diffusion coefficient in terms of the thermodynamic properties of the system. Their result can be anticipated as follows.

Consider a two-component system comprising molecules A and B and let the gradient of concentration C_A of A be maintained in an equilibrium condition by the application of a force F_A per g mol of A in the direction of increasing x. This is purely a hypothetical operation but it can be realized in the case of large molecules, much different in density from the solvent, by a centrifugal field. The generalized form of the condition for this thermodynamic equilibrium is

$$F_A = \partial \mu_A / \partial x, \tag{10.37}$$

where μ_A is the chemical potential of component A. The rate of transfer of A due to the force F_A is

$$\frac{F_A C_A}{\sigma_A \eta} = \frac{C_A}{\sigma_A \eta} \frac{\partial \mu_A}{\partial x}, \tag{10.38}$$

where $\sigma_A \eta$ is a resistance coefficient. But in the equilibrium condition (10.38) is also the rate of transfer by diffusion relative to a section through which there is no bulk-flow and so we have

$$\mathscr{D}_A \frac{\partial C_A}{\partial x} = \frac{C_A}{\sigma_A \eta} \frac{\partial \mu_A}{\partial C_A} \frac{\partial C_A}{\partial x}, \tag{10.39}$$

and hence

$$\mathscr{D}_A = \frac{C_A}{\sigma_A \eta} \frac{\partial \mu_A}{\partial C_A}. \tag{10.40}$$

On applying the same treatment to the labelled molecules in a system of uniform chemical composition we have

$$F_A^* = \frac{\partial \mu_A^*}{\partial x} = \frac{RT}{C_A^*} \frac{\partial C_A^*}{\partial x}, \tag{10.41}$$

because of the ideality of the system, where asterisks denote properties of labelled molecules. Instead of (10.40), therefore, we obtain

$$\mathscr{D}_A^* = \frac{RT}{\sigma_A \eta}, \tag{10.42}$$

and finally

$$\mathscr{D}_A = \mathscr{D}_A^* C_A (\partial \mu_A / \partial C_A)/RT, \tag{10.43}$$

if we assume the resistance coefficient $\sigma_A \eta$ to depend only on chemical composition. Carman and Stein (1956) used these concepts in discussing their measurements of self-diffusion coefficients in mixtures of ethyl iodide and n-butyliodide.

Pattle *et al.* (1967) questioned the validity of this assumption. It implies that the mobility of a few tracer molecules moving relative to all the others is the same as the corresponding mobility in mutual diffusion where all the molecules of one kind are moving in one direction and the rest in the other. They propose as an alternative a 'one-process' theory which postulates no bulk-flow but instead introduces 'enhancement factors' relating mobilities in mutual and tracer diffusion. The factors are chosen so that the theory fits the experimentally observed values of the different diffusion coefficients and the associated thermodynamic data. Patel *et al.* agree that the 'two-process' theory incorporating bulk-flow can be modified by the insertion of similar arbitrary parameters. The subject needs further careful experimental investigation of the kind carried out by Pattle *et al.* in the rubber–benzene system.

Kirkwood, Baldwin, Dunlop, Gosting, and Kegeles (1960) discussed the various frames of reference described in § 10.3 (p. 205) with reference to the transport equations of irreversible thermodynamics. They were particularly concerned to test Onsager's reciprocal relations and to measure diffusion coefficients in systems in which changes of volume occur on mixing. In this connection Fujita and Gosting (1956) presented mathematical solutions of Onsager's flow equations for diffusion in a three-component system with interacting flows. They developed a general procedure for calculating the four diffusion coefficients from experimental data.

Bearman (1961) examined the absolute reaction rate theory (Glasstone, Laidler, and Eyring (1941)) and the equations of Hartley and Crank (1949) and of Gordon (1937) from the viewpoint of statistical thermodynamics. He found the different approaches basically equivalent. In his criticism of the idea of intrinsic diffusion coefficients Bearman seems to confuse the mass-fixed frame of reference (§ 10.3.1, p. 207) with that moving with the bulk-flow (§ 10.4, p. 209). Partly for this reason the term 'bulk-flow' may be preferable to 'mass-flow' as originally used by Hartley and Crank (1949). Wright (1972) in a series of papers dealing with gaseous diffusion at 'strictly uniform pressure' obtains expressions similar in algebraic form to those of this section. He sees this as quite natural since the two discussions are expressing the same basic concepts in different terms.

Tyrrell (1971) considers that all the phenomenological descriptions of diffusion based on Fick's original concepts are deceptively too simple. He discusses critically the problems of the definition of diffusion coefficients and suggests that understanding of diffusion processes is most likely to advance through the application of non-equilibrium thermodynamics and through mechanical analogies involving frictional coefficients (Mills 1963; Tyrell 1963).

10.6. Methods of measurement

Crank and Park (1968) reviewed the more useful experimental techniques for measuring diffusion coefficients and their concentration dependence. In this section the emphasis is on the associated mathematical analysis.

Most of the earlier methods assumed constant diffusion coefficients. When such methods are applied to systems in which this is not true, a mean value is obtained. Both steady-state and transient methods are used, sometimes in combination, as in the time-lag method of § 4.3.3 (p. 51). Various methods based on the analysis of concentration–distance curves are reviewed by Alexander and Johnson (1949, Chapter X). A variation is to observe the over-all rate of uptake or loss of diffusing substance by a specimen of known size and shape, and to compare this with the calculated rate of uptake expressed as a function of Dt. In some cases, special tables have been constructed to facilitate the calculations. Stefan's tables (1879), to which reference was made earlier in § 4.3.8 (p. 63) refer to the diffusion of solute from a column of solution into a column of water. They give the amount of diffusing substance contained in successive layers of equal height. These tables and others of a similar nature are reproduced and discussed by Jost (1952). In all these methods, the difficulties lie in the experimental techniques rather than in the subsequent mathematics. Adequate accounts are already available (Barrer 1951; Jost 1952). It suffices to say here, following the discussion of the alternative definitions of diffusion coefficients, that in the early measurements it is usually assumed that the total volume of the system remains constant as diffusion proceeds and so the mutual diffusion coefficient D^V is measured. A notable exception is that of Clack (1916; 1921), who introduced a correction for mass-flow and obtained in effect the coefficient D_A^B of solute with respect to stationary solvent. A number of examples are discussed in more detail by Hartley and Crank (1949).

This is a convenient point at which to refer to a method of measurement suggested by Taylor (1953, 1954) based on observations of the dispersion of soluble matter in solvent flowing slowly through a small-bore tube. The distribution of concentration is found to be centred on a point which moves with the mean speed of flow U and is symmetrical in spite of the asymmetry of flow. Taylor (1953) shows that the distribution is determined by a longitudinal diffusion coefficient k which is related to the molecular diffusion coefficient D by the relation

$$k = \frac{a^2 U^2}{48D},$$

where U, the mean speed of flow, is defined as $\frac{1}{2}u_0$, where u_0 is the maximum velocity on the axis and a is the radius of the tube.

Two useful experimental conditions are as follows.

(i) Material of mass M concentrated at $x = 0$ when time $t = 0$. The solution for this is

$$C = \tfrac{1}{2}Ma^{-2}\pi^{-\frac{3}{2}}(kt)^{-\frac{1}{2}}\exp(-x_1^2/4kt),$$

where $x_1 = x - \frac{1}{2}u_0 t$.

(ii) Material of constant concentration C_0 is allowed to enter the tube at a uniform rate at $x = 0$, starting at $t = 0$. Initially the tube is filled with solvent only ($C = 0$). The solution for this case is

$$C/C_0 = \tfrac{1}{2} + \tfrac{1}{2}\operatorname{erf}\left\{\frac{x_1}{2(kt)^{\frac{1}{2}}}\right\}, \qquad x_1 < 0,$$

$$C/C_0 = \tfrac{1}{2} - \tfrac{1}{2}\operatorname{erf}\left\{\frac{x_1}{2(kt)^{\frac{1}{2}}}\right\}, \qquad x_1 > 0.$$

In either case k, and hence D, may be deduced by comparing the appropriate mathematical solution with an observed concentration distribution. The mathematical analysis rests on the assumption that radial differences in concentration are smoothed out quickly by molecular diffusion compared with the time necessary for appreciable effects to appear owing to convective transport. Also longitudinal diffusion is neglected.

In a second paper, Taylor (1954) suggested necessary conditions to be

$$\frac{4L}{a} \geqslant \frac{Ua}{D} \geqslant 6.9,$$

where L is the length of tube over which appreciable changes in concentration occur. In later papers Philip (1963) criticizes some aspects of this work and a similar analysis by Aris (1956), and suggests a more generalized approach. Carrier (1956) deals with the effluent from a tube when the intake is oscillatory. This is a method which may be of particular interest in physiological systems (Philip 1969b).

Two more recent methods rely on less familiar mathematical solutions. Pasternak, Schimscheimer, and Heller (1970) described a 'dynamic', isobaric permeation experiment in which atmospheric pressure is maintained on both sides of a membrane. One face is in contact with the penetrant and a carrier gas, flowing at a constant rate past the other face, sweeps away the penetrant which diffuses through the membrane to a recording system. Thus the permeation rate is observed continuously. The mathematical statement of an experiment is

$$C = 0, \qquad x = l, \qquad t \geqslant 0, \tag{10.44}$$

$$C = C_i, \qquad x = 0, \qquad t = 0, \tag{10.45}$$

$$C = C_f, \qquad x = 0, \qquad t > 0, \tag{10.46}$$

$$C = C_i(l-x)/l, \qquad 0 \leqslant x \leqslant l, \qquad t = 0, \tag{10.47}$$

$$C = C_f(l-x)/l, \qquad 0 \leqslant x \leqslant l, \qquad t = \infty. \tag{10.48}$$

Thus there is a change from one steady state to another with the concentration at $x = l$ always zero. Pasternak $et\ al.$ use Holstein's solution (4.27) modified to allow for their initial condition (10.47). It is

$$F = \frac{DC_i}{l} + \frac{4D(C_f - C_i)}{l\pi^{\frac{1}{2}}} \left(\frac{l^2}{4Dt}\right)^{\frac{1}{2}} \sum_0^\infty \exp\left\{-(2n+1)^2 l^2/(4Dt)\right\} \tag{10.49}$$

where F is the flux at $x = l$. Taking the first term of the series in (10.49) we have

$$\Delta F = F - \frac{DC_i}{l} = \Delta F_\infty \frac{4}{\pi^{\frac{1}{2}}} \left(\frac{l^2}{4Dt}\right)^{\frac{1}{2}} \exp\left\{-l^2/(4Dt)\right\}, \tag{10.50}$$

where $\Delta F_\infty = D(C_f - C_i)/l$ is the change in flux from the start to the finish of the experiment. They then write (10.50) in the form

$$\Delta F/\Delta F_\infty = (4/\pi^{\frac{1}{2}}) X \, e^{-X^2}, \tag{10.51}$$

where $X^2 = l^2/(4Dt)$. Then D is easily derived from a comparison of the master plot of $\Delta F/\Delta F_\infty$ from (10.51) with the corresponding experimental plot against t.

Evnochides and Henley (1970) used a frequency-response technique to measure the diffusion coefficient and the solubility simultaneously. A polymer sample is exposed to a permeating gas whose pressure is varied sinusoidally. The phase angle and amplitude of the weight changes are measured as a function of the frequency of the sinusoidal pressure change. Initially, the main pressure of the gas is P_m and the concentration inside the film is uniform and equal to KP_m. At $x = \pm L$ we have the concentration

$$C_s = KP = K\{P_m + A_P \sin(\omega t + \varepsilon)\}$$

where the amplitude $A_P = \frac{1}{2}(P_1 - P_0)$. Defining $C_m = KP_m$ and $C_0 = KP_0$ and

$$\gamma = (C - C_m)/(C_m - C_0) \tag{10.52}$$

we require solutions of

$$\partial\gamma/\partial t = D\,\partial^2\gamma/\partial x^2 \tag{10.53}$$

subject to

$$\gamma = 0, \qquad t = 0, \qquad -L \leqslant x \leqslant L, \tag{10.54}$$

$$\gamma = \sin(\omega t + \varepsilon), \qquad t \geqslant 0, \qquad x = \pm L. \tag{10.55}$$

The full solution is given by Carslaw and Jaeger (1959).

It is

$$\gamma = A_\gamma \sin(\omega t + \varepsilon + \phi)$$

$$+ 4\pi D \sum_{n=0}^{\infty} \frac{(-1)^n (2n+1)\{4L^2\omega \cos\varepsilon - D(2n+1)^2\pi^2 \sin\varepsilon\}}{16L^4\omega^2 + D^2\pi^4(2n+1)^4}$$

$$\times \cos\frac{(2n+1)\pi x}{2L} \exp\left\{-\frac{D(2n+1)^2\pi^2 t}{4L^2}\right\}. \tag{10.56}$$

At large times a periodic steady state is reached

$$\gamma = A_\gamma \sin(\omega t + \varepsilon + \phi), \tag{10.57}$$

where

$$A_\gamma = \mathrm{mod}\left\{\frac{\cosh \zeta x(1+\mathrm{i})}{\cosh \zeta L(1+\mathrm{i})}\right\} = \frac{\cosh 2\zeta x + \cos 2\zeta x}{\cosh 2\zeta L + \cos 2\zeta L}, \tag{10.58}$$

$$\phi = \arg\left\{\frac{\cosh \zeta x(1+\mathrm{i})}{\cosh \zeta L(1+\mathrm{i})}\right\}, \qquad \zeta = (\omega/2D)^{\frac{1}{2}} \tag{10.59}$$

Defining M_t, as the mass of gas in the film at time t we have

$$M_t = 2A\rho \int_0^L C\,\mathrm{d}x, \tag{10.60}$$

where A and ρ are the area and density of the film. Taking $M_m = WC_m$ and

$M_0 = WC_0$, where W is the mass of the film, eqns (10.52) and (10.57) with appropriate substitutions lead after integration of (10.60) to

$$M_t - M_m = A \sin(\omega t + \varepsilon + \psi), \tag{10.61}$$

where

$$A = \frac{\sqrt{(2)}KWA_p(\sin^2 2\zeta L + \sinh^2 2\zeta L)^{\frac{1}{4}}}{2\zeta L(\cos 2\zeta L + \cosh 2\zeta L)}, \tag{10.62}$$

$$\psi = \tan^{-1}\left\{\frac{\sin 2\zeta L - \sinh 2\zeta L}{\sin 2\zeta L + \sinh 2\zeta L}\right\}. \tag{10.63}$$

Evnochides and Henley conclude that a single experiment in which both A and ψ are measured suffices to determine both K and ζ. Both occur in (10.62) but the phase angle is independent of the solubility.

10.6.1. *Analysis of steady-state flow*

In a series of papers Ash, Barrer, and others have studied extensively the theory of steady-state flow through membranes. From measurements of the flux through the membrane they have been concerned to extract information about the concentration profile, the amount of diffusant within the membrane, and the dependence of the diffusion coefficient on concentration and the positional coordinate.

Ash and Barrer (1971) have developed a general theory of one-dimensional diffusion in slabs, hollow cylinders, and spherical shells when D is a separable function of concentration and the space coordinate. For v-dimensional diffusion, where $v = 1$ corresponds to flow through unit area of a slab, $v = 2$ through a cylinder of unit length, and $v = 3$ through a spherical shell, the general equation can be written

$$\frac{\partial C}{\partial t} = \frac{1}{r^{v-1}} \frac{\partial}{\partial r}\left(r^{v-1} D \frac{\partial C}{\partial r}\right), \tag{10.64}$$

and in the steady state $\partial C/\partial t = 0$. At the boundaries R_1 and $R_2 (R_2 > R_1)$ it is assumed that

$$C(R_1, t) = C_1 \quad \text{and} \quad C(R_2, t) = C_2, \tag{10.65}$$

where C_1 and C_2 are constant and $C_1 > C_2$. Thus flow occurs in the direction of r increasing and is termed 'forward flow'. When C_1 and C_2 are interchanged we have 'reverse' flow.

Take D to be given by

$$D = D_0\phi(C)f(r), \tag{10.66}$$

where $\phi(0) = 1$, $f(R_1) = 1$. The steady-state flow in the direction of r

increasing is

$$F = -\omega_v r^{v-1} D_0 \phi(C) f(r) \, dC/dr \qquad (10.67a)$$

$$= \omega_v D_0 \int_{C_2}^{C_1} \frac{\phi(u) \, du}{N_v(R_1, R_2)}, \qquad (10.67b)$$

where

$$N_v(x, y) = \int_x^y \frac{dr}{r^{v-1} f(r)}$$

and $\omega_1 = 1$, $\omega_2 = 2\pi$, $\omega_3 = 4\pi$. The expression (10.67b) for F follows from (10.67a) by integrating with respect to r, and remembering that F is independent of r. We note in passing that $|F|$ is independent of the direction of flow for the type of D assumed in (10.66) (Hartley 1948). Integration of (10.67a) with respect to r from R_1 to r yields

$$\frac{N_v(R_1, r)}{N_v(R_1, R_2)} \int_{C_2}^{C_1} \phi(u) \, du = \int_{C(r)}^{C_1} \phi(u) \, du. \qquad (10.68)$$

If $\phi(C)$ and $f(r)$ are known, the concentration profile through the membrane, in principle, may be determined from (10.68). Interchanging C_1 and C_2 in (10.68) gives the corresponding relationship for $C(r)$ in 'reverse' flow. By adding the two relationships we obtain

$$\int_{C(r)}^{C_1} \phi(u) \, du + \int_{\bar{C}(r)}^{C_2} \phi(u) \, du = 0. \qquad (10.69)$$

In the particular case of constant D, (10.69) becomes

$$C(r) + \bar{C}(r) = C_1 + C_2,$$

so that the 'reverse' distribution is readily found if the 'forward' distribution is known. It is likely, however, that in practice $\phi(C)$ and $f(r)$ will not be known in advance. Ash and Barrer (1971) propose the following procedure for determining $C(r)$ in such cases.

We keep C_2 and R_1 constant and make a series of measurements of $F = F(C_1, R_2)$ for different values of C_1 and R_2. From (10.68) we have

$$\frac{N_v(R_1, r)}{N_v(R_1, R_2)} = \frac{\int_{C_2}^{C_1} \phi(u) \, du - \int_{C_2}^{C(r)} \phi(u) \, du}{\int_{C_2}^{C_1} \phi(u) \, du} = \frac{F(C_1, R_2) - F(C(r), R_2)}{F(C_1, R_2)}$$

by using (10.67b). We must keep in mind that $C(r)$ is the concentration at the point r, $R_1 < r < R_2$ in an experiment for which the surfaces R_1 and R_2 are at concentrations C_1 and C_2 respectively, but that $F(C(r), R_2)$ denotes the flow that would be observed in an experiment for which the concentration at the surface R_1 were $C(r)$. Similarly, $F(C_1, r)$ signifies the flow that would be observed through a membrane with surfaces at R_1 and r kept at concentrations C_1 and C_2 respectively. With this nomenclature

$$F(C_1, r) = \omega_v D_0 \int_{C_2}^{C_1} \phi(u)\, du / N_v(R_1, r),$$

and $F(C_1, R_2)$ is given by (10.67b).

It follows that

$$\frac{N_v(R_1, r)}{N_v(R_1, R_2)} = \frac{F(C_1, R_2)}{F(C_1, r)} = \frac{F(C_1, R_2) - F(C(r), R_2)}{F(C_1, R_2)}. \qquad (10.70)$$

We now draw two master plots based on experimental data: the first of F against C_1 for constant C_2, R_1 and R_2; the second of F against R_2 for C_1, C_2 and R_1 held constant (Figs. 10.1(a) and 10.1(b)).

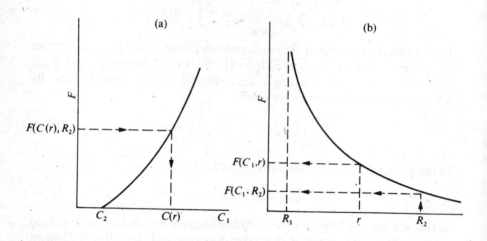

Fig. 10.1.

We select a value of r in the range $R_1 < r < R_2$. From the plot in Fig. 10.1(b) we read the values of $F(C_1, r)$ and $F(C_1, R_2)$ and find their ratio. The value of $F(C(r), R_2)$ follows from (10.70) and from Fig. 10.1(a) the value of $C(r)$ for the selected value of r. Proceeding in this way for various selected values of r the profile of $C(r)$ against r is determined. To deal with reverse flow we write

$$\int_{C(r)}^{C_1} \phi(u)\, du + \int_{\tilde{C}(r)}^{C_1} \phi(u)\, du = \int_{C_2}^{C_1} \phi(u)\, du$$

by using (10.69). Hence

$$F(C(r), R_2) - F(\bar{C}(r), R_2) = F(C_1', R_2). \qquad (10.71)$$

This relation may be substituted into (10.70) to obtain

$$\frac{N_v(R_1, r)}{N_v(R_1, R_2)} = \frac{\bar{F}(C_1, R_2)}{\bar{F}(C_1, r)} = \frac{F(\bar{C}(r), R_2)}{\bar{F}(C_1, R_2)}. \qquad (10.72)$$

The master plots allow $\bar{C}(r)$ to be found as a function of r. The amount of diffusant in the membrane is

$$M = \int_{R_1}^{R_2} \omega_v r^{v-1} C(r)\, dr. \qquad (10.73)$$

Once $C(r)$ has been determined as a function of r the integration of (10.73) can be performed numerically or graphically. The procedures for finding $C(r)$ and M require no knowledge of $\phi(C)$ or of $f(r)$. These functions can be obtained, however, at least in theory, from the master plots.

For C_2 and R_1 held constant (10.67b) may be written

$$F(C_1, R_2)N_v(R_1, R_2) = \omega_v D_0 \int_{C_2}^{C_1} \phi(u)\, du.$$

Partial differentiation with respect to R_2 and C_1 respectively gives

$$\frac{F(C_1, R_2)}{R_2^{v-1} f(R_2)} + \frac{\partial F(C_1, R_2)}{\partial R_2} N_v(R_1, R_2) = 0, \qquad (10.74)$$

and

$$\frac{\partial F(C_1, R_2)}{\partial C_1} = \frac{\omega_v D_0 \phi(C)}{N_v(R_1, R_2)}. \qquad (10.75)$$

On combining (10.74) and (10.75) and re-arranging we see that

$$D_0 \phi(C_1) f(R_2) = -\frac{F(C_1, R_2)}{R_2^{v-1} \omega_v} \frac{\partial F(C_1, R_2)/\partial C_1}{\partial F(C_1, R_2)/\partial R_2}. \qquad (10.76)$$

All the quantities needed to find the functional dependence of D can be obtained graphically or numerically from the master plots in Figs. 10.1(a) and 10.1(b).

The particular cases of $D = D_0 \phi(C)$ or $D = D_0 f(r)$ only can be extracted from the equations developed above. They are treated separately by Ash and Barrer (1971), who give some additional relationships.

10.6.2. *Time-lag methods*

In § 4.3.3 (p. 51) the approach to steady-state flow through a membrane was analysed and the time-lag L related to the diffusion coefficient by (4.26) for cases of constant D.

Paul and Dibenedetto (1965) have shown how the time-lag technique can be extended to the situation in which the pressure or concentration on the 'downstream' side of the membrane increase with time, as the diffusant emerges into a limited volume.

Frisch (1957) has obtained expressions for the time lag which apply also to systems with concentration dependent diffusion coefficients without explicitly solving the diffusion equation. He considers $C = C_0$, $x = 0$, $C = 0$, $x = l$, $t \geqslant 0$; $C = 0$, $0 < x < l$, $t = 0$.

We denote the non-steady rate of flow through the surface $x = l$ by $F(t)$, where

$$F(t) = -\left(D_c \frac{\partial C}{\partial x}\right)_{x=l} \tag{10.77}$$

The total flow through this surface in time t is given by

$$Q(t) = \int_0^t F(t)\, \mathrm{d}t. \tag{10.78}$$

To find $Q(t)$ we first integrate both sides of the diffusion eqn (7.5) over x from x to l and obtain

$$\int_x^l \frac{\partial C}{\partial t}(z, t)\, \mathrm{d}z + F(t) + D_c \frac{\partial C}{\partial x} = 0.$$

Integrating again over x from 0 to l and re-arranging we obtain

$$F(t) = \frac{1}{l}\left\{\int_0^{C_0} D(u)\, \mathrm{d}u - \int_0^l \int_x^l \frac{\partial C}{\partial t}(z, t)\, \mathrm{d}z\, \mathrm{d}x\right\}.$$

Finally, by integrating over t from 0 to t and changing the order of integration we find

$$Q(t) = \frac{1}{l}\left\{t\int_0^{C_0} D(u)\, \mathrm{d}u - \int_0^l \int_x^l C(z, t)\, \mathrm{d}z\, \mathrm{d}x\right\}. \tag{10.79}$$

The asymptote $Q_s(t)$ to $Q(t)$ is

$$Q_s(t) = F(t - L) \tag{10.80}$$

since its gradient is the steady-state flow rate F and the time lag L is its intercept on the t-axis. Comparison of (10.79) with (10.80) using the appropriate form of (10.67b) shows that

$$L = \frac{\int_0^l \int_x^l C_s(z)\,dz\,dx}{\int_0^{C_0} D(u)\,du},$$ (10.81)

where $C_s(x)$ is the concentration distribution through the membrane in the steady state. Integration by parts leads to the form

$$L = \int_0^l x C_s(x)\,dx \bigg/ \int_0^{C_0} D(u)\,du.$$ (10.82)

The appropriate form of (10.68) gives

$$\int_{C_s}^{C_0} D(u)\,du = \frac{x}{l}\int_0^{C_0} D(u)\,du.$$ (10.83)

In principle, $C_s(x)$ can be derived from (10.83) if $D(u)$ is known or (10.82) can be rewritten as

$$L = \frac{l^2 \int_0^{C_0} w D(w)\{\int_0^{C_0} D(u)\,du\}\,dw}{\{\int_0^{C_0} D(u)\,du\}^3}$$ (10.84)

Provided the functional form of $D(C) = D_C$ is known, the measurement of L for various values of C_0 allows the constant parameters of D, and hence D as a function of C, to be determined.

It can be shown (Frisch 1958; Pollack and Frisch 1959) that when $\ln \int_0^C D(u)\,du$ is a convex function of C, the time lag is governed by the inequality

$$\frac{1}{6} \leqslant \frac{L(C_0)}{l^2 C_0}\int_0^{C_0} D(u)\,du \leqslant \frac{1}{2}.$$

Barrie and Machin (1967) stress that the relative variation of L with C is always considerably smaller than that of $D(C)$ with C. This is particularly true when $D(C)$ decreases with C increasing. They show graphs of $L(C)$ and $D(C)$ illustrating these general statements for various concentration dependent diffusion coefficients.

Ash, Baker, and Barrer (1968) have suggested two extensions of (10.82) and (10.84) which may be more convenient in practice when D is a function of concentration only. The first is based on the appropriate special case of (10.76) which, in conjunction with the master plots of Figs. 10.1(a) and (b), allows the concentration distribution to be deduced from measurements of the steady-state flow F. We can determine $C_s(x)$ in (10.82) from the relation

$$\frac{x}{l} = \frac{F(C_0) - F\{C_s(x)\}}{F(C_0)}. \tag{10.85a}$$

Furthermore, from (10.85a) we have

$$dx = \frac{-l}{F(C_0)} dF\{C_s(x)\}, \tag{10.85b}$$

and by substituting (10.85a) and (10.85b) in (10.82) we obtain

$$L = \frac{l \int_0^{F(C_0)} C_F[F(C_0) - F\{C_s(x)\}] dF\{C_s(x)\}}{\{F(C_0)\}^3},$$

where $C_F = C[F\{C_s(x)\}]$ and is the value of the concentration which corresponds to $F\{C_s(x)\}$ on the appropriate master plot. Frisch (1957) illustrates his method for the cases of D constant and $D(C) = D_0(1 + \beta C)$, and also considers more general boundary conditions. Barrer and Ferguson (1958) and also Meares (1958, 1965) quote the expression for L given in (10.86) below for an exponential dependence of D on C.

Frisch (1958) showed that from measurements of L and the steady-state flow F at different pressures, both the parameters in the $D(C)$ functional relationship and the solubility coefficient can be found without resource to a separate measurement of solubility.

Meares (1965) carried out a single permeation experiment for a system in which the diffusion coefficient was known to have the form

$$D = D_0 e^{\beta C}.$$

One face of the membrane is in contact with vapour at a constant pressure P_1. Frisch's expression (10.84) leads in this case to the relationship

$$\frac{4D_0 L}{l^2} = \frac{4 e^{\beta C_1} - 1 + e^{2\beta C_1}(2\beta C_1 - 3)}{(e^{\beta C_1} - 1)^3}. \tag{10.86}$$

The limiting slope as t approaches zero, of the graph expressing Holstein's solution (4.27a) gives a value for D_0, since at this time the concentration is still very small in the greater part of the sheet. The steady-state rate of permeation through a sheet of area A into a volume V in which the pressure is $p(t)$ is given by

$$\frac{dp}{dt} = \frac{ART}{Vl} \int_0^{C_1} D_0 e^{\beta C} dC = \frac{ARTD_0}{Vl\beta}(e^{\beta C_1} - 1). \tag{10.87}$$

When D_0 has been determined, βC_1 follows from (10.86) and the individual values of β and C_1 are calculated from (10.87).

If Henry's law is obeyed the solubility coefficient S is also obtained from C_1 and the pressure on the ingoing face. Meares (1965) found the method successful in three quite different systems.

Frisch (1959) has generalized his early results to include cases in which D may depend explicitly not only on concentration but also on time and x. He considers the v-dimensional, diffusion problem defined by

$$\frac{\partial C}{\partial t}(r, t) = \frac{1}{r^{v-1}} \frac{\partial}{\partial r} \left\{ r^{v-1} \frac{\partial \phi}{\partial r}(C, r, t) \right\}, \quad v = 1\ 2, 3, \qquad (10.88)$$

in $R_1 < r < R_2, t > 0$, where

$$C(R_1, t) = C_1 \quad \text{or} \quad \phi(C_1, R_1, t) = \phi_1(t)$$

$$C(R_2; t) = C_2 \quad \text{or} \quad \phi(C_2, R_2, t) = \phi_2(t)$$

$$C(r, 0) = C_0(r). \qquad (10.88a)$$

Here ϕ is a 'potential of diffusion' defined in general by

$$F = -D \operatorname{grad} C = -\operatorname{grad} \phi,$$

and for spherically symmetric systems by

$$F = -D\, \partial C / \partial r = -\partial \phi / \partial r. \qquad (10.89)$$

Integration of (10.89) gives

$$\phi_1(t) - \phi_2(t) = \int_{R_1}^{R_2} F(r, t)\, dr,$$

in general and

$$\phi_1(t) - \phi_2(t) = -\int_{C_1}^{C_2} D(u, t)\, du$$

when $D = D(C, t)$.

The case, $v = 1$, corresponds to one-dimensional diffusion through a flat membrane of thickness l, $v = 2$ to radial diffusion through infinite concentric cylinders and $v = 3$ to radial diffusion between concentric spheres.

We assume that as $t \to \infty$ a steady state is attained and $\phi \to \phi_s$, the solution of

$$\frac{1}{r^{v-1}} \frac{\partial}{\partial r} \left(r^{v-1} \frac{\partial \phi_s}{\partial r} \right) = 0$$

with

$$\phi_s(R_1) = \phi_1^s = \lim_{t \to \infty} \phi_1(t).$$

$$\phi_s(R_2) = \phi_2^s = \lim_{t \to \infty} \phi_2(t).$$

Following the method outlined above Frisch arrives at the relation

$$L = \frac{\int_{R_1}^{R_2} r^{1-\nu}\, dr \int_r^{R_2} \{C_s(z) - C_0(z)\} z^{\nu-1}\, dz}{\phi_1^s - \phi_2^s}$$

$$- \frac{\int_0^\infty \{\phi_1(\tau) - \phi_2(\tau) - (\phi_1^s - \phi_2^s)\}\, d\tau}{\phi_1^s - \phi_2^s} \tag{10.90}$$

We note that the expression (10.90) is generally true for all equations of flow which conform to the conservation of mass condition. Ash *et al.* (1968) stress this point and quote expressions analogous to (10.90) for the time lag measured at any plane in a membrane. When $D = D(C)$, a function of concentration only, the second term on the right side of (10.90) vanishes identically. This follows since for $D = D(C)$, ϕ is independent of t and hence $\phi_1(t) = \phi_1^s$, $\phi_2(t) = \phi_2^s$. The first term on the right side then reduces for the particular case $\nu = 1$ and $C_2 = C_0 = 0$ to (10.82) and hence to (10.84).

Thus in general $L(l)$ can be written as the sum of a 'Fickian' time lag $L_F(l)$ and a 'non-Fickian' or 'time-dependent' contribution L_T which is independent of l. Frisch (1962a) wrote

$$L(l) = L_F(l) + L_T = \alpha(C_1)l^2 + L_T,$$

where $\alpha(C_1)$ is independent of l. Thus, if time lags can be measured for films of different thicknesses but which are otherwise of identical physical properties a plot of $L(l)$ against l^2 should be a straight line of ordinate intercept L_T, while the slope gives a value of $D(C_1, \infty)$ obtained from steady-state measurements. So far, we have considered only the time lag measured at the 'downstream' boundary, e.g. at $x = l$, for the plane membrane. Frisch (1962b) used his technique to obtain expressions for $L(x)$ and in particular $L(0)$ measured at the 'upstream' face of the membrane. The 'upstream–downstream' difference is shown to be

$$\Delta L = L(l) - L(0) = (1/F) \int_0^l \{C_s(x) - C_0\}\, dx. \tag{10.91}$$

Furthermore, we may refer to a time lag measured by an experiment in which $C_0 = C_2$ as an 'absorption time lag' L_a. The condition $C_0 = C_1$, leading to a 'desorption time lag' L_d, is also possible. On inserting this value for C_0 in the Fickian part of (10.90) and taking $\nu = 1$ it follows that the absorption–desorption time-lag difference for a Fickian membrane is given by

$$\delta L(l) \equiv \delta L = L_a - L_d = l(C_1 - C_2)/2F. \tag{10.92}$$

The double difference $\delta\, \Delta L$ is defined as

$$\delta\, \Delta L = \Delta L_a - \Delta L_d = l(C_1 - C_2)/F. \tag{10.93}$$

Barrer (1969) gives a list of time lags and related quantities. Petropoulos and Roussis (1967) made extensive studies of the properties of these time lags and their differences in attempts to detect and characterize different types of non-Fickian behaviour. They distinguish between systems in which the diffusion and solubility parameters are time-dependent or distance-dependent. They introduce the gradient of chemical potential into the diffusion equation which becomes

$$\frac{\partial C}{\partial t} = \frac{\partial}{\partial x}\left(D_T S \frac{\partial a}{\partial x}\right) = \frac{\partial}{\partial x}\left(P_T \frac{\partial a}{\partial x}\right), \tag{10.94}$$

with boundary conditions

$$a(0, t) = a_0, \qquad a(l, t) = a_l, \qquad a(x, 0) = a_1, \tag{10.95}$$

and where D_T is the 'thermodynamic diffusion coefficient', and P_T the corresponding permeability. We assume the a_0, a_l, a_1 to be constants. Comparison of (10.94) with (7.5) shows that

$$D = D_T S \, \partial a / \partial C. \tag{10.96}$$

By applying the method of Frisch to (10.94) we find that L_T, the non-Fickian time-lag increment, is given by

$$L_T = \int_0^\infty dt \int_{a_l}^{a_0} \{P_T(a, \infty) - P_T(a, t)\} \, da \bigg/ \int_{a_l}^{a_0} P_T(a, \infty) \, da. \tag{10.97}$$

Petropoulos and Roussis introduced expressions for $S(a, t)$ and $D_T(a, t)$ which they thought to be appropriate to diffusion in polymers showing evidence of time-dependent behaviour. They used

$$S(a, t)/S(a, \infty) = 1 - u(a) \exp\{-\beta_1(a)t\}, \tag{10.98}$$

where

$$u(a) = 1 - S(a, 0)/S(a, \infty), \qquad u(a_1) = 0;$$

$$D_T(a, t)/D_T(a, \infty) = 1 - w(a) \exp\{-\beta_2(a)t\}, \tag{10.99}$$

where

$$w(a) = 1 - D_T(a, 0)/D_T(a, \infty).$$

All these functions of a may alternatively be expressed in terms of $C^\alpha = aS(a, \alpha)$, the corresponding concentration in the fully relaxed membrane after theoretically infinite time. Substitution of (10.98) and (10.99) in (10.97) remembering (10.96) yields, after integration with respect to t,

$$L_T = \int_{C_l^\alpha}^{C_0^\alpha} D(C^\alpha, \alpha)\left\{\frac{u}{\beta_1} + \frac{w}{\beta_2} - \frac{uw}{\beta_1 + \beta_2}\right\} dC^\alpha \bigg/ \int_{C_l^\alpha}^{C_0^\alpha} D(C^\alpha, \alpha) \, dC^\alpha$$

$$= L_{T_1} + L_{T_2} + L_{T_3}, \tag{10.100}$$

where all functions in (10.100) are to be taken as functions of C^∞, i.e. $u = u(C^\alpha)$ etc. Petropoulos and Roussis suggested that u and w will normally be much less than unity, so that $L_T = L_{T_1} + L_{T_2}$, approximately, which is the sum of the contributions from the time-dependence of S and D_T respectively. They proceed to examine some examples in detail.

Time-lag expressions for a system in which the solubility and thermo-dynamic diffusion coefficient are distance-dependent are obtained by the same authors for

$$S = S(a, x) = S(a, 0)\phi_1(x), \tag{10.101}$$

$$D_T = D_T(a, x) = D_T(a, 0)\phi_2(x), \tag{10.102}$$

where $\phi_1(0) = \phi_2(0) = 1$ and $\phi_1(x), \phi_2(x)$ are non-negative, analytic, and continuous functions in $0 \leqslant x \leqslant l$.

The time-dependent systems we have considered tend towards a Fickian, purely concentration-dependent system in the final steady state of permeation. We have studied the increment in the time lag which results because the system is not Fickian throughout the permeation process. In contrast, the distance-dependence with which we are now concerned persists for all time and so the time-lag increment has a different significance. Any experimentally measured time lag for a non-homogeneous sheet can be interpreted as if it were homogeneous by applying the Frisch method to (7.5). This will yield an apparent diffusion coefficient $\tilde{D} = \tilde{D}(\tilde{C})$ which is an average function for the sheet and where $\tilde{C}$ is the averaged concentration variable obtained from the corresponding equilibrium sorption measurements; i.e.

$$\tilde{C} = a\tilde{S} = \frac{a}{l}\int_0^l S \, dx, \tag{10.103}$$

where $\tilde{S}$ is the apparent solubility coefficient. The purely concentration-dependent parameters $\tilde{D}$ and $\tilde{S}$ define an 'equivalent Fickian system'. We wish to see how the expression for the time lag for a non-homogeneous sheet differs from the standard expression for the equivalent, purely concentration-dependent system.

Petropoulos and Roussis transformed (10.94) into

$$\frac{\partial C}{\partial t} = \frac{\partial}{\partial x}\left(D_e \frac{\partial \tilde{C}}{\partial x}\right),$$

$$\tilde{C}(0, t) = \tilde{C}_0, \tilde{C}(l, t) = \tilde{C}_l, \tilde{C}(x, 0) = \tilde{C}_1, \tag{10.104}$$

where $\tilde{C}_0, \tilde{C}_l, \tilde{C}_1$ are constants, $\tilde{C}_0$ and $\tilde{C}_l$ being in equilibrium with the respective a_r. In view of (10.101) and (10.102), D_e in (10.104) is given by

$$D_e = D_e(\tilde{C}, x) = D_T S(\partial a/\partial \tilde{C}) = D_e(\tilde{C}, 0)\phi(x), \tag{10.105}$$

where

$$\phi(x) = \phi_1(x)\phi_2(x). \tag{10.106}$$

The procedure developed by Frisch (1957) can be applied to (10.104) and yields the result

$$L(l) = \frac{1}{F\Psi(l)\Phi(l)} \int_0^l \phi_1(x)\Phi(x)\{\tilde{C}(x, \infty) - \tilde{C}_1\} \, dx, \tag{10.107}$$

where $\tilde{C}(x, \infty)$ is given by the solution of

$$\int_{\tilde{C}_l}^{\tilde{C}} \tilde{D}(z) \, dz = \left\{1 - \frac{\Phi(x)}{\Phi(l)}\right\} \int_{\tilde{C}_l}^{\tilde{C}_0} \tilde{D}(z) \, dz. \tag{10.108}$$

The subsidiary functions in (10.107) are

$$\Phi(x) = \int_0^x \frac{dz}{\phi(z)}, \qquad \Psi(x) = \int_0^x \phi_1(z) \, dz, \tag{10.109}$$

and the steady-state flow F is now given by

$$F = \{\Phi(l)\}^{-1} \int_{\tilde{C}_1}^{\tilde{C}_0} D_e(\tilde{C}, 0) \, d\tilde{C} \equiv l^{-1} \int_{\tilde{C}_1}^{\tilde{C}_0} \tilde{D}(z) \, dz, \tag{10.110}$$

from which we find

$$\tilde{D}(\tilde{C}) = lD_e(\tilde{C}, 0)/\Phi(l). \tag{10.111}$$

Corresponding expressions for time-lag differences were given by Petropoulos and Roussis (1967) and some special cases were examined. Any time lag can be split in this case into $L = L_F + L_h$ by analogy with the time-dependent case $L = L_F + L_T$.

Ash et al. (1968) quote a particular result thought to be relevant for a membrane made by compaction of powder under pressure. The middle of the membrane is less compressed than the outer layers. For a diffusion coefficient D defined by

$$D = D_0/\{1 - \alpha u(1 - u)\}, \tag{10.112}$$

where $u = x/l$ and α is a coefficient, they quote the result

$$L_l = \frac{l^2}{D_0}\left(\frac{1}{6} - \frac{\alpha}{20} + \frac{\alpha^2}{280}\right) \bigg/ \left(1 - \frac{1}{6}\alpha\right). \tag{10.113}$$

If the form of (10.112) is assumed or known for D then (10.113) allows α to be determined from a measurement of L_l.

A few time-lag relationships permit distinction between time- and distance-dependent anomalies in their most general form. For example:

(i) Always, we have $\Delta L_T^a(a_0, a_l) = 0$,

$$\Delta L_h^a(a_0, a_l) \neq 0 \quad \text{for all } a_0 \neq a_l.$$

(ii) All non-Fickian components of time lag are unaffected by flow reversal in time-dependent systems but not in distance-dependent cases except when the solubility and diffusion coefficient are symmetrical about the mid plane of the membrane for all activities.

Other results are given by Petropoulos and Roussis (1967) and are extended (1969*a*) to cases in which activity and distance factors are not separable.

Ash, Barrer, and Nicholson (1963) and later Paul (1969) used Frisch's method to study the effect of an immobilizing reaction on the diffusion time lag. Paul's diffusion equation is modified to allow for some molecules being trapped in non-diffusing holes according to a Langmuir-type isotherm. It has the form

$$\left(1 + \frac{K}{(1+\alpha C)^2}\right)\frac{\partial C}{\partial t} = D\frac{\partial^2 C}{\partial x^2}, \tag{10.114}$$

where C is the concentration of molecules free to diffuse with a diffusion coefficient D, and α and K are parameters of the immobilization isotherm. Paul gives general expressions for both the absorption and desorption time lags. Commonly used conditions are those of zero concentration at the downstream face of a plane sheet and a constant concentration C_0, maintained at the upstream face. In addition, $C = 0$ initially throughout the sheet for absorption and $C = C_0$ initially for desorption. For these cases, Paul obtained for the time lag for absorption L_a

$$6L_aD^2/l^2 = 1 + Kf(y), \tag{10.115}$$

where

$$f(y) = 6y^{-3}\{\tfrac{1}{2}y^2 + y - (1+y)\ln(1+y)\},$$

and for the desorption lag L_d,

$$3L_dD^2/l^2 = -1 + Kg(y), \tag{10.116}$$

where

$$g(y) = 3y^{-3}(1+y)^{-1}\{\tfrac{3}{2}y^2 + y - (1+y)^2\ln(1+y)\}.$$

Paul discussed the effect of various parameters on the time lags and their implications for models of diffusion in glassy polymers. Petropoulos (1970) examined the physical assumptions underlying earlier treatments.

10.6.3. *Analysis of concentration–distance curves*

There are a number of optical methods for observing how either the refractive index or its gradient depends on distance measured in the direction of diffusion at a given time (Crank and Park 1968). Concentration–distance curves can also be obtained in the case of two metals interdiffusing (Barrer

1951). If two infinite media are brought together at $t = 0$, e.g. two long columns of liquid or two metal bars, the diffusion coefficient and its concentration-dependence can readily be deduced from the concentration distribution observed at some known subsequent time. The conditions of the experiment are

$$C = C_\infty, \qquad x < 0, \qquad t = 0, \tag{10.117}$$

$$C = 0, \qquad x > 0, \qquad t = 0, \tag{10.118}$$

where C is the concentration of the component in which we are interested, and $x = 0$ is the position of the initial interface between the two components at time $t = 0$. Assume for the moment that there is no over-all change of volume on mixing and that C is measured as mass per unit volume of the system. Then we may use the Boltzmann variable $\eta = x/2t^{\frac{1}{2}}$ and as in § 7.2.1 (p. 105) we obtain the ordinary differential equation

$$-2\eta \frac{dC}{d\eta} = \frac{d}{d\eta}\left(D\frac{dC}{d\eta}\right). \tag{10.119}$$

On integration with respect to η (10.119) becomes

$$-2\int_0^{C_1} \eta \, dC = \left[D\frac{dC}{d\eta}\right]_{C=0}^{C=C_1} = \left(D\frac{dC}{d\eta}\right)_{C=C_1} \tag{10.120}$$

since $D \, dC/d\eta = 0$ when $C = 0$. Here C_1 is any value of C between 0 and C_∞. Finally, by rearrangement of (10.120) and introducing x and t, we have

$$D_{C=C_1} = -\frac{1}{2t}\frac{dx}{dC}\int_0^{C_1} x \, dC. \tag{10.121}$$

Since $(D \, dC/d\eta)_{C=C_\infty} = 0$ also, it follows from (10.120) that

$$\int_0^{C_\infty} x \, dC = \int_0^{C_\infty} \eta \, dC = 0, \tag{10.122}$$

and in order that the boundary conditions shall be satisfied the origin from which x is measured must be such that (10.122) is satisfied. In other words, the plane, $x = 0$, must be chosen so that the two shaded areas in Fig. 10.2(a) are equal. In a constant volume system (10.122) is a conservation of mass condition and it is clearly satisfied if x is measured from the initial position of the boundary between the two components at time $t = 0$. In this case, therefore, the procedure is to plot the concentration-distance curve for a known time as in Fig. 10.2(b), to locate the plane $x = 0$ by use of (10.122), and then to evaluate D at various concentrations C_1 from (10.121). The integrals can be obtained by using a planimeter or by counting squares and the gradients dx/dC by drawing tangents. In Fig. 10.2(b) the area representing $\int_0^{C_1} x \, dC$ is shown shaded. Diffusion coefficients in metal systems were

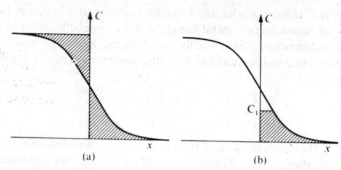

Fɪɢ. 10.2. Evaluation of D from a concentration–distance curve using (10.121).

obtained by Matano (1932–3) using this method, and since then it has been widely used. Alternative ways of using (10.121) have been suggested.

(i) One development has been found useful in dilute solutions (Eversole, Peterson, and Kindswater 1941). On integrating by parts (10.121) becomes

$$D_{C=C_1} = -\frac{1}{2t}\frac{dx}{dC}\left(x_1 C_1 - \int_x^{x_1} C\,dx'\right).$$

$$= -\frac{1}{4t}\frac{x_1}{(\ln C_1)^{\frac{1}{2}}}\frac{dx}{d(\ln C)^{\frac{1}{2}}}\left(1 + \frac{1}{x_1 C_1}\int_{x_1}^x C\,dx'\right), \quad (10.123)$$

where x_1 is the value of x at which $C = C_1$ and the derivatives are to be measured at the point $x = x_1$. The determination of the gradients at various concentrations is made simpler by this modification if the graph of x against $(\ln C)^{\frac{1}{2}}$ is nearly linear as it is in some systems.

A corresponding analysis can be applied to the penetration of diffusing substance into an effectively semi-infinite sheet, provided the concentration–distance curve can be obtained. An example is the diffusion of dye molecules from a well-stirred solution into a cellulose sheet (Crank 1948a). Eqn (10.121) still holds if x is measured from the surface of the sheet.

(ii) Graphical or numerical methods of evaluating the diffusion coefficient from (10.121) have certain disadvantages. They entail the measurement of slopes and areas under curves and it is clear from the behaviour of the concentration–distance curves in Fig. 10.2 that considerable uncertainty arises for concentrations near the limiting values. Often the values of diffusion coefficients for very small concentrations of one component are of great importance and so it is desirable to be able to calculate them as accurately as the experimental data permit. A method of improving the the accuracy of the calculations near the extremes of the concentration-range has been suggested by Hall (1953). He examines some experimental data of da Silva and Mehl (1951) for the copper–silver system. Using the nomenclature of the previous section, he takes C to be the concentration of copper and C_{γ} to be 100 per cent copper. Fig. 10.3, taken from Hall's paper, is a probability plot of C/C_{γ} against $x/t^{\frac{1}{2}}$, i.e. the variable u which is used as ordinate is given by

$$\tfrac{1}{2}(1 + \operatorname{erf} u) = C/C_{\gamma} . \tag{10.124}$$

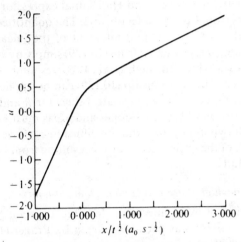

FIG. 10.3. Probability plot of a concentration distribution. a_0 is the side of the unit cell.

Other relationships involving error functions can of course be used. Hall finds (10.124) convenient for purposes of calculation. We avoid his use of the function erfc for the left-hand side of (10.124) because it is not standard notation. The point of interest is that when plotted in this way the concentration–distance curve becomes linear at the two ends of the concentration range. Now a straight line on this plot corresponds to an equation of the type

$$u = h\eta + k, \tag{10.125}$$

so that, from (10.124) and (10.125), we have

$$C/C_{\gamma} = \tfrac{1}{2}\{1 + \operatorname{erf}(h\eta + k)\}. \tag{10.126}$$

It readily follows that

$$\frac{d\eta}{dC} = \frac{\pi^{\frac{1}{2}}}{hC_{\gamma}} \exp(u^2), \tag{10.127}$$

and

$$\int_0^C \eta \, dC' = \frac{hC_{\gamma}}{\pi^{\frac{1}{2}}} \int_{-\infty}^{\eta} \eta' \exp\{-(h\eta' + k)^2\} \, d\eta'$$

$$= -\frac{C_{\gamma}}{2h\pi^{\frac{1}{2}}} \exp(-u^2) - \frac{kC}{h}. \tag{10.128}$$

If now we write (10.121) in terms of η and substitute from (10.127) and (10.128) we obtain

$$D(C) = \frac{1}{h^2} + k\frac{\pi^{\frac{1}{2}}}{h^2}(1 + \text{erf } u)\exp(u^2).\tag{10.129}$$

The differences between (10.129) and Hall's final expression are because the η used here differs from his λ by a factor of 2. The quantities h and k are obtained respectively as the slope and intercept of the linear portion of the probability plot, and D is evaluated from (10.129) simply by substituting these values and that of u which through (10.124) corresponds to the C/C_∞ for which D is required. The relationship (10.129) can be applied at each end of the concentration range with appropriate values for h and k. In the intermediate curved portion of Fig. 10.3 slopes and areas must be measured on a plot of the type shown in Fig. 10.2 and the diffusion coefficient deduced from (10.121). Evidence of the improved accuracy resulting from the use of (10.129) is advanced by Hall.

10.6.4. *Systems in which there is a volume change on mixing*

The interdiffusion of two components forming a system in which volume changes occur on mixing has been considered by Prager (1953) as follows. The rates at which the concentrations of the two components C_A and C_B change at a point are given by

$$\frac{\partial C_A}{\partial t} = \frac{\partial}{\partial x}\left(\mathscr{D}_A\frac{\partial C_A}{\partial x}\right) - \frac{\partial}{\partial x}(vC_A),\tag{10.130}$$

$$\frac{\partial C_B}{\partial t} = \frac{\partial}{\partial x}\left(\mathscr{D}_B\frac{\partial C_B}{\partial x}\right) - \frac{\partial}{\partial x}(vC_B),\tag{10.131}$$

where $\mathscr{D}_A$ and $\mathscr{D}_B$ are the intrinsic diffusion coefficients introduced in § 10.4, and v is the velocity of the mass-flow assumed to be in the x direction and dependent only on the x coordinate and the time t. There is a relationship between C_A and C_B at constant temperature and pressure which is

$$V_BC_B + V_AC_A = 1,\tag{10.132}$$

where the Vs are the partial molal volumes of the two components, and also we have

$$(\partial C_B/\partial C_A)_{P,T} = -V_A/V_B.\tag{10.133}$$

Using (10.132) and (10.133) in (10.131) we find

$$-\frac{V_A}{V_B}\frac{\partial C_A}{\partial t} = -\frac{\partial}{\partial x}\left(\mathscr{D}_B\frac{V_A}{V_B}\frac{\partial C_A}{\partial x}\right) - C_B\frac{\partial v}{\partial x} + \frac{V_A}{V_B}\frac{\partial C_A}{\partial x},\tag{10.134}$$

and combining (10.134) with (10.130) yields

$$\frac{\partial v}{\partial x} = V_A \frac{\partial}{\partial x}\left(\mathscr{D}_A \frac{\partial C_A}{\partial x}\right) - V_B \frac{\partial}{\partial x}\left(\mathscr{D}_B \frac{V_A}{V_B}\frac{\partial C_A}{\partial x}\right). \tag{10.135}$$

Integration by parts from $-\infty$ to x transforms (10.135) into

$$v = V_A(\mathscr{D}_A - \mathscr{D}_B)\frac{\partial C_A}{\partial x} + \int_{-\infty}^{x}\frac{C_B}{C_A}\left(\mathscr{D}_A + \mathscr{D}_B\frac{V_A C_A}{V_B C_B}\right)\left(\frac{\partial V_B}{\partial C_A}\right)\left(\frac{\partial C_A}{\partial x}\right)^2 \mathrm{d}x', \tag{10.136}$$

where v and $\partial C_A/\partial x$ have been assumed zero at $x = -\infty$. Substituting (10.136) into (10.130) and (10.131) we find

$$\frac{\partial C_A}{\partial t} = \frac{\partial}{\partial x}\left(D^V \frac{\partial C_A}{\partial x}\right) - \frac{\partial}{\partial x}\left\{C_A \int_{-\infty}^{x}\frac{D^V}{V_B C_A}\left(\frac{\partial V_B}{\partial C_A}\right)\left(\frac{\partial C_A}{\partial x}\right)^2 \mathrm{d}x'\right\}, \tag{10.137}$$

$$\frac{\partial C_B}{\partial t} = \frac{\partial}{\partial x}\left(D^V \frac{\partial C_B}{\partial x}\right) - \frac{\partial}{\partial x}\left\{C_B \int_{-\infty}^{x}\frac{D^V}{V_B C_A}\left(\frac{\partial V_B}{\partial C_A}\right)\left(\frac{\partial C_A}{\partial x}\right)^2 \mathrm{d}x'\right\}, \tag{10.138}$$

where D^V is related to $\mathscr{D}_A$ and $\mathscr{D}_B$ by

$$D^V = \mathscr{D}_A V_B C_B + \mathscr{D}_B V_A C_A, \tag{10.139}$$

which is eqn (10.32). We should note that we are here defining D^V by (10.139). It has the significance of a mutual diffusion coefficient as defined in § 10.2 (p. 205) only if there is no volume change on mixing. The second terms on the right-hand sides of (10.137) and (10.138) arise because of the volume changes on mixing, and they vanish when such changes do not occur, i.e. when $(\partial V_A/\partial C_A)_{P,T} = 0$, in which case (10.137) and (10.138) reduce to the usual diffusion equations.

If the initial distribution is such that

$$C_A = 0, \qquad x < 0, \qquad t = 0, \tag{10.140}$$

$$C_A = C_x, \qquad x > 0, \qquad t = 0, \tag{10.141}$$

then we can make the Boltzmann substitution (1894) even if $(\partial V_A/\partial C_A)_{P,T}$ is not zero. Thus if we suppose C_A to be a function of $\eta = x/2t^{\frac{1}{2}}$ only, eqn (10.137) becomes

$$-2\eta\frac{\mathrm{d}C_A}{\mathrm{d}\eta} = \frac{\mathrm{d}}{\mathrm{d}\eta}\left(D^V \frac{\mathrm{d}C_A}{\mathrm{d}\eta}\right) - \frac{\mathrm{d}}{\mathrm{d}\eta}\left\{C_A \int_{-\infty}^{\eta}\frac{D^V}{V_B C_A}\left(\frac{\partial V_B}{\partial C_A}\right)\left(\frac{\mathrm{d}C_A}{\mathrm{d}\eta}\right)^2 \mathrm{d}\eta'\right\}, \tag{10.142}$$

with boundary conditions

$$C_A = 0, \qquad \eta = -\infty; \qquad C_A = C_x, \qquad \eta = \infty. \tag{10.143}$$

If the concentration distribution $C_A(\eta)$ is known from experiment, (10.142) can be solved for D^V to give

$$D^V_{C_A = C_1} = -2\frac{d\eta}{dC_1}\left\{\int_0^{C_1} \eta\, dC_A + C_1 \int_0^{C_1} \frac{\int_0^{C_1} \eta\, dC_A}{C_A V_B}\left(\frac{\partial V_B}{\partial C_A}\right)dC_A\right\}. \qquad (10.144)$$

The first term on the right-hand side of (10.144), when written in terms of x and t, is the expression (10.121) for calculating D from the concentration distribution when there is no volume change on mixing. The second term is a small correction term in which we have substituted for D^V from (10.121) written in terms of η as an approximation.

It is of interest to examine what diffusion coefficient is obtained by the Matano (1932–3) procedure, as described in § 10.6.3 (p. 230) when there is a volume change on mixing. Continuing to denote by η the distance coordinate measured from the initial position of the boundary between the two components, Matano introduces a new coordinate η_1 measured from a new origin chosen so that $\int_0^{C_\infty} \eta_1\, dC_A = 0$. If the new origin be at $\eta = -\delta$ then we have $\eta_1 = \eta + \delta$ and hence

$$\int_0^{C_\infty} (\eta - \delta)\, dC_A = 0, \qquad (10.145)$$

so that

$$\delta = \frac{1}{C_\infty}\int_0^{C_\infty} \eta\, dC_A. \qquad (10.146)$$

Furthermore, Matano deduces a diffusion coefficient from the expression

$$D_{C_A = C_1} = -2\frac{d\eta_1}{dC_A}\int_0^{C_1} \eta_1\, dC_A. \qquad (10.147)$$

Now

$$\int_0^{C_1} \eta_1\, dC_A = \int_0^{C_1} (\eta + \delta)\, dC_A = \int_0^{C_1} \eta\, dC_A + C_1 \delta, \qquad (10.148)$$

and hence the diffusion coefficient calculated in this way is

$$D_{C_A = C_1} = -2\frac{d\eta}{dC_A}\left(\int_0^{C_1} \eta\, dC'_1 + C_1 \delta\right), \qquad (10.149)$$

since $d\eta_1/dC_A = d\eta/dC_A$ and where δ is given by (10.146). Comparing (10.149) with (10.144) shows that this procedure does not yield exactly the coefficient D^V related to the intrinsic diffusion coefficients by (10.139) but only an approximation to it. It assumes

$$\int_0^{C_1} \frac{\int_0^{C_1} \eta\, dC_A}{C_A V_B}\left(\frac{\partial V_B}{\partial C_A}\right)dC_A = \text{constant}, \qquad (10.150)$$

the constant being δ, the displacement of the origin. Prager (1953) suggests that a better approximation is to treat V_B and $(1/C_A)(\partial V_B/\partial C_A)$ as constants, in which case (10.144) becomes

$$D_{C_A = C_1} = -2\frac{d\eta}{dC_A}\left(\int_0^{C_1} \eta \, dC_A\right) + BC_1 \int_0^{C_1} \left(\int_0^{C_1} \eta \, dC_A\right) dC_A, \quad (10.151)$$

where B is given by

$$B = -\left(\int_0^{C_\infty} \eta \, dC_A\right) \Bigg/ \left\{C_\infty \int_0^{C_\infty} \left(\int_0^{C_\infty} \eta \, dC_A\right) dC_A\right\}. \quad (10.152)$$

This procedure, like Matano's, requires no data on the partial molal volumes. If such data are available it is of course possible, though more laborious, to use the complete equation (10.144) to evaluate D^V.

We can now see what is the physical significance of the diffusion coefficient deduced by the Matano procedure when there is a volume change on mixing. Returning to eqn (10.142) and writing ϕ for the flow velocity due to volume change, we have

$$-2\eta \frac{dC_A}{d\eta} = \frac{d}{d\eta}\left(D^V \frac{dC_A}{d\eta}\right) - \frac{d}{d\eta}(C_A\phi), \quad (10.153)$$

and on integrating from $C_A = 0$ to $C_A = C_\infty$, this becomes

$$-2\int_0^{C_\infty} \eta \, dC_A = \left[D^V \frac{dC_A}{d\eta}\right]_0^{C_\infty} - [C_A\phi]_0^{C_\infty}. \quad (10.154)$$

Since $dC_A/d\eta = 0$ at both ends of the range of integration, we see that

$$C_\infty\phi_\infty = -2\int_0^{C_\infty} \eta \, dC_A, \quad (10.155)$$

where ϕ_∞ is the value of ϕ when $C_A = C_\infty$. Thus by choosing the origin of η such that the right-hand side of (10.155) is zero, we select a frame of reference such that, provided $C = 0$ at one end of the system, there is no flux of component A at either end due to volume change. It must be emphasized that the removal of the effect of volume change is complete only at the ends of the system, and so the diffusion coefficient deduced by using the Matano procedure is not the mutual diffusion coefficient. In fact, it is not a diffusion coefficient which is readily related to those obtained by other methods.

An alternative treatment of a system in which a volume change occurs makes use of the frame of reference fixed with respect to the total mass of the system. Thus in terms of the variables C_A^M and ξ_M introduced in § 10.3, eqn (10.121) (p. 231) becomes

$$D_{C_A = C_1}^M = -\frac{1}{2t}\frac{d\xi_M}{dC_A^M}\int_0^{C_1} \xi_M \, dC_A^M, \quad (10.156)$$

and according to the Matano procedure the origin of ξ_M is to be chosen such that

$$\int_0^{1/V_B^0} \xi_M \, dC_A^M = 0. \qquad (10.157)$$

The upper limit of integration for C_A^M comes from (10.16) since when $\xi_M = \infty$, $C_B^M = 0$. Choosing the origin of ξ_M so that (10.157) holds ensures that the boundary conditions are satisfied as in § 10.6.3 (p. 231). Thus, provided concentrations and distance scales are expressed in the new units as described in § 10.3, the coefficient D^M is obtained from (10.156) and (10.157).

By combining (10.23) and (10.35) we have a relationship between D^M and the intrinsic diffusion coefficients, $\mathscr{D}_A$ and $\mathscr{D}_B$, which takes the place of (10.139) in Prager's method.

10.6.5. *Sorption method*

(i) *Constant diffusion coefficient.* A different approach is to deduce the diffusion coefficient from observations of the over-all rate of uptake of component A by a plane sheet of component B. Such a method has been used, for example, to determine the diffusion coefficient of direct dyes in cellulose sheet (Neale and Stringfellow 1933) and of oxygen in muscle (Hill 1928). In each case the diffusion coefficient was assumed constant.

We shall describe the method in terms of the uptake of vapour by a plane sheet, first on the assumptions that the diffusion coefficient is constant and the sheet does not swell, and then for cases in which the sheet swells and the diffusion coefficient is concentration-dependent. Its application to other systems will be obvious. The experimental procedure is to suspend a plane sheet of thickness l in an atmosphere of vapour maintained at constant temperature and pressure, and to observe the increase in weight of the sheet and hence the rate of uptake of vapour. This can be done most conveniently by hanging the sheet on a spring of known stiffness. The appropriate solution of the diffusion equation may be written

$$\frac{M_t}{M_\infty} = 1 - \frac{8}{\pi^2} \sum_{m=0}^{\infty} \frac{1}{(2m+1)^2} \exp\{-D(2m+1)^2\pi^2 t/l^2\}, \qquad (10.158)$$

if the uptake is considered to be a diffusion process controlled by a constant diffusion coefficient D. Here M_t is the total amount of vapour absorbed by the sheet at time t, and M_∞ the equilibrium sorption attained theoretically after infinite time. The application of (10.158) is based on the assumption that immediately the sheet is placed in the vapour the concentration at each surface attains a value corresponding to the equilibrium uptake for the vapour pressure existing, and remains constant afterwards. The sheet is considered to be initially free of vapour. The value of t/l^2 for which $M_t/M_\infty = \frac{1}{2}$, conveniently written $(t/l^2)_{\frac{1}{2}}$, is given by

$$\left(\frac{t}{l^2}\right)_{\frac{1}{2}} = -\frac{1}{\pi^2 D} \ln \left\{ \frac{\pi^2}{16} - \frac{1}{9}\left(\frac{\pi^2}{16}\right)^9 \right\} \qquad (10.159)$$

approximately, the error being about 0·001 per cent. Thus we have

$$D = 0·049/(t/l^2)_{\frac{1}{2}}, \qquad (10.160)$$

and so, if the half-time of a sorption process is observed experimentally for a system in which the diffusion coefficient is constant, the value of this constant can be determined from (10.160). The extension of this method to less simple systems is as follows.

(ii) *Sorption by a swelling sheet.* In deriving (10.158) the thickness l of the sheet is assumed to remain constant as diffusion proceeds. In practice it often happens, however, that the sheet swells and the thickness increases as the vapour enters. Eqn (10.158) can still be used in such cases, provided we take a frame of reference fixed with respect to the substance of the sheet, and concentration and thickness are measured in the units discussed in § 10.3 (p. 207). Thus we take the basic volume of the sheet to be its volume in the absence of vapour and use the unit of length ξ_B such that unit ξ_B contains, per unit area, unit basic volume of the substance of the sheet B. Then the thickness of the sheet, measured in these units, is constant and equal to the original unswollen thickness, and the diffusion coefficient deduced from (10.158) by substituting the original thickness for l is that for the diffusion of vapour relative to stationary sheet (denoted by D_A^B in § 10.3). If there is no over-all volume change on mixing, i.e. if the increase in volume of the sheet is equal to the volume of vapour sorbed at the vapour pressure existing in the experiment, the coefficient obtained by this sorption method is related to the mutual diffusion coefficient D^V, deduced by the Matano procedure, by eqn (10.28), i.e.

$$D_A^B = D^V(1 - \text{volume fraction of vapour})^2. \qquad (10.161)$$

(iii) *Concentration-dependent diffusion coefficients.* Clearly from (10.158) and (10.160) the value of t/l^2 for which M_t/M_∞ has any given value, and in particular the value of $(t/l^2)_{\frac{1}{2}}$, is independent of M_∞ when the diffusion coefficient is constant. Fig. 10.4 shows a set of curves obtained experimentally by Park (see Crank and Park 1949) for the uptake of chloroform by a polystyrene sheet, each curve corresponding to a different vapour pressure and hence a different M_∞. It is evident that $(t/l^2)_{\frac{1}{2}}$ decreases considerably the greater the value of the final uptake M_∞ and therefore the diffusion coefficient is not constant but increases as the concentration of chloroform is increased. The problem is to deduce quantitatively how the diffusion coefficient is related to concentration, given the half-times of sorption experiments carried out for a number of different vapour pressures.

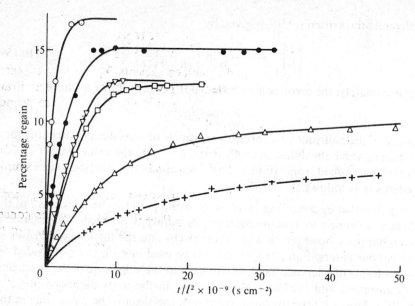

FIG. 10.4. Rate of uptake of chloroform by polystyrene sheet at 25 °C.

Application of (10.160) to each of the curves of Fig. 10.4 yields some mean value $\bar{D}$, say, of the variable diffusion coefficient averaged over the range of concentration appropriate to each curve. The method devised by Crank and Park (1949) depends on the fact that, for any experiment, $\bar{D}$ provides a reasonable approximation to $(1/C_0) \int_0^{C_0} D \, dC$, where 0 to C_0 is the concentration range existing in the sheet during that experiment. This was shown to be so (Crank and Henry 1949a) by evaluating numerical solutions of the diffusion equation for a number of variable diffusion coefficients. By applying (10.160) to the sorption-time curves so calculated, values of $\bar{D}$ were obtained and compared with corresponding values of $(1/C_0) \int_0^{C_0} D \, dC$. Thus by deducing a value of $\bar{D}$ from each of the experimental curves of Fig. 10.4 using (10.160), and assuming the approximate relationship

$$\bar{D} = (1/C_0) \int_0^{C_0} D \, dC, \qquad (10.162)$$

a graph showing $\bar{D} C_0$ as a function of C_0 can be drawn as in Fig. 10.5 and numerical or graphical differentiation of the curve with respect to C_0 gives a first approximation to the relationship between D and C. Numerical data are given in Table 10.1.

In many cases this first approximation may be sufficiently accurate, but successively better approximations can be obtained as follows. Sorption–time curves are calculated numerically for the D–C relationship obtained as

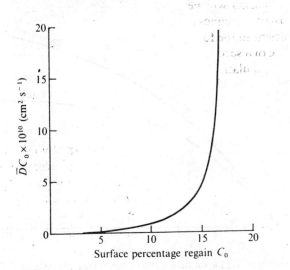

FIG. 10.5. Graph of $\bar{D}C_0$ as a function of C_0.

the first approximation, there being one calculated curve for each experimental curve of Fig. 10.4. The $\bar{D}$ values derived by applying (10.160) to the calculated curves are shown in Table 10.1, column 6. Comparison of these calculated $\bar{D}$ values with the experimental ones and therefore with the first estimate of $(1/C_0)\int_0^{C_0} D \, dC$ shows the errors involved in use of (10.162) for this particular type of diffusion coefficient. The correct relationship between $\bar{D}$ and $(1/C_0)\int_0^{C_0} D \, dC$ for the D–C relationship given by the first approximation can be plotted, and from this improved values of $(1/C_0)\int_0^{C_0} D \, dC$ can be read off for the experimental values of $\bar{D}$. By repeating the differentiation, a second approximation to the diffusion coefficient D is obtained. The process can be repeated till the calculated and experimental values of $\bar{D}$ agree to the accuracy desired. Results are shown in Table 10.1.

In this form the method can be used whatever the relationship between D and C but the calculations involved in evaluating successive approximations are tedious and laborious.

(iv) *Exponential and linear diffusion coefficients.* In many systems the diffusion coefficient is found to depend either linearly or exponentially on concentration, and so for these cases correction curves have been produced showing the difference between $(1/C_0)\int_0^{C_0} (D/D_0) \, dC$ and $\bar{D}/D_0$, where D_0 is the value of D at zero concentration of vapour or whatever the diffusing substance is. The correction curves are shown in Fig. 10.6. By using the appropriate curve, the diffusion coefficient–concentration relationship can be deduced as readily from sorption experiments as from steady-state measurements. In each case, differentiation of $\int D \, dC$ to obtain D is the

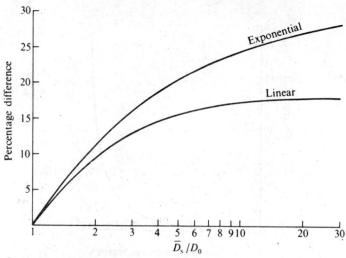

FIG. 10.6. Correction curves for sorption method. Percentage difference is

$$(\bar{D}/D_0) - (1/C_0) \int_0^{C_0} (D/D_0)\, dC$$

expressed as a percentage of $\bar{D}/D_0$. These curves can be applied to both half-times and initial gradients.

only mathematical operation involved. Park (1950) used a method of differentiation which is considerably more accurate than direct graphical differentiation if D is an exponential function of C. Denoting $\int_0^{C_0} (D/D_0)\, dC$ by I, he plotted $\log I$ against C_0 to obtain a curve which approximates to a straight line. Then $d \log I/dC_0$ is easily obtained by graphical differentiation and D follows from the expression

$$D/D_0 = dI/dC_0 = 2.303 I\, d \log I/dC_0. \qquad (10.163)$$

The method for exponential and linear diffusion coefficients is therefore as follows:

(1) plot sorption–time curves for different vapour pressures as in Fig. 10.4;

(2) calculate $\bar{D}$ for each curve using (10.160);

(3) extrapolate $\bar{D}$ to zero concentration to give D_0;

(4) read off from the appropriate correction curve of Fig. 10.6 the value of $(1/C_0) \int_0^{C_0} (D/D_0)\, dC = I/C_0$ for each $\bar{D}/D_0$;

(5) differentiate I using (10.163) or otherwise to obtain D/D_0 and hence D.

A second approximation may be necessary due to the uncertainty in the first estimate of D_0. It is, of course, necessary to have some idea of the type of diffusion coefficient involved before the choice of correction curve can be made. For example, if the initial choice is exponential, but the final result is nothing like an exponential function, and if use of the linear correction curve also fails, then the method of successive approximations has to be carried out as described above in § 10.6.5 (iii) (p. 239)

For diffusion coefficients varying exponentially with concentration Hansen (1967) calculated correction factors, F_a and F_d, which when applied to $\bar{D}$ give $D(C_0)$, where C_0 is the surface concentration for absorption or the initial uniform concentration for desorption, i.e.

$$D(C_0) = F_a\bar{D} \quad \text{or} \quad F_d\bar{D}.$$

His computed results are reproduced in Table 10.2, where exp K is the ratio of the diffusion coefficients at the extremes of the concentration range in any experiment and $T_{\frac{1}{2}} = D_0 t_{\frac{1}{2}}/(L')^2$, where L' is the half-thickness of a free sheet or the full thickness of a film attached on one surface to an impermeable backing. After deducing $\bar{D}$ from the half-times measured for several concentration ranges, D_0 is estimated by extrapolation. Values of exp K, and hence of the correction factors F can then be estimated and an approximation to a $D(C)$ relationship obtained. This can be improved iteratively. Hansen finds his results agree reasonably well with those of Table 10.1.

Fels and Huang (1970) described a method of obtaining the diffusion coefficient and free volume parameters (Fujita 1961) from desorption data. They obtained numerical solutions of the equation

$$\frac{\partial C}{\partial t} = \frac{\partial}{\partial x}\left\{(1-v_P)^2 \exp\left(\frac{v_P}{a+bv_P}\right)\frac{\partial C}{\partial x}\right\},$$

where a and b involve free volume parameters and the diffusion coefficient, and v_P is the volume fraction of the penetrant.

(v) *Use of a single sorption experiment.* Duda and Vrentas (1971) develop a new technique for deducing the concentration-dependence from a single sorption curve, based on the method of moments (§ 7.4, p. 129). They also use only the first two moments, but a novel feature is that they express their concentration profile in terms of a variable related to the amount of penetrant in a given part of the polymer rather than the space coordinate as in (7.152). In the nomenclature of § 7.4 (p. 129) their profile can be written

$$\frac{c-\psi}{1-\psi} = \left(\frac{m}{M}\right)^{\frac{1}{3}}\left\{A_0 + A_1\left(\frac{m}{M}\right)^{\frac{1}{3}}\right\},$$

where A_0, A_1 are undetermined functions of time T, $\psi(T) = c(0, T)$ and

$$m = \int_0^X \{c(X', T) - \psi(T)\}\,dX', \qquad M = \int_0^1 \{c(X', T) - \psi(T)\}\,dX'.$$

Such a form is capable of describing the concentration profile over the entire duration of the sorption experiment. After considerable manipulation Duda and Vrentas arrive at a fairly simple set of two algebraic equations and one differential equation from which the concentration-dependence $F(c)$ can be obtained when the sorption–time curve and its derivative are available from experimental data. Duda and Vrentas tested the accuracy of the method for exponential diffusion coefficients varying by as much as five-fold and found about 5 per cent accuracy over 80 per cent of the concentration range. There was a serious loss of accuracy at low concentrations but complementary studies of desorption could circumvent the difficulty. In practice, the need to carry out numerical differentiation may prove a serious drawback. But this and other ways of extracting information from single sorption, desorption or permeation curves deserve further study. Duda and Vrentas include the effects of volume changes on mixing and phase volume change. They also derive expressions for the concentration distribution corresponding to a prescribed diffusion coefficient.

10.6.6. *Sorption–desorption method*

A method which uses both sorption and desorption data, is very quick and simple in cases to which it is applicable. If $\bar{D}_s$ is calculated using (10.160) from the half-time for sorption and $\bar{D}_d$ from the half-time for desorption over the same concentration range, then $\frac{1}{2}(\bar{D}_s + \bar{D}_d)$ is a better approximation to $(1/C_0)\int_0^{C_0} D\,dC$ than either $\bar{D}_s$ or $\bar{D}_d$ separately. Often it is a very good approximation and has been used without correction in some instances (Prager and Long 1951; Kokes *et al.* 1952). If the range in D is small enough for this to give the accuracy required, the method is particularly simple since there is no need even to extrapolate to obtain D_0 at zero concentration. Kokes, Long, and Hoard (1952) have applied the sorption–desorption method to successive small concentration ranges so that the approximation $\frac{1}{2}(\bar{D}_s + \bar{D}_d)$ can be used with more confidence than it could if applied to the complete concentration range. Should it be necessary to obtain higher accuracy, the correction curves of Fig. 10.7 can be used to obtain better estimates of $(1/C_0)\int_0^{C_0} D\,dC$. These are then differentiated as in the sorption method to obtain D.

10.6.7. *Use of initial rates of sorption and desorption*

In § 10.6.5 (p. 240) we deduced an average diffusion coefficient from the half-time of a sorption curve by using (10.160). It is also possible to deduce an average diffusion coefficient from the initial gradient of the sorption curve when plotted against the square root of time. Thus, in the early stages, for a constant diffusion coefficient D and a sheet of thickness l, we have (§ 4.3.2, p. 48)

$$\frac{M_t}{M_\infty} = \frac{4}{\pi^{\frac{1}{2}}}\left(\frac{Dt}{l^2}\right)^{\frac{1}{2}} \tag{10.164}$$

If the initial gradient, $R = d(M_t/M_\infty)/d(t/l^2)^{\frac{1}{2}}$, is observed in a sorption

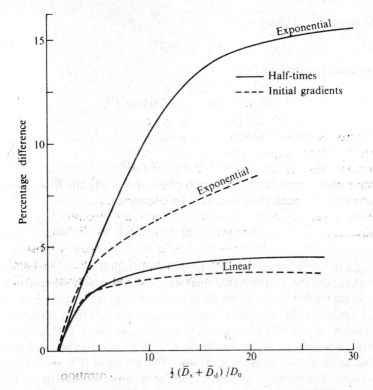

FIG. 10.7. Correction curve for sorption–desorption method. Percentage difference is

$$(1/C_0)\int_0^{C_0}(D/D_0)\,\mathrm{d}C - \tfrac{1}{2}(\bar{D}_s+\bar{D}_d)/D_0$$

expressed as a percentage of $\tfrac{1}{2}(\bar{D}_s+\bar{D}_d)/D_0$. (N.B. The differences here are opposite in sign to those of Fig. 10.6.)

experiment in which D is concentration-dependent, then the average diffusion coefficient $\bar{D}$ deduced from (10.164) is

$$\bar{D} = \frac{\pi}{16}R^2. \tag{10.165}$$

This, too, provides an approximation to $(1/C_0)\int_0^{C_0} D\,\mathrm{d}C$, and the sorption method can proceed as above but starting with the new values of $\bar{D}$ given by (10.165) as original data. Initial rates of desorption can be used similarly. Correction curves for exponential and linear diffusion coefficients are shown for the sorption method in Fig. 10.6 and for sorption–desorption in Fig. 10.7.

If the sorption curve when plotted against $(t/l^2)^{\frac{1}{2}}$ is approximately linear as far as $M_t/M_\infty = \tfrac{1}{2}$, and this is often true in practice, then it is easy to see that (10.160) and (10.165) yield roughly the same diffusion coefficient. Thus for a linear sorption curve we have

$$R = \tfrac{1}{2}/\sqrt{(t/l^2)_{\frac{1}{4}}}, \qquad (10.166)$$

and so from (10.165) we find

$$\bar{D} = \frac{\pi}{64} \Big/ (t/l^2)_{\frac{1}{4}} = 0{\cdot}049/(t/l^2)_{\frac{1}{4}}, \qquad (10.167)$$

which is the same as (10.160).

10.6.8. *Use of final rates of sorption and desorption*

In the later stages of diffusion in a plane sheet only the first term in the series of (10.158) need be considered, and we have

$$\frac{d}{dt}\{\ln(M_\infty - M_t)\} = -\frac{D\pi^2}{l^2}. \qquad (10.168)$$

During a sorption experiment the final concentration, C_0 say, is approached throughout the sheet; during the final stages of desorption the concentration everywhere tends to zero. If the diffusion coefficient is constant, both experiments should yield the same value on applying (10.168). Otherwise, the application of (10.168) to a sorption experiment gives $D(C_0)$, while from desorption we obtain D_0, the value at $C = 0$. By this method, we avoid the extrapolation needed to estimate D_0 in the method of § 10.6.5 (iii). Frensdorff (1964) used both methods and examined the extent of the final linear portions of desorption curves by computing numerical solutions of the diffusion equation for $D(C)$ given by

$$\frac{D(C)}{D(0)} = 1 + A'\frac{C}{C_0},$$

where C_0 is the initial equilibrium concentration. His results are reproduced in Fig. 10.8

10.6.9. *A step-function approximation to the diffusion coefficient*

Prager (1951) described an alternative method of deducing the diffusion coefficient-concentration relationship from sorption data. The principle of the method is to approximate to the actual concentration-dependence by a step function. For such a function the differential equations can be solved analytically (Chapter 13) and the heights of the individual steps computed. The step function is then smoothed out by an averaging process.

The concentration range to be studied is divided into a number of intervals and we assume that the diffusion coefficient $D(C)$ has a constant (although as yet unknown) value in each. For a diffusion experiment covering only the first concentration interval D is constant throughout and may be calculated if M_t is known as a function of t, using the solution (10.164). For a diffusion experi-

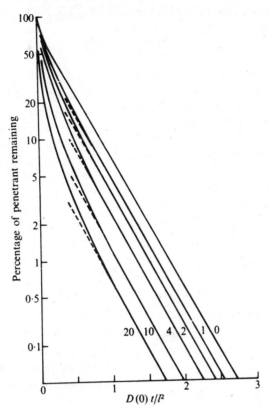

Fig. 10.8. Calculated desorption curves for a linear diffusion coefficient $D(C)/D(0) = 1 + A'C/C_0$, where C_0 is the initial concentration. Numbers on the curves are values of A', and l is the thickness of the sheet. (From Frensdorff 1964).

ment covering the first two intervals, $D(C)$ is given by a step function whose value in the first interval is now known from the first calculation. The value in the second interval may be calculated as described below using the equations of § 13.2.2 (p. 290). This procedure may be continued so as to calculate the value of $D(C)$ in the third interval, its value in the first two being known, and so on, until the entire concentration range has been covered. The method as described here is based on the treatment of diffusion with discontinuous boundaries given in §§ 13.2.2 and 13.2.4 (p. 290 and 296). An alternative treatment is given in Prager's original paper (1951).

Before the concentration in the centre of the film attains an appreciable value we may consider diffusion to take place into a semi-infinite medium. If we have a two-step diffusion coefficient such that for concentrations greater than C_X, $D = D_1$, and for concentrations less than C_X, $D = D_2$, then the diffusion process can be described by the equations of § 13.2.2 if

D_1 and D_2 are known. In particular, $\partial M_t/\partial t^{\frac{1}{2}}$ is given using the nomenclature of that section by

$$\frac{\partial M_t}{\partial t^{\frac{1}{2}}} = -\frac{2AD_1^{\frac{1}{2}}}{\pi^{\frac{1}{2}}}.$$

Here we have the opposite problem, i.e. given $\partial M_t/\partial t^{\frac{1}{2}}$ from an experiment and D_2 having been calculated from the previous experiment, to find D_1. If M_t is measured as a function of $t^{\frac{1}{2}}$ this means essentially that $AD_1^{\frac{1}{2}}$ is measured. Hence if we estimate D_1, k is determined from (13.12) on p. 291 and a new D_1 follows from (13.15) of the same section. Successive estimates of D_1 are made till agreement is obtained between estimated and final values of D_1. The graphs shown in Figs. 13.4, 13.5 and 13.6 of Chapter 13 can be used to help the calculation.

The method is readily extended to a three-step diffusion coefficient and then to higher numbers of steps. Thus if

$$D = D_1, \qquad C_1 > C > C_X, \qquad (10.169)$$

$$D = D_2, \qquad C < C_Y, \qquad (10.170)$$

$$D = D_3, \qquad C_X > C > C_Y, \qquad (10.171)$$

the process is described by the equations of § 13.2.4 (p. 296). Again $AD_1^{\frac{1}{2}}$ is measured and D_2, D_3 are known from previous experiments. We estimate D_1 and calculate k_1 from (13.32) and k_2 from (13.38). Finally we check k_1, k_2, D_1 in (13.39), repeating if necessary as before. The number of equations of the type (13.38) and (13.39) increases with the number of steps in the diffusion coefficient but the method still holds.

Fujita and Kishimoto (1952) analysed Prager's data by the method of moments, § 7.4.

10.6.10. *Sorption by sheets initially conditioned to different uniform concentrations*

Barrer and Brook (1953) used a method of measuring the concentration-dependence of the diffusion coefficient based on a series of sorption experiments. The concentration C_0 at the surface of the specimen is kept constant throughout the series but the initial uniform concentration, C_1, through the sheet, is different for each experiment. Denoting again by M_t the amount of diffusing substance taken up by the sheet in time t, we may write for the initial stages of sorption,

$$M_t = K(C_1)t^{\frac{1}{2}}. \qquad (10.172)$$

Here K is different for each experiment, being a function of C_1, the initial concentration through the sheet. A curve showing K as a function of C_1 must cut the axis of C, at the point $C_1 = C_0$ because then the initial rate of sorption, and hence K, is zero, there being no concentration gradient. Thus the experimental curve of K against C_1, when extrapolated, must always pass through $C_1 = C_0$ when $K = 0$. Furthermore, as the interval $C_0 - C_1$ is decreased progressively in successive experiments the sorption process can more and more nearly be described by the diffusion equation and its solution for a constant diffusion coefficient, i.e. by

$$M_t = 4(C_0 - C_1)(D_0 t/\pi)^{\frac{1}{2}}, \qquad (10.173)$$

where D_0 is the value of the diffusion coefficient at the concentration C_0. Hence, as C_1 approaches C_0, the curve of K against C_1 approaches asymptotically a tangent of slope $-4D_0^{\frac{1}{2}}/\pi^{\frac{1}{2}}$. This differs by a factor of two from the expression given by Barrer and Brook, because they consider the amount entering through one face of the sheet, whereas we include both. If the experimental curve is sufficiently well defined by the data obtained from the sorption experiments, the tangent can be drawn and D_0 deduced from its gradient using (10.173). By repeating the series of experiments for different values of the surface concentration C_0, the relationship between the diffusion coefficient and concentration is obtained. If, as in some zeolites, D decreases as C increases, then the curve of K against C_1 approaches the tangent from above as in Fig. 10.9. If, however, D increases as C increases, which is often

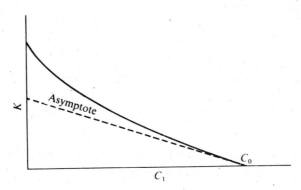

FIG. 10.9. Curve of K against C_1 approaching asymptote from above because D decreases as C increases.

the case in solvent–polymer systems, then the tangent is approached from below as in Fig. 10.10. Thus the curves of K and C_1 show at a glance the nature of the concentration-dependence of D. If it is practicable to carry out desorption experiments as well as sorption, that is to include values of C_1 which are higher than the surface concentration C_0, the curve of K against C_1 will cross the C_1 axis at $C_1 = C_0$, as in Fig. 10.11. This should allow the tangent to be drawn more accurately since extrapolation is avoided.

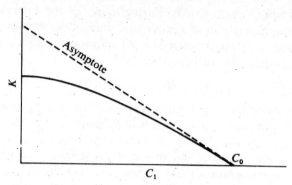

FIG. 10.10. Curve of K against C_1 approaching asymptote from below because D increases as C increases.

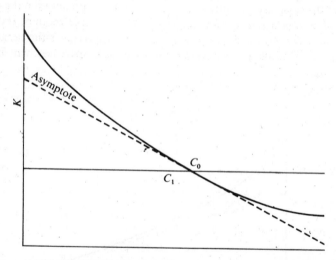

FIG. 10.11. Curve of K against C_1 using data from both sorption and desorption experiments.

10.7. Weighted-mean diffusion coefficients

In § 10.6.7 (p. 244) we looked upon the average diffusion coefficient deduced from the initial rate of sorption by (10.165) as as approximation to $(1/C_0) \times \int_0^{C_0} D \, dC$. Subsequently, numerical solutions showed that the initial rates of sorption and desorption are much more closely controlled by weighted-mean diffusion coefficients of the form

$$\bar{D}_a = p C_0^{-p} \int_0^{C_0} C^{p-1} D(C) \, dC \qquad (10.174a)$$

for sorption and

$$\bar{D}_d = qC_0^{-q} \int_0^{C_0} (C_0 - C)^{q-1} D(C)\, dC \qquad (10.174b)$$

for desorption.

On the basis of his original calculations for diffusion coefficients increasing with concentration over ranges up to 200-fold in one or other of the ways shown by curves, 1, 3, 4 in Fig. 9.4, Crank suggested the values $p = 1.67$, $q = 1.85$.

Kishimoto and Enda (1963) found good agreement between $D(C)$ obtained for the polymethyl acrylate–benzene system from steady-state measurements and by using (10.174a, b) with these values of p and q. Barrie and Machin (1971) found that their experimental data also supported the same values. They pointed out, however, that the correspondence principle between sorption and desorption (see § 9.3.1, p. 173) implies that for very small concentration ranges p and q should approach the same limiting value. By using Lin Hwang's (1952) series solution they concluded that as $C_0 \to 0$, $p \to 2/(\pi - 2)$. Wilkins's (1963) solutions for small ε presented in Table 7.6 confirm this limiting value. Also, we may regard his solutions for $\varepsilon = \infty$ as limiting cases for linear diffusion coefficients of very large range. It is easy to deduce that for D increasing $1.62 < p < 1.75$ and $1.75 < q < 1.89$. These ranges include the values of 1.67 and 1.85 which fit experimental measurements over a wide intermediate range of $D(C)$.

Kishimoto and Enda (1963) proposed a convenient approximate way of using (10.174b). They assume

$$D(C) = D_0 + k_1 C + k_2 C^2 + \dots , \qquad (10.175)$$

where k_1, k_2 etc. are unknown coefficients. By introducing (10.175) into (10.174b) and integrating they find that $\bar{D}_d$ may be written

$$\bar{D}_d = D_0 + k_1' C + k_2' C^2 + \dots , \qquad (10.176)$$

where

$$k_1 = 2.85 k_1'; \qquad k_2 = (3.85 \times 2.85/2!) k_2' = 5.486 k_2';$$

$$k_3 = (4.85 \times 3.85 \times 2.85/3!) k_3' = 8.871 k_3'. \qquad (10.177)$$

Higher coefficients are readily calculated. The coefficients D_0 and k' are obtained by fitting the expression (10.176) to the experimentally observed values of $\bar{D}_d$ and suitably chosen values of C_0. The $D(C)$ plot follows readily from (10.175) and (10.177).

10.8. Radiotracer methods

A number of experimental arrangements have been proposed for observing the diffusion of labelled molecules. If the total concentration of labelled and unlabelled molecules is kept constant throughout the system, self-diffusion coefficients are measured. The same principles have been applied in the presence of an overall concentration gradient.

In the 'thin-smear' method, Moore and Ferry (1962, 1968) use a radio-actively-labelled penetrant in trace amounts so that its concentration is very small throughout the experiment. Initially, penetrant is spread in a thin layer on one surface of a polymer disc and the variation of activity with time at the opposite face is observed. In general, this activity is due to the inte-grated effect of β-particles emanating from different depths in the polymer. The diffusion coefficient is deduced by comparing the measured activity change with the corresponding theoretical expression

$$\frac{N(t)}{N(\infty)} = 1 - 2\left(\frac{e^{-\tau}}{1+\varepsilon} - \frac{e^{-4\tau}}{1+4\varepsilon} + \frac{e^{-9\tau}}{1+9\varepsilon} - \cdots \right), \qquad (10.178)$$

where $N(t)$ and $N(\infty)$ are the activities just outside the inert surface of the disc at time t and at equilibrium, $\tau = \pi^2 Dt/h^2$, $\varepsilon = (\pi/\mu h)^2$, D is the diffusion coefficient, h the disc thickness, and μ the absorption coefficient for the β-particles in the polymer, assuming an exponential law. Eqn (10.178) takes into account radiation contributions from below the inert surface. If, however, the absorption of the β-particles is sufficiently high we have the case of $\varepsilon = 0$ and the activity is simply proportional to the concentration at the inert surface.

In the twin-disc method (Moore and Ferry 1962), a disc containing unlabelled penetrant is place on top of an identical disc containing labelled penetrant. In this case, the activity is taken to measure the concentration of radioactive penetrant at the upper surface of the sample and the appropriate mathematical solution is

$$\frac{N(t)}{N(\infty)} = 1 - \frac{4}{\pi}(e^{-\tau} - \tfrac{1}{3}e^{-9\tau} + \tfrac{1}{5}e^{-25\tau} - \cdots). \qquad (10.179)$$

Jackson, Oldland, and Pajaczkowski (1968) used an arrangement which can be described as follows. The radioactive source occupies the region $h < x < k$; diffusion takes place into a polymer sheet in $k < x < L$ and into a cover film in $0 < x < h$. They used the appropriate particular form of the solution (4.56) to calculate the concentration distribution

$$\frac{C}{C_\infty} = 1 + \frac{2L}{\pi(k-h)} \sum_1^\infty \frac{1}{n} \exp(-n^2\tau)\left(\sin\frac{n\pi k}{L} - \sin\frac{n\pi h}{L}\right)\cos\frac{n\pi x}{L}, \qquad (10.180)$$

where $\tau = \pi^2 Dt/L^2$. This is substituted into the expression for the activity $N(t)$, at the surface $x = 0$,

$$N(t) = H \int_0^L Cf(x)\,dx, \qquad (10.181)$$

where $f(x)$ is the fraction of β-rays reaching the surface from nuclei at distance x; H is a constant of the apparatus. Jackson *et al.* assume $f(x) = \exp(-\mu x)$ and obtain

$$\frac{N(t)}{N(\infty)} = 1 + \left\{ \frac{2L\mu^2}{\pi(k-h)} \right\} \sum_1^\infty A(n) \exp(-n^2 t) \sin\frac{n\pi k}{L} \sin\frac{n\pi h}{L}, \quad (10.182)$$

where

$$A(n) = \frac{1-e^{-\mu L}}{\mu^2 n + n^3 \pi^2 / L^2}.$$

If the activity is observed at the surface $x = L$, eqn (10.182) still holds with the modification that

$$A(n) = \frac{\cos n\pi - e^{-\mu L}}{\mu^2 n + n^3 \pi^2 L^2}.$$

Expressions (10.181) and (10.182) are based on the assumption that the diffusion coefficient is the same throughout the system though there are, in fact, three distinct layers. Jackson *et al.* (1968) examined another extreme model. The middle layer, in which the polymer is initially confined, is now considered as a perfectly stirred liquid. The penetrant concentration just inside the polymer is K times that in the stirred liquid at the interface between them. Also, the cover film is now taken to be of zero thickness and so forms the surface $x = 0$. The mathematical solutions of § 4.3.5 (p. 56) apply, and when substituted in (10.181) result in the expression, for the surface $x = L$,

$$\frac{N(t)}{N(\infty)} = 1 + 2\mu^2(1+\alpha) \sum_1^n B(n) \left\{ \mu^2 + \frac{q_n^2}{(L-k)^2} \right\}^{-1} \frac{\exp(-q_n^2 T)}{\cos q_n}, \quad (10.183)$$

where q_n are the non-zero positive roots of $\tan q_n = -\alpha q_n$,

$$\alpha = k/KL, \quad T = Dt/(L-k)^2, \quad \text{and} \quad B(n) = (1+\alpha+\alpha^2 q_n^2)^{-1}.$$

Park (1954) who applied a similar technique to the self-diffusion of tricresyl phosphate in polystyrene found it necessary to construct a β-ray absorption curve under the conditions actually used in the diffusion experiments. He used two layers of equal thickness and so numerical and graphical methods were needed to obtain the diffusion coefficient.

The β-particle absorption techniques described above are difficult to apply when the penetrant is volatile. Park (1961) therefore used a radioactive exchange process in a sealed equilibrium sorption system. The mathematical solutions required are again those of § 4.3.5 (p. 56).

11

NON-FICKIAN DIFFUSION

11.1. Glassy polymers

THE diffusion behaviour of many polymers cannot be described adequately by a concentration-dependent form of Fick's law with constant boundary conditions, especially when the penetrant causes extensive swelling of the polymer. Generally, this is the case with so-called glassy polymers which are said to exhibit 'anomalous' or 'non-Fickian' behaviour. In rubbery polymers, on the other hand, diffusion is generally Fickian. The essential distinction is that polymers in the rubbery state respond rapidly to changes in their condition. For example, a change in temperature causes an almost immediate change to a new equilibrium volume. The properties of a glassy polymer, however, tend to be time-dependent; for example, the stress may be slow to decay after such a polymer has been stretched. Deviations from Fickian behaviour are considered to be associated with the finite rates at which the polymer structure may change in response to the sorption or desorption of penetrant molecules.

Anomalous effects may be directly related to the influence of the changing polymer structure on solubility and diffusional mobility, or they may result from the internal stresses exerted by one part of the medium on another as diffusion proceeds.

Polymers usually have a wide spectrum of relaxation times associated with structural changes, but all of them decrease as temperature or penetrant concentration is increased and motion of the polymer segments enhanced. At a given concentration, the change from the glassy to the rubbery state is said to occur at the glass transition temperature. A sorption process, for example, will be influenced by those segmental motions which occur at about the same rate or slower than the motivating diffusion process. In rubbery polymers, well above their glass temperature, the polymer chains adjust so quickly to the presence of the penetrant that they do not cause diffusion anomalies.

Alfrey, Gurnee, and Lloyd (1966) proposed a useful classification according to the relative rates of diffusion and polymer relaxation. Three classes are distinguished:

    (i) Case I or Fickian diffusion in which the rate of diffusion is much less than that of relaxation;

    (ii) Case II diffusion, the other extreme in which diffusion is very rapid compared with the relaxation processes;

(iii) Non-Fickian or anomalous diffusion which occurs when the diffusion and relaxation rates are comparable.

The significance of the terms Case I and Case II is that they are both simple cases in the sense that the behaviour of each can be described in terms of a single parameter. Case I systems are controlled by the diffusion coefficient. In Case II the parameter is the constant velocity of an advancing front which marks the innermost limit of penetration of the diffusant and is the boundary between swollen gel and glassy core. Case II is also a second extreme or limiting case with respect to the shape of the sorption–time curve. If we denote the amount sorbed at time t by Kt^n, with K and n constants, then Case II systems are characterized by $n = 1$ and Case I systems by $n = \frac{1}{2}$.

Non-Fickian systems lie between Case I and Case II in that n takes an intermediate value between $\frac{1}{2}$ and 1, or changes sigmoidally from one to the other. Also, non-Fickian behaviour needs two or more parameters to describe the interacting diffusion and relaxation effects inherent in it.

11.2. Characteristic features

Extensive reviews are available of the experimentally observed diffusion anomalies in various polymer systems (Rogers 1965; Alfrey *et al.* 1966; Park 1968; Stannet, Hopfenberg, and Petropoulos 1972; Hopfenberg and Stannet 1973). These contain many references to original papers. We shall confine attention in this chapter to a qualitative description of the main characteristic features. It provides a background for the discussion in § 11.3 below of the mathematical models which have been advanced in partial explanation of the anomalies. The different types of behaviour are summarized in Fig. 11.1, taken from Rogers (1965).

(i) *Fickian: Case I.* The characteristic features of Fickian diffusion, controlled by a concentration-dependent diffusion coefficient, are listed in § 9.4 (p. 179). The term pseudo-Fickian has been used (Rogers 1965) to describe sorption–desorption curves of the same general shape and disposition, but for which the initial portion persists for a shorter time.

(ii) *Sigmoid.* Curves of the general shape of Fig. 11.1(b) have been observed experimentally in many systems (Park 1968). The sorption curves are sigmoid in shape with a single point of inflexion often at about 50 per cent equilibrium sorption. The initial rate of desorption exceeds that of sorption, but desorption soon becomes slower and the curves cross.

(iii) *Two-stage sorption.* Bagley and Long (1955) carried out a series of 'interval' sorptions of acetone by cellulose acetate. The polymer was first brought into equilibrium with acetone vapour at one vapour pressure and then the sorption–(time)$^{\frac{1}{2}}$ relationship resulting from a small increase in vapour pressure was observed. An initial rapid uptake, a linear function of

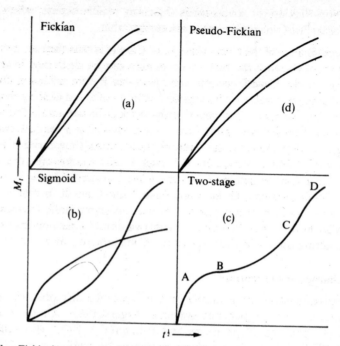

FIG. 11.1. 'Non-Fickian' or 'anomalous' sorption and desorption curves compared with Fickian-type curves. (Rogers 1965).

(time)$^{\frac{1}{2}}$, leads to a quasi-equilibrium acetone uptake as in Fig. 11.1c, followed by a slow approach to a final true equilibrium. Bagley and Long (1955) advanced the simple interpretation that a quasi-equilibrium is reached rapidly first at the polymer surface and then by simple diffusion throughout the polymer sheet. The second stage of sorption is associated with an increase in surface concentration which occurs slowly compared with the diffusion process and is the rate-determining factor for sorption. The concentration is virtually uniform throughout the sheet and increases at a rate independent of the thickness. The sigmoid curve as in Fig. 11.1(b) is a special case in which surface concentration changes and diffusion flow occur at comparable rates.

An experimental observation of the acetone distribution in a sheet by Long and Richman (1960) confirmed this interpretation and showed that the surface concentration C_s is well represented by

$$C_s = C_i + (C_e - C_i)(1 - e^{-\beta t}), \tag{11.1}$$

where C_i and C_e are the surface concentrations achieved instantaneously and at final equilibrium, t is time and β a constant. Park (1961) has shown that the expected relation between the diffusion coefficient calculated from the first stage and the self-diffusion coefficient is satisfied provided the activity and volume fraction from the quasi-equilibrium isotherm are used and not those from the final equilibrium isotherm.

(iv) *Advancing boundaries: Case II.* When a liquid penetrant diffuses into a polymer sheet or filament, sharp boundaries can often be seen under a microscope. The topic is reviewed by Park (1968) and Rogers (1965) with references to original papers. Hartley (1946) noted three boundaries. The inner boundary marks the limit of penetration of the liquid, while the outer boundary shows the limit of the swollen gel. The third, intermediate, boundary is usually thought to lie between polymer in the elastic rubbery state and glassy polymer. When the polymer sheet is in contact with vapour rather than liquid penetrant similar phenomena are observed but in many cases the middle boundary will be missing if the whole system is in the glassy state.

Often the positions of the boundaries are proportional to $t^{\frac{1}{2}}$. Cases in which each boundary moves with a constant velocity and also intermediate power laws have been observed by Hartley (1946) and Kwei and Zupko (1969). Wang, Kwei, and Frisch (1969) point out that an alternative expression for the depth of penetration is $at + bt^{\frac{1}{2}}$, where a and b are constants.

Robinson (1950) obtained interferometric fringe photographs of chloroform penetrating cellulose acetate between parallel glass plates. In particular, the concentration was seen to change very rapidly in the region of the inner boundary, implying the existence of large internal stresses due to the interaction of the swelling polymer and the central unattacked core of the sheet. Not only can the orientation of the polymer molecules be changed by these stresses (Dreschel, Hoard, and Long (1953)), but also Hermans (1948) and Alfrey *et al.* (1966) have observed anisotropic dimensional changes and transverse cracking of the unswollen core.

Related effects in permeation experiments were observed by Park (1951). Meares (1958) found a rapid initial flux of allyl chloride vapour through a membrane of polyvinyl acetate, which quickly fell to a low value before the more normal gradually increasing flow occurred. Meares attributed the initial behaviour to the increase in diffusion coefficient brought about by the stress exerted by the swollen surface layers on the solvent-free region beneath. Similar overshooting in sorption experiments has been observed (Rogers 1965).

11.3. Mathematical models

A number of authors have proposed mathematical models of the penetrant-polymer diffusion phenomena. No single model successfully predicts all the experimental observations. Nevertheless, each one accounts for some of the features described in § 11.2 (p. 255), and collectively they contribute

towards an elucidation, albeit incomplete, of the complex mechanisms
involved. It is only possible in this book to describe briefly the salient features
of some of the models. Details are given in published papers. Essentially
they are all concerned with the combination of penetrant diffusion and polymer relaxation proceeding at different relative rates according to the conditions of the experiment.

11.3.1. *History-dependence*

Crank (1953) explored the suggestion that part of the movement of the
polymer molecules may be effectively instantaneous, and part relatively
slow compared with the diffusion of penetrant. An increase in penetrant
concentration is assumed to lead to an immediate increase in diffusion coefficient, followed by a slow drift towards an equilibrium value as a result of the
relaxation process characteristic of the glassy state. This means that the value
of the diffusion coefficient in any element of the system depends on the
concentration history of the element. We assume the surface concentration
to be held constant and that the fast and slow changes in the diffusion coefficient proceed independently of each other. If the concentration changes
infinitely rapidly then we can expect the instantaneous component D_i of the
diffusion coefficient to satisfy

$$D_i = D_0 \, e^{aC}, \tag{11.2}$$

where C is the penetrant concentration and D_0 and a are constants. At a
point that remains at a given concentration C the diffusion coefficient is
assumed to increase further with time until a final equilibrium value D_e is
obtained given by

$$D_e = D_0 \, e^{bC}, \tag{11.3}$$

where b is larger than a. This approach to D_e is assumed to be a first-order
change such that

$$\frac{\partial D}{\partial t} = \alpha(D_e - D), \tag{11.4}$$

where α is also a function of concentration given by

$$\alpha = \alpha_0 \, e^{\sigma C} \tag{11.5}$$

The over-all expression for the diffusion coefficient D becomes

$$\left(\frac{\partial D}{\partial t} \right)_x = \left(\frac{\partial D_i}{\partial C} \right) \left(\frac{\partial C}{\partial t} \right)_x + \alpha(D_e - D). \tag{11.6}$$

Calculations based on this equation produce sorption–desorption curves
exhibiting many features of the sigmoid and psuedo-Fickian types of curves.
With five adjustable parameters available we must not expect to draw detailed

conclusions from a comparison of calculated curves with experimental data. Nevertheless, the orders of magnitude obtained were reasonable. This history-dependent model cannot explain two-stage behaviour which must indicate some influence of solubility.

11.3.2. *Two-stage theories*

We have seen in § 11.2(iii) that a good description of two-stage sorption is afforded by solutions of the concentration-dependent form of the diffusion equation with a time-dependent surface concentration satisfying eqn (11.1). It is, however, unlikely in general that slow relaxation of the polymer is important only on the surface. In the history-dependent model, internal relaxation influenced only the diffusion coefficient. In the second stage of two-stage sorption the concentration is thought to be uniform throughout the sheet as in § 11.2(iii), and Newns (1956) has developed the other extreme hypothesis that internal relaxation influences only the solubility. Following Flory (1953), Newns considers that in the quasi-equilibrium at the end of the first stage the elastic forces in the swollen polymer network have increased the chemical potential of the sorbed penetrant to such an extent that no further sorption occurs. On standing, the elastic forces slowly relax, the chemical potential decreases, and more penetrant is absorbed to establish equilibrium again. This process continues to a final, true equilibrium when the elastic forces have decayed away completely or reached equilibrium. The hypothetical hydrostatic pressure Π exerted by the polymer network at a time during the second stage when the concentration is C_q is given by

$$\Pi = (RT/\overline{V}) \ln (p/p_q), \qquad (11.7)$$

where $\overline{V}$ is the partial molar volume of the penetrant, the pressure p is the actual vapour pressure of the penetrant, and p_q the pressure that would correspond to a concentration C_q if a final equilibrium had been reached. The internal stress due to the presence of the penetrant is $\gamma \Delta v$, where Δv is the fractional volume change and γ an elastic constant. By balancing the opposing stresses Newns obtained an expression for the ratio of the amount sorbed during the first stage, to that at final equilibrium, and predicted correctly several features of the two-stage sorption curve.

Petropoulose and Roussis (1969b) obtained numerical solutions of eqn (10.94) (p. 227) containing 'thermodynamic' parameters, and so incorporated relaxation effects both on the surface and in the body of the polymer.

11.3.3. *Strain-dependent models*

In § 11.2(iv) evidence of internal stresses arising from the differential swelling of an outer swollen layer and an internal core of a sheet or filament was described. Attempts have been made to relate anomalous diffusion behaviour to these stresses.

Crank (1953) assumed that only the diffusion coefficient is affected by the stresses and not the surface concentration or internal solubility. The essential features of the mathematical model are clear from Fig. 11.2. Sorption experiments (Park 1953) suggest that the cross-sectional area A of the sheet can be assumed to remain effectively constant throughout its thickness.

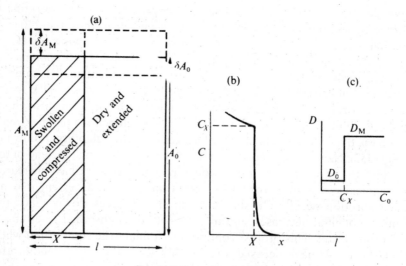

FIG. 11.2. From Crank 1953.

The diffusion coefficient in the unstressed polymer is approximated by the step-wise form in Fig. 11.2(c). The associated concentration distribution is shown in Fig. 11.2(b), where C_0 is the constant surface concentration. We assume the mechanical compression to be uniform in the region $0 < x < X$, and that the stress-free swelling in this region corresponds to that for a mean concentration $\frac{1}{2}(C_0 + C_X)$. The stress-free swelling in the central core $X < x < l$ is assumed zero. Internal stresses are assumed to have a negligible effect when $X = l$. The sheet is initially of area A_0 and finally A_1, i e. A_1 is the stress-free swelling for a final uniform concentration C_0. The increases in area are small, and so in the absence of stress they are taken to be proportional to mean concentrations. In the swollen region, $0 < x < X$, the mean stress-free area A_M is given by

$$A_M - A_0 = (A_1 - A_0)C_M/C_0, \tag{11.8}$$

when the mean concentration is C_M. The compression is

$$\delta A_M = A_M - A.$$

At the stage shown in Fig. 11.2(a), the swollen region, $0 < x < X$, is compressed by δA_M and the dry region $X < x < l$ extended by δA_0. The assumption of constant area throughout the thickness means that

$$A = A_0 + \delta A_0 = A_M - \delta A_M. \tag{11.9}$$

By equating total forces of compression and extension we obtain

$$\lambda_0 \delta A_0 (l - X)/A_0 = \lambda_M \delta A_M X/A_M, \tag{11.10}$$

where λ_0 and λ_M are Young's moduli for the dry and swollen polymer respectively. The diffusion coefficients are assumed to be related to δA_M and δA_0 by

$$D_1 = D_M(1 - \mu_M \delta A_M/A_M), \tag{11.11}$$

$$D_2 = D_0(1 + \mu_0 \delta A_0/A_0), \tag{11.12}$$

where D_1 and D_2 are the diffusion coefficients in the swollen and dry regions when the area is A. Crank (1953) considerably simplified the mathematics by further assuming that the ratio of the diffusion coefficients for the fully compressed and unstressed states is the same for swollen and dry polymer regions. It then follows from (11.11) and (11.12) that D_1/D_2 remains constant as diffusion proceeds and so

$$D = D(t)D(C). \tag{11.13}$$

Calculations based on this model reproduced an experimental sorption curve for methylene chloride in polystyrene and, to some extent, the change in area with time. The model also predicts qualitatively the experimental observation that in the early stages a thin sheet should absorb more penetrant than a thicker one. But according to the model, plots of M_t/M_∞ against $t^{\frac{1}{2}}/l$ should all coincide for different thicknesses l, and this is not true.

Another view of the effect of internal stresses has been discussed by Frisch, Wang, and Kwei (1969) and Wang, Kwei, and Frisch (1969). They include in the basic transport equation, in addition to the gradient of chemical potential, a second term deriving from the partial stress of the penetrant. In this case, the thermodynamics of irreversible processes leads to an expression for the flux of penetrant

$$F = -BC\left(\frac{\partial \mu}{\partial x} - \frac{1}{C}\frac{\partial S}{\partial x}\right), \tag{11.14}$$

where B is a mobility coefficient, μ the chemical potential, and S the partial stress tensor in one dimension. On making the assumption that S is proportional to the total uptake of penetrant so that

$$S(x) = s\int_0^x C(x', t)\,\mathrm{d}x', \tag{11.15}$$

where s is a constant, we arrive at the generalized diffusion equation

$$\frac{\partial C}{\partial t} = \frac{\partial}{\partial x}\left\{D(\dot{C}, x, t)\frac{\partial C}{\partial x} - B(C, x, t)sC\right\}. \tag{11.16}$$

If the stress gradient predominates so that $\partial\mu/\partial n \ll \partial S/\partial x$, eqn (11.16) is seen to lead to Case II diffusion (§ 11.2(iv), p. 257). Wang *et al.* (1969) point out that the solution of the linear form of (11.16) (for which D and $v = Bs$ are constants) for diffusion into a penetrant-free semi-infinite medium $x \geqslant 0$, with $x = 0$ maintained at a constant concentration C_0, is

$$C(x, t) = \tfrac{1}{2}C_0\left[\exp(xv/D)\,\mathrm{erfc}\left\{\frac{x+vt}{2\sqrt{(Dt)}}\right\} + \mathrm{erfc}\left\{\frac{x-vt}{2\sqrt{(Dt)}}\right\}\right]. \tag{11.17}$$

When x is small compared with vt, inspection of the function erfc reveals that the second term on the right side of (11.17) is much larger than the first. Wang *et al.* therefore make the approximation that except for very small t the movement of any plane X with a given concentration is such that

$$X = 2k\sqrt{(Dt)} + vt, \tag{11.18}$$

where k is a constant. They then take an expression of this approximate form to describe the movement of the sharp boundary associated with a discontinuous diffusion coefficient (§ 13.2, p. 287). They apply Danckwerts' moving frames of reference (§ 13.4, p. 298) and obtain solutions of the linear form of (11.16) in which, of course, the sharp boundary moves according to (11.18). Kwei, Wang, and Zupko (1972) find their model predicts successfully the rates of sorption and penetration of acetone in poly(vinyl chloride). The two special cases of $D \ll v$ and $D \gg v$ correspond to Fickian and Case II diffusion respectively.

Alfrey *et al.* (1966) used approximate mathematical models to calculate swelling stresses and fracturing in glassy-polymer discs and cylinders in which two-stage diffusion is occurring.

11.3.4. *Irreversible thermodynamical model*

Frisch (1964, 1966) has discussed diffusion in glassy polymers in the language of irreversible thermodynamics. In view of the many excellent general accounts of this subject which are available (de Groot 1961; de Groot and Mazur 1962; Prigogine 1967; Haase, 1969] we shall attempt only to convey the essence of Frisch's ideas in the simplest terms.

In addition to the usual thermodynamic variables such as concentration, temperature and pressure, Frisch introduces a set of unspecified internal parameters which are used to describe the gradual 'freezing' or relaxing of internal degrees of freedom as the system passes through a 'glass-like' transition region. Above the transition temperature the system is in internal chemical equilibrium with respect to these parameters, but below the transition region it is not. Thus, the Gibbs free energy per unit mass G is a function of the mass-fraction concentrations c_j, temperature T, pressure p, and the proposed internal parameters ξ_α ($\alpha = 1, 2, ..., N$). Systems with a glass-like transition are defined to have a characteristic temperature $T_\xi(c_j, p)$ such that $\partial G/\partial \xi_\alpha = 0$ for $T > T_\xi$ but is not zero for all α when $T < T_\xi$. Below T_ξ these free energy derivatives depend on ξ_α, which themselves depend on the history of the system as well as its composition, i.e. $\xi_\alpha = \xi_\alpha(t, c_j, p)$ where t is time.

At constant T, p etc. the Gibbs–Duhem relation is given by

$$\sum_j c_j \, d\mu_j = 0, \qquad T > T_\xi,$$

$$= \sum_\alpha \xi_\alpha \, dA_\alpha, \qquad T < T_\xi, \tag{11.19}$$

where $A_\alpha = \partial G/\partial \xi_\alpha$.

We discuss the diffusion flux with respect to a mass-fixed section. A basic idea in irreversible thermodynamics is that the flux J_i of one component i depends not only on its own gradient of chemical potential but also on those of the other components in the system. Thus we write

$$J_i = \sum_{j=0}^{q} L_{ij} X_j, \qquad T > T_\xi, \tag{11.20}$$

where in the one-dimensional case the forces X_j are given by

$$X_j = -\partial \mu_j/\partial x. \tag{11.21}$$

The L_{ij} are the so-called Onsager coefficients and for $T > T_\xi$ they are functions of T, p, c_j only, i.e. $L_{ij} = L_{ij}(T, p, c_j)$. Correspondingly, we define

$$J_i = \sum_{j=0}^{q} \bar{L}_{ij} X_j, \qquad T < T_\xi, \tag{11.22}$$

but now $\bar{L}_{ij}$ are time-dependent since they are also functions of ξ_α, i.e. $\bar{L}_{ij} = \bar{L}_{ij}(T, p, c_j, \xi_\alpha)$. Reciprocal relations

$$L_{ij} = L_{ji}, \qquad \bar{L}_{ij} = \bar{L}_{ji}, \tag{11.23}$$

are to be satisfied and also

$$\sum_{i=0}^{q} J_i = 0.$$

Following Frisch we consider as an example a two-component system in one dimension, with component 0 as the polymer and component 1, the penetrant. Thus $q = 1$, giving

$$J_0 + J_1 = 0, \qquad L_{00} = L_{01} = -L_{10} = L_{11}, \tag{11.24}$$

so that (11.20) and (11.22) with (11.21) become

$$J_1 = -J_0 = -L_{11}\frac{\partial}{\partial x}(\mu_1 - \mu_0), \qquad T > T_\xi$$

$$= -\bar{L}_{11}\frac{\partial}{\partial x}(\mu_1 - \mu_0), \qquad T < T_\xi \qquad (11.25)$$

In this example, we have only one independent concentration variable since $c_1 = 1 - c_0$. Thus, using (11.19),

$$\frac{\partial}{\partial x}(\mu_1 - \mu_0) = \frac{\partial}{\partial c_1}(\mu_1 - \mu_0)\frac{\partial c_1}{\partial x}$$

$$= \left(\frac{1}{1-c_1}\right)\frac{\partial c_1}{\partial x}\frac{\partial \mu_1}{\partial x}, \qquad T > T_\xi$$

$$= \left(\frac{1}{1-c_1}\right)\frac{\partial c_1}{\partial x}\left\{\frac{\partial \mu_1}{\partial c_1} + \sum_\alpha \xi_\alpha \frac{\partial A_\alpha}{\partial c_1}\right\}, \qquad T < T_\xi, \quad (11.26)$$

where

$$\frac{\partial A_\alpha}{\partial c_1} = \frac{\partial^2 G}{\partial \xi_\alpha \partial c_1} = \frac{\partial \mu_1}{\partial \xi_\alpha}. \qquad (11.27)$$

Finally, the diffusion flux is given by

$$J_1 = -D\,\partial c_1/\partial x, \qquad T > T_\xi,$$

$$= -\bar{D}\,\partial c_1/\partial x, \qquad T < T_\xi, \qquad (11.28)$$

where the diffusion coefficients are

$$D = -\frac{Lu \cdot}{1-c_1}\frac{\partial \mu_1}{\partial c_1}, \qquad T > T_\xi,$$

$$\bar{D} = -\frac{\bar{L}u}{1-c_1}\left(\frac{\partial \mu_1}{\partial c_1} + \sum_\alpha \frac{\partial \mu_1}{\partial \ln \xi_\alpha}\right), \qquad T < T_\xi. \qquad (11.29)$$

Thus, even if there is no change of mechanism across T_ξ, i.e. $L_{11} = \bar{L}_{11}$, $\bar{D}$ will differ from D and will be history-dependent.

Frisch (1964) developed the approach further in other reference frames and in a consideration of boundary conditions he showed that the time-dependent surface concentration observed experimentally by Long and Richman (see eqn (11.1)) can be deduced.

11.4. Conclusion

No single one of the models discussed in (11.3) can account satisfactorily for all the known aspects of anomalous diffusion. A synthesizing model is still needed to include the combined effects of history, orientation, and stress on the diffusion process within the polymer system and on the surface concentration. Furthermore, a composite model needs to be of such a form that it can be used to predict diffusion behaviour under given conditions and be useful for the interpretation of experimental data.

12

DIFFUSION IN HETEROGENEOUS MEDIA

12.1.

PRACTICAL materials are frequently heterogeneous in structure. We distinguish two types: particulates, in which discrete particles of one phase are dispersed in a continuum of another; and laminates, comprising layers of different properties sandwiched together. Barrer (1968) has reviewed some studies of both types.

12.2. Laminates

We consider some properties of laminates in which the layers A, B, C ... are normal to the direction of flow.

(i) *Steady state.* In the steady state, the flow through the ith lamina is the same as for any other lamina and is given by

$$F = -\omega_v r^{v-1} D_i \, \partial C_i / \partial r, \tag{12.1}$$

where for a plane sheet, hollow cylinder, and spherical shell, $v = 1, 2$, and 3 respectively and $\omega_v r^{v-1}$ is the surface area with $\omega_1 = 1$, $\omega_2 = 2\pi$, and $\omega_3 = 4\pi$ for unit area of a plane sheet, unit length of a cylinder, and for a whole spherical shell. If D_i is independent of concentration, and the solubility is proportional to the pressure in all laminae, the integration of (12.1) with respect to r gives

$$F = \frac{\omega_v D_i \Delta C_i}{I_v(R_{i-1}, R_i)}, \tag{12.2}$$

where ΔC_i is the concentration drop across the ith lamina and

$$I_v(x, y) = \int_x^y \frac{dr}{r^{v-1}}.$$

The permeability is given by $P_i = D_i \sigma_i$ where σ_i is the solubility coefficient in the ith lamina, i.e. $C_i = \sigma_i p_i$ and so (12.2) may be written

$$\Delta p_i = \frac{F I_v(R_{i-1}, R_i)}{\omega_v P_i}, \tag{12.3}$$

where Δp_i is the pressure drop across the ith lamina which would correspond to ΔC_i. It follows immediately that

$$\frac{I_v(R_0, R_1)}{P_1} + \frac{I_v(R_1, R_2)}{P_2} + \dots + \frac{I_v(R_{n-1}, R_n)}{P_n}$$

$$= \frac{\omega_v}{F} \sum_{i=1}^{n} \Delta p_i = \frac{\omega_v}{F} \Delta p, \qquad (12.4)$$

where Δp is the total pressure drop across the n-layer laminate. The over-all permeability P of the composite is defined by

$$P = \frac{F I_v(R_n, R_0)}{\omega_v \Delta p} \qquad (12.5)$$

We arrive at the three formulae

$$\frac{l_1}{P_1} + \frac{l_2}{P_2} + \dots + \frac{l_n}{P_n} = \frac{l}{P}, \quad \text{(slab)};$$

$$\frac{\ln(R_1/R_0)}{P_1} + \frac{\ln(R_2/R_1)}{P_2} + \dots + \frac{\ln(R_n/R_{n-1})}{P_n} = \frac{\ln(R_n/R_0)}{P}, \quad \text{(cylinder)};$$

$$\frac{(1/R_0)-(1/R_1)}{P_1} + \frac{(1/R_1)-(1/R_2)}{P_2} + \dots + \frac{(1/R_{n-1})-(1/R_n)}{P_n}$$

$$= \frac{(1/R_0)-(1/R_n)}{P}, \quad \text{(spherical shell)}. \qquad (12.6)$$

Here $l_i \equiv R_i - R_{i-1}$ and $l = R_n - R_0$. Also on the left sides but not on the right of eqn (12.6) we may substitute $P_i = D_i \sigma_i$, $i = 1, 2, \dots, n$. The corresponding relationships between diffusion coefficients follow by putting $\sigma = 1$ and hence $P = D$ everywhere in (12.6).

Barrer (1968) draws attention to an important property of a laminate in which the diffusion coefficient of each lamina is a function of concentration, and the solubility of the diffusant is not proportional to the pressure. Integration of (12.1) now leads to

$$\frac{\omega_v}{F} \Delta C = \frac{I_v(R_{i-1}, R_i)}{\bar{D}_i}, \qquad (12.7)$$

where

$$\bar{D}_i = \frac{1}{\Delta C_i} \int_{\lambda_{2i}}^{\lambda_{2i-1}} D_i \, dC_i \qquad (12.7a)$$

in which $\lambda_{2i-1}, \lambda_{2i}$ are the values of C_i just within the two surfaces R_{i-1}

and R_i. On summation, (12.7) becomes

$$\frac{\dot{\omega}_v}{F} \sum_{i=1}^{n} \Delta C_i = \sum_{i=1}^{n} \frac{I_v(R_{i-1}, R_i)}{D_i}. \tag{12.8}$$

Finally, we may define an overall integral diffusion coefficient for the laminate $\bar{D}$ by the expression

$$\frac{\omega_v}{F} \sum_{i=1}^{n} \Delta C_i = \frac{I_v(R_0, R_n)}{\bar{D}}. \tag{12.9}$$

The magnitude of the sum on the left side may depend on the sequence of the laminae. Ash, Barrer, and Palmer (1965) have investigated the distribution of concentration through the laminate in the case where each D_i is constant, and Henry's law holds in each lamina. In (12.2) above they substitute $\Delta C_i = \lambda_{2i} - \lambda_{2i-1}$, where the λs are as used in (12.7a) above. Then we can write

$$\frac{F}{\omega_v} H_i = (\lambda_{2i-1} - \lambda_{2i}) \prod_{j=0}^{i-1} k_j, \tag{12.10}$$

where

$$H_i = \frac{I_v(R_{i-1}, R_i)}{D_i} \prod_{j=0}^{i-1} k_j \tag{12.11}$$

and where $\lambda_{2i} = k_i \lambda_{2i+1}$ and $k_0 = 1$. After summing over all values of i and re-arranging, we find

$$F = \omega_v \lambda_1 \Big/ \sum_{i=1}^{n} H_i. \tag{12.12}$$

Also from (12.10) we find

$$\lambda_{2i} = \lambda_1 \left(\sum_{i=1}^{n} H_i - H_i \right) \Big/ \left(\prod_{j=0}^{i-1} k_j \right) \sum_{i=1}^{n} H_i. \tag{12.13}$$

By integrating (12.1) from r to R_{i-1} and re-arranging we see that

$$C_i(r) = \frac{F I_v(r, R_{i-1})}{\omega_v D_i} + \lambda_{2i-1}, \tag{12.14}$$

which can be rewritten by substituting for λ_{2i-1} from the equivalent of (12.13) if need be.

(ii) *Time lag.* The time lag as defined in § 4.3.3 (p. 51) for a single lamina has also been calculated for laminates by methods which avoid the need to obtain complete solutions of the transient diffusion equation. One of these is due to Jaeger (1950b) and employs Laplace transforms. He expresses the time lag in terms of eight subsidiary quantities related to certain determinants. Four sets of boundary conditions are considered at the two outer faces of the laminate:

(i) constant temperature, zero temperature;
(ii) constant heat flux, zero temperature;
(iii) constant flux, zero flux;
(iv) temperature proportional to time, zero flux.

For each combination he tabulates numerical values for the time lag for heat flow through two different building walls each composed of six layers.

Barrie, Levine, Michaels, and Wong (1963) applied Jaeger's procedure to laminates AB and ABA. Ash, Barrer, and Petropoulos (1963) used Frisch's method described in § 10.6.2 (p. 222) to obtain time lags for repeated laminates ABAB.... Ash, Barrer, and Palmer (1965) deal also with the laminate ABCD... and with composite hollow cylinders and spherical shells. Barrer (1968) quotes a number of the formulae for particular cases. They are all cumbersome and here we give only the general formula for the ν-dimensional case derived by Ash et al. (1965).

$$\frac{L}{\nu \sum_{i=1}^{n} H_i} = \sum_{i=1}^{n} \left(\frac{1}{D_i} \left\{ R_i^{\nu} I_{\nu}(R_{i-1}, R_i) - \frac{1}{2}(R_i^2 - R_{i-1}^2) \right\} \sum_{i=1}^{n} H_i \right.$$

$$- \frac{1}{D_i^2} \left(\prod_{j=0}^{i-1} k_j \right) \left[R_i^{\nu} \{ I_{\nu}(R_{i-1}, R_i) \}^2 - R_i^2 I_{\nu}(R_{i-1}, R_i) \right.$$

$$\left. + \int_{R_{i-1}}^{R_i} \frac{dr}{r^{\nu-3}} \right] \right) + \sum_{i=1}^{n} \left[H_i \sum_{\beta=i+1}^{n} \left(1 \left[\frac{R_\beta^{\nu} - R_{\beta-1}^{\nu}}{\prod_{j=0}^{\beta-1} k_j} \sum_{\beta} H_\beta \right. \right. \right.$$

$$\left. \left. \left. - \frac{1}{D_\beta} \left\{ R_\beta^{\nu} I_{\nu}(R_{\beta-1}, R_\beta) - \frac{1}{2}(R_\beta^2 - R_{\beta-1}^2) \right\} \right] \right) \right] \qquad (12.15)$$

Ash et al. consider two special systems in more detail: (a) a lamina of substance B is inserted in a medium A and the position of B is varied; (b) the lamina B is inserted centrally in A and its thickness varied. Results are presented graphically for plane sheets, cylinders and spheres. Kubín and Špaček (1967) consider an ABA type of plane membrane separating two compartments, each of the same finite volume, which contain solutions differing in concentrations of a common solute.

12.3. Disperse phase in a continuum

Holliday (1963) gives examples of practical systems made up of one continuous polymer phase and one dispersed phase. The following parameters are required to describe a composite material of this kind:

(i) The geometry of the dispersed phase (shape, size, and size distribution, concentration and its distribution, orientation, topology);

(ii) the composition, state of matter, and other properties of both the disperse phase and the continuous phase.

Essentially the same mathematical equations describe the phenomena of electrical and heat conductivity, and the diffusion of penetrant. The many theoretical formulae that have been proposed to predict electrical and thermal properties of these composites are easily adapted to permeability studies. Several authors have summarized some of the papers on the subject and given extensive lists of references. In this account, the methods described have been selected to represent the different types of mathematical approach that have met with some success in correlating experimental and theoretical results. For a formulation of diffusive flow in heterogeneous systems in terms of irreversible thermodynamics, we refer to Barrer (1968) and to Kedem and Katchalsky (1963), who consider multi-component systems and electrolytes.

12.4. The mathematical problem of a two-phase system

We consider a number of discrete particles of material A embedded in a continuous medium of material B. In general, the particles constituting phase A may vary in size and shape, and their distribution in the medium phase B may be random or regular both in respect of spacing, arrangement and orientation. We denote by D_a and D_b the diffusion coefficients, assumed constant, of materials A and B respectively and by C_a, C_b the corresponding concentrations. These are normally functions of three space variables, but not of time, when steady-state problems are being studied.

The steady-state concentration in each phase satisfies Laplace's equation

$$\nabla^2 C_i = 0, \qquad i = a, b. \tag{12.16}$$

The two conditions which the concentration must satisfy are that both the concentration and the flux of the diffusing substance must be continuous everywhere. In particular, on any surface separating two phases we have

$$C_a = C_b; \qquad D_a \, \partial C_a / \partial n = D_b \, \partial C_b / \partial n, \tag{12.17}$$

where $\partial/\partial n$ denotes the directional derivative normal to the interface. Conditions (12.17) hold over the whole surface of each of the particles constituting phase A. Additional boundary conditions will be specified over the outermost surface or surfaces of phase B, according to the nature of the problem under consideration. In stating (12.17) we have assumed that no contact resistance exists at the interface. The appropriate modification is easily made if necessary.

12.4.1. *Arrays of spheres, cylinders, and ellipsoids*

Earlier workers investigated successively more complicated approximations to the general problem. The first paper usually quoted is by Maxwell (1873), who solved the problem of a suspension of spheres so sparsely distributed in the continuum that any interaction between them is negligible. The volume fraction v_a of the dispersed phase is therefore small. By applying the conditions (12.17) to appropriate harmonic functions, he obtained an expression for the effective diffusion coefficient of the composite medium, which can be written in the form

$$\frac{D - D_b}{D + 2D_b} = v_a \frac{D_a - D_b}{D_a + 2D_b} \tag{12.18}$$

The coefficient D is the diffusion coefficient of a hypothetical homogeneous medium exhibiting the same steady-state behaviour as the two-phase composite. Burger (1919) extended the treatment to ellipsoids and Eucken (1932) included particles of different sizes. Fricke (1924) derived an expression for spheroids suspended in a continuum which he expressed as a generalized form of Maxwell's equation (12.18), namely

$$\frac{D - D_b}{D + xD_b} = v_a \frac{D_a - D_b}{D_a + xD_b}, \tag{12.19}$$

where x is a function of D_a/D_b and the ratio a/b is the ratio of the axis of symmetry of the spheroid to the other axis. For the case of spheres, $x = 2$, we have Maxwell's formula. In general

$$x = \frac{\beta D_a - (D_a - D_b)}{(D_a - D_b) - \beta D_b}, \tag{12.20}$$

where

$$\beta = \tfrac{1}{3}\left[\frac{2}{1 + \tfrac{1}{2}M\{D_a/D_b - 1\}} + \frac{1}{1 + (1 - M)\{D_a/D_b - 1\}}\right]\left(\frac{D_a}{D_b} - 1\right),$$

$$M(a < b) = \frac{(\phi - \tfrac{1}{2}\sin 2\phi)}{\sin^3 \phi}\cos \phi, \quad \text{where} \quad \cos \phi = a/b,$$

and

$$M(a > b) = \frac{1}{\sin^2 \phi'} - \frac{1}{2}\frac{\cos^2 \phi'}{\sin^2 \phi'}\ln\left(\frac{1 + \sin \phi'}{1 - \sin \phi'}\right),$$

$$\text{where } \cos \phi' = b/a.$$

The influence of the geometric factors of the suspended particles is expressed solely in the parameter β.

Fricke presented graphs of β and x against D_a/D_b for different values of a/b and concluded that for a constant volume concentration the diffusion coefficient of the suspension is independent of the size of the suspended particles, and also nearly independent of their form, when the difference between the diffusion coefficient of the two phases is not very large. This is especially true for suspensions of prolate spheroids ($a > b$) which have a smaller diffusion coefficient than the suspending medium.

Comparison with experimental data for the blood of a dog, for which the values $D_a = 0$, $b/a = 4.25$, and $x = 1.05$ are considered appropriate, shows excellent agreement for volume concentrations ranging from 10 per cent to 90 per cent. On the other hand, Hamilton and Crosser (1962) attempted to correlate Fricke's expression (12.19) with experimental data relating to mixtures of rubber and particles of aluminium or balsa in the form of spheres, cylinders, parallelepipeds, or discs. They found that (12.19) predicts all the experimental data more satisfactorily if x is not calculated from (12.20), but from an empirical relationship

$$x + 1 = 3/\psi. \tag{12.21}$$

The sphericity of a particle ψ is defined as the ratio of the surface area of a sphere of volume equal to that of the particle, to the surface area of the particle. In these terms Fricke's values of $x + 1$ would be $3/\psi^2$ for prolate ellipsoids and $3/\psi^{1.5}$ for oblate ellipsoids.

The general conclusion is that if the continuous phase has the higher diffusion coefficient there is little shape-effect. In the opposite case, a marked shape-effect is noticeable if the ratio of the diffusion coefficients exceeds 100, and the expression (12.19) with x given by (12.21) accounts satisfactorily for the variation over a wide range of variables.

Starting from solutions of Laplace's equation expressed in Legendre functions, Rayleigh (1892) obtained an improved approximation for identical spheres arrayed on a single cubic lattice. His work was developed further by Runge (1925) and de Vries (1952a,b) to include both body-centred and face-centred cubic lattices and cylinders.

Another way of stating the nature of Maxwell's approximation is that he assumed any one spherical particle to be embedded in a medium whose properties were those of pure phase B, unmodified by the presence of the spheres of phase A. Subsequently, other assumptions were made, notably that the surrounding properties are those of the equivalent homogeneous medium. Barrer (1968) quotes several references and formulae and a table from de Vries (1952a,b) comparing calculated results for spheres by several authors. These are seen to be consistent only when D_a/D_b is small. Barrer suggests that the formulae may account reasonably well for plastics containing impermeable filler particles, though the question of non-spherical shapes is not resolved. Higuchi and Higuchi (1960) discuss a number of theoretical relationships for both steady and transient flow, particularly in relation to pharmaceutical problems.

Keller (1963) obtained an asymptotic solution for a dense cubic array of spheres of infinite diffusivity embedded in a medium with diffusion coefficient D. His expression for the effective diffusion coefficient D^∞ is

$$D^\infty/D = -\tfrac{1}{2}\pi \ln ((\pi/6)-f)+ \dots , \tag{12.22a}$$

with $(\pi/6)-f \ll 1$, where f is the volume fraction occupied by the spheres. The singularity at $f = \pi/6$ occurs when the spheres touch each other. The corresponding result for a square array of cylinders is

$$\frac{D^\infty}{D} = \frac{\pi^{\frac{3}{2}}}{2\{(\pi/4)-f\}^{\frac{1}{2}}}+ \dots , \tag{12.22b}$$

with $(\pi/4)-f \ll 1$. Keller (1963) also established the following theorem. Let D^∞ and D^0 denote the effective diffusion coefficients of the two composite media obtained by imbedding in a medium of diffusivity D a square lattice of circular cylinders with diffusion coefficients ∞ and 0, respectively. Then

$$D^\infty/D = D/D^0. \tag{12.23a}$$

A similar theorem for a rectangular lattice relates D^∞ in the x-direction with D^0 in the y-direction. That theorem also holds for non-circular objects provided they are symmetric in the x- and y-axes. It states that

$$D_x^\infty/D = D/D_y^0, \tag{12.23b}$$

where the subscripts x and y indicate the directions in which the effective coefficients are measured. Theorem (12.23a) holds for objects which are symmetric in the x- and y-axes and also in the $45°$ lines $x = \pm y$.

12.4.2. Series–parallel formulae

Some papers have explored the idea that neighbouring parts of a composite medium may be considered to be in series or parallel. We have seen in § 12.2 (p. 267) that a laminate composed of n sheets of thicknesses $l_1, l_2 \dots l_n$ and diffusion coefficients $D_1, D_2 \dots D_n$ placed in series has an effective diffusion coefficient D, given by

$$\frac{l_1}{D_1}+\frac{l_2}{D_2}+ \dots \frac{l_n}{D_n} = \frac{l}{D},$$

where l is the total thickness of the laminate. The effective diffusion coefficient for a composite consisting of n sheets in parallel is given by

$$l_1 D_1 + l_2 D_2 + \dots l_n D_n = lD.$$

Both expressions are based on the assumption that the flow is uni-directional. If it is not, the relationships are approximations.

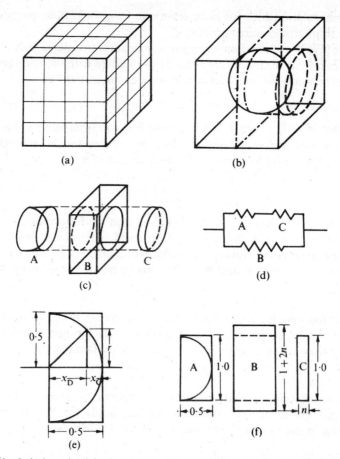

FIG. 12.1. Analytical conductivity is determined by considering the fluid–solid suspension as divided into cubes. (After Jefferson *et al.* 1958).

(i) *Uniform spheres on a regular lattice.* Jefferson, Witzell, and Sibbett (1958) consider the composite medium to be divided into a number of cubes as in Fig. 12.1(a), each having a sphere of the discontinuous phase at its centre; Fig. 12.1(b) shows half a typical cube containing a hemisphere. Jefferson *et al.* consider this to comprise the three components A, B, and C as in Fig. 12.1(c). The cylinder A has length and radius both equal to the radius of the sphere; the cylinder C has the same radius but its length is the difference between the side of the cube and the radius of the sphere; B is a hollow cylinder of length equal to half the side of the cube, outer diameter twice its length, and inner diameter that of the sphere. The three components are considered to form a series–parallel diffusion circuit, as in Fig. 12.1(d). The underlying assumption is that the diffusion flow through the composite is everywhere uni-directional. This implies first, that the net flow through components A and C is the same; and second, that concentrations are constant over all planes perpendicular to the flow In particular, the concentration difference between the ends of B is the sum of the differences across A and C respectively.

In reality, these assumptions are not strictly true, even in the most favourable case of a plane sheet of composite material with opposite faces maintained at different concentrations so that flow would be uni-directional in a homogeneous sheet. In the composite, it will be two- or three-dimensional to an extent depending on the geometry of the discontinuous phase and its distribution, and the relative properties of the two phases. We present Jefferson *et al.*'s treatment with slightly modified nomenclature. They were concerned with the corresponding thermal problem.

The diffusion flow through the composite model q_s of Figs. 12.1(c) and 12.1(d) is given by $q_s = q_A + q_B$. Denote by D_A, A_A, l_A the diffusion coefficient, cross-sectional area, and length of component A, and by ΔC_A the concentration difference between its ends, with corresponding nomenclature for components B and C. Then

$$q_A = -\frac{D_A A_A \Delta C_A}{l_A} = q_C = -\frac{D_C A_C \Delta C_C}{l_C}.$$

Since the total change in concentration across the half cube is ΔC_S, where $\Delta C_S = \Delta C_A + \Delta C_C$ and also $A_A = A_C$, we find

$$-q_A = \frac{\Delta C_S A_A}{(l_A/D_A)+(l_C/D_C)} = \frac{\Delta C_S A_A D_{AC}}{l_A+l_C}, \tag{12.24}$$

where D_{AC} is the equivalent diffusion coefficient of A and C taken together.

By considering A and C together in parallel with B we have for the total flow q_S through the half-cube,

$$-q_S = -(q_A+q_B) = \frac{D_{AC} \Delta C_S A_A}{l_A+l_C}+\frac{D_B \Delta C_S A_B}{l_B}$$

$$= \frac{D_S A_S \Delta C_S}{l_A+l_C}, \tag{12.25}$$

where D_S is the effective diffusion coefficient for the whole system and A_S the cross-sectional area of the cubic element.

Remembering that $l_B = l_A + l_C$, we have on substituting for D_{AC} from (12.24)

$$D_S = D_B \frac{A_B}{A_S}+\left\{\frac{(l_A+l_C)D_A D_C}{l_A D_C+l_C D_A}\right\}\frac{A_A}{A_S}. \tag{12.26}$$

If now we take the sphere to be of unit diameter and component C to be of length n then the geometry of the model gives (Fig. 12.1f)

$$A_A = \pi/4, \qquad A_S = (1+2n)^2, \qquad A_B = (1+2n)^2 - \pi/4;$$

$$l_A = l_B = 0.5, \qquad l_C = n. \tag{12.27}$$

It remains to determine the diffusion coefficient D_A. With reference to Fig. 12.1(e), and by applying the series formulae similar to the first of (12.24) to a cylindrical element of A lying between r and $r+\delta r$;

$$-q_A = \frac{D_A A_A \Delta C_A}{l_A} = \int_{r=0}^{r=0.5} \frac{2\pi \Delta C_A r \, dr}{(x_D/D_D)+(x_C/D_C)},$$

where D_D is the diffusion coefficient of the spheres forming the discrete phase, and D_C is already defined for component C which can be identified as the continuous phase.

Substituting $x_D^2 = (0.5)^2 - r^2$ and $x_C = 0.5 - x_D$, and performing the integration we find

$$D_A = \frac{2D_D D_C}{D_D - D_C} \left\{ \frac{D_D}{D_D - D_C} \ln\left(\frac{D_D}{D_C}\right) - 1 \right\}. \tag{12.28}$$

The parameter, n, is related to the volume fraction v_D of the discontinuous phase by the expression

$$n = 0.403 v_D^{-1/3} - 0.5. \tag{12.29}$$

Eqns (12.26), (12.27), (12.28), and (12.29), when combined, yield an expression for the effective diffusion coefficient of an array of uniform spheres arranged on a regular lattice in a continuum, in terms of their volume fraction and the diffusion properties of the two phases.

(ii) *Random distribution of discrete phase.* Tsao (1961) extended the use of the series–parallel formulae to composites in which the discontinuous phase consists of randomly distributed particles of irregular size and shape. A unit cube of such a composite is shown in Fig. 12.2(a). It is assumed that the surfaces parallel to the xy- and xz-planes are sealed, so that the over-all direction of diffusion is along the x-axis. The shaded regions represent the discontinuous phase d and the rest is the continuous phase c. Tsao illustrates, by Fig. 12.2(b), three porosity parameters. For a line in the cube parallel to the x-axis, the one-dimensional porosity P_1 is defined to be the fraction of the unit length occupied by the discrete phase. For a plane parallel to the yz-plane in Fig. 12.2(b), the two-dimensional porosity P_2 is defined as the fractional area occupied by phase d. Finally, the three-dimensional porosity P_3 is the volume fraction occupied by phase d.

Tsao slices the two-phase composite material into many thin layers, parallel to the yz-plane as in Fig. 12.2(a). Each layer is so thin that P_2 may be

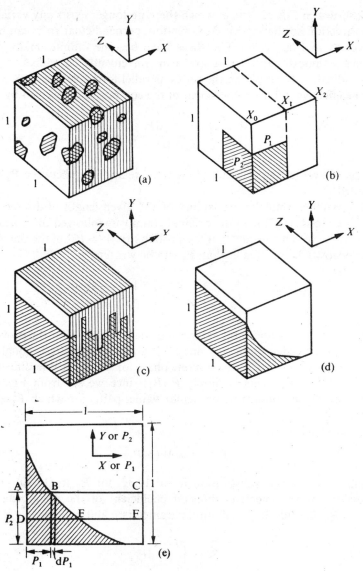

FIG. 12.2. From Tsao (1961).

assumed constant within it. Since both the parallel and series formulae are independent of the order in which the separate components are placed, we can re-arrange the elements of the discontinuous phase within each slice and then re-arrange the slices in any order we find convenient. All this assumes that the diffusion may be considered uni-directional in the x-direction. In this way Tsao proceeds through Fig. 12.2(c) to the arrangement of discrete phase shown in Fig. 12.2(d), in which there no longer exists any variation in the z-direction so that a unit cross-section, perpendicular to z, can be considered, as in Fig. 12.2(e). The slices have become infinitesimally thin to produce a smooth curve whose axes can conveniently be relabelled P_1 and P_2 as indicated. By applying the series–parallel formulae it is easy to show that the effective diffusion coefficient of the composite D_e is given by

$$\frac{1}{D_e} = \int_0^1 \frac{dP_1}{D_c + (D_d - D_c)P_2}. \tag{12.30}$$

In order to evaluate eqn (12.30), a functional relationship between P_1 and P_2 is needed.

Tsao assumed that the probability of any given length of discrete phase occurring along any line in the original composite followed the normal distribution curve. This means that the probability $W(>P'_1)$ of the one-dimensional porosity being greater than P'_1 can be written

$$W(>P'_1) = \int_{P'_1}^1 \frac{1}{\sigma\sqrt{(2\pi)}} \exp\left\{ -\frac{1}{2}\left(\frac{P_1 - \mu}{\sigma}\right)^2 \right\} dP'_1, \tag{12.31}$$

where μ is the mean and σ the standard deviation of P_1. But the probability of P'_1 being exceeded along any line through the composite is equal to the fractional area occupied by the discrete phase for which the one-dimensional porosity equals P'_1. This is simply $P_2(P'_1)$ since we see from Fig. 12.2(e) that for all P_2 less than this particular value, paths for which P_1 exceeds P'_1 exist.

Thus we may write

$$P_2(P_1) = W(>P_1) \tag{12.32}$$

as defined in (12.31). Substitution from (12.32) for P_2 in (12.30) gives an expression for the effective diffusion coefficient of the composite D_e in terms of D_d, D_c and the distribution parameters μ and σ. Since

$$P_3 = \int_0^1 P_2 \, dP_1, \tag{12.33}$$

it follows that the mean value $\mu = P_3$, the volume fraction of the discrete phase. Tsao decided that the standard deviation σ of P_1 had to be determined by experiment in addition to the volume fraction, in order to fix the two parameters μ and σ in his probability function.

Cheng and Vachon (1969) applied Tsao's basic idea in a different way. They re-arranged the particles of the discontinuous phase to form a centralized symmetrical area as in Fig. 12.3(a), which they then transformed to

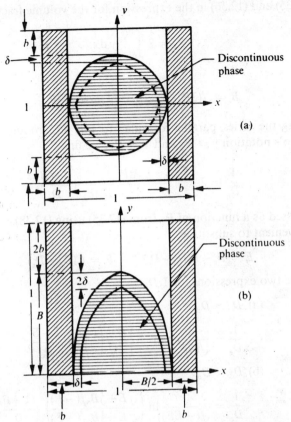

FIG. 12.3. Phase distribution for a two-phase mixture as represented by: (a) the x-axis bisecting the square; (b) the x-axis coinciding with the lower border of the square. (From Cheng and Vachon 1969).

Fig. 12.3(b) by dropping the x-axis to coincide with the lower edge of the square. Following Tsao, they assumed the discontinuous phase to be dispersed according to a normal distribution so that the curve in Fig. 12.3(b) is of the form

$$y = C_1 \exp(-C_2 x^2), \qquad (12.34)$$

where C_1 and C_2 are parameters. Having in mind that $-\frac{1}{2} \leqslant x \leqslant \frac{1}{2}$, they neglected all higher terms in the expansion of the exponential function in (12.34) and wrote

$$y = B + Cx^2. \qquad (12.35)$$

From the symmetry of Fig. 12.3(a) we see that

$$B = -4/C. \qquad (12.36)$$

Using (12.35) and (12.36) in the expression for the volume fraction,

$$P_3 = 2 \int_0^{B/2} y \, dx,$$

(12.37)

we find

$$B = \sqrt{(3P_3/2)}, \quad C = -4\sqrt{2}/\{\sqrt{(3P_3)}\}.$$

(12.38)

By applying the series–parallel formulae to Fig. 12.3(b) and remembering that in Tsao's notation $y = P_2$ and $x = P_1$, we find

$$\frac{1}{D_e} = 2 \int_0^{B/2} \frac{dP_1}{D_c + (D_d - D_c)P_2} + \frac{1-B}{D_c}.$$

(12.39)

P_2 is expressed as a function of P_1 from (12.35) using (12.38).

It is convenient to substitute

$$E = D_c - B(D_d - D_c), \quad F = C(D_d - D_c)$$

(12.40)

and so write two expressions for $1/D_e$:

(a) $D_c > D_d$.

$$\frac{1}{D_e} = \frac{2}{\sqrt{(EF)}} \tan^{-1} \left\{ \frac{B}{2} \sqrt{\left(\frac{F}{E} \right)} \right\} + \frac{1-B}{D_c},$$

(12.41)

(b) $D_c < D_d$.

$$\frac{1}{D_e} = \frac{1}{\sqrt{(-EF)}} \ln \left(\frac{\sqrt{E} + \frac{1}{2}B\sqrt{(-F)}}{\sqrt{E} - \frac{1}{2}B\sqrt{(-F)}} \right) + \frac{1-B}{D_c}.$$

(12.42)

In this case F is a negative quantity.

Since only one parameter P_3, the volume fraction, is needed to determine B and C through (12.38) it appears that Cheng and Vachon have removed the need to measure a second parameter experimentally as Tsao suggested. What they have done is to specify one particular set of normal distributions by their assumption of symmetry in Fig. 12.3(a). In fact, in terms of B and C the normal distribution (12.34) is

$$y = B \exp (Cx^2/B),$$

which is identical with that used by Tsao to derive (12.31) if we identify $x = P_1 - \mu$ and $B/C = 2\sigma^2$. The only difference is that Tsao normalized his integral by the factor $1/(\sigma\sqrt{2\pi})$, whereas Cheng and Vachon chose $B = -4/C = (3P_3/2)^{\frac{1}{2}}$. Clearly they have also chosen $\sigma = \frac{1}{4}\sqrt{(3P_3)}$ by their method of rearranging the discontinuous phase in Figs. 12.3(a) and 12.3(b).

Cheng and Vachon suggest how their method can be applied to some three- and multi-phase systems for certain ratios of the diffusion coefficients of the component phases. They also compare results obtained by their own and other formulae with experimental data for two- and three-phase systems. Fidelle and Kirk (1971) measured thermal conductivities of three different composites covering a broad range of filler concentration and heat conductivity ratios. The composites were carborundum plaster, alumina plaster, and alumina polyester. Over-all, the best predictor of their experimental data is the Cheng–Vachon formula. That theirs is better than the three-dimensional computer results obtained by Fidelle and Kirk (1971) perhaps suggests that the assumption that the spheres are on a regular lattice and not randomly distributed, as in Cheng and Vachon's treatment, is more serious than that of uni-directional flow. The uni-directional, approximate treatment has the advantage of leading to a simple explicit expression for the effective diffusion coefficient.

12.4.3. *Numerical solutions*

Several authors have obtained numerical solutions of Laplace's equation for a region in which identical uniform particles of one phase are arranged on a regular lattice in a continuum of a different phase.

(i) Fidelle and Kirk (1971) used finite-difference methods analogous to those described in Chapter 8 to compute numerical solutions for Jefferson's model of spherical particles § 12.4.2(i) (p. 274). They took account of the three-dimensional nature of the problem but did not reproduce the spherical interfaces exactly. They used the best approximation possible with a rectangular grid of points.

(ii) Bell and Crank (1974) studied a regular two-dimensional array of identical, rectangular blocks, impenetrable by the diffusing substance, i.e. having zero diffusion coefficient. They first showed that for any region with repeated symmetry, as in Fig. 12.4(a), it suffices to study the steady-state

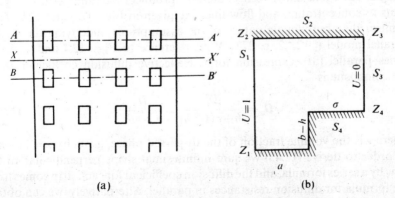

FIG. 12.4. (a) Aligned rectangular blocks. (b) Basic element.

problem in a rectangular region with a re-entrant corner as in Fig. 12.4(b), provided the number of repeating basic units is an integer power of two. The blocks need not be rectangular. The argument applies in general, provided symmetry is identified. A staggered arrangement, as in Fig. 12.5(a), can be studied by solving the problem for the rectangle with two re-entrant corners, as in Fig. 12.5(b).

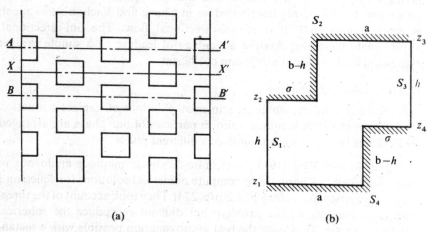

(a)   (b)

FIG. 12.5. (a) Staggered rectangular blocks. (b) Basic element.

Bell and Crank obtained finite-difference solutions for the regions shown in Figs. 12.4(b) and 12.5(b). The singularities at the re-entrant corners (see §§ 8.10.2 and 8.12, pp. 153 and 159) were treated by the methods of Whiteman and Papamichael (1972). An alternative approach by Symm (1973) using integral equations and Green's functions produces the same results. Typical plots of concentration and flow-lines are given in Figs. 12.6 and 12.7. They confirm the approximate nature of the assumptions underlying the series–parallel model (§ 12.4.2, p. 273). With reference to Figs. 12.4 and 12.5, the series–parallel (SP) expression for the effective diffusion coefficient D_e of the composite is

$$D_e = \frac{Dh}{h+\sigma(1-h)} = \frac{D}{1+v/h},$$ (12.43)

where v is the volume fraction of the dispersed phase, given by $v = (1-h)\sigma$. In order to derive (12.43) we sum infinitesimal strips perpendicular to the flow by a series formula, and the diffusion coefficient for each strip comes from the formula for diffusion resistances in parallel. Alternatively, we can obtain an approximation based on the summation of strips lying parallel to the flow

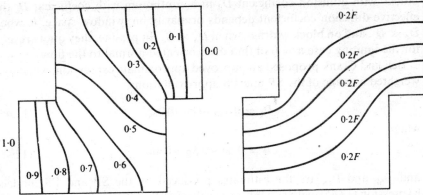

FIG. 12.6. Lines of constant concentration.　　FIG. 12.7. Flow lines (where F is the total flow through the region).

rather than across it. For this problem we obtain the apparently trivial expression

$$D_e = Dh \qquad (12.44)$$

which, for convenience, we refer to as the parallel–series or PS approximation. Fig. 12.8 shows how D_e/D varies with the 'window' width h and that the numerical solution lies between the SP and PS approximations. These and other calculated results, taken together with Keller's reciprocal relation (12.23b) led Bell and Crank to the general conclusion that for rectangular

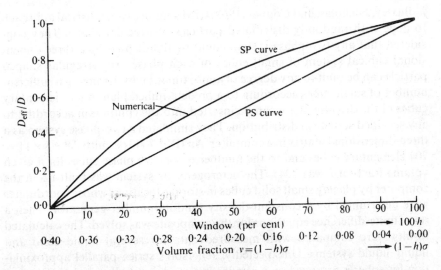

FIG. 12.8.

blocks with diffusion coefficient D_d in a continuum with coefficient D_c the effective diffusion coefficient depends primarily on 'window space' h, when $D_d \ll D_c$, and on block width σ, when $D_d \gg D_c$. In passing, they quote results for the limiting case $\sigma = 0$ of thin discs oriented normal to the flow.

Bell and Crank proposed an improved approximate expression, which is a weighted average of the SP and PS approximations,

$$D_e = \theta D_{SP} + (1 - \theta)D_{PS}, \tag{12.45}$$

where

$$\theta = 0.56 - 0.5\sigma + 0.4h, \tag{12.46}$$

and D_{SP} and D_{PS} are the estimates produced by the SP and PS models. Expression (12.45), with (12.46), gives values of D_e differing by only 2 per cent or 3 per cent from the numerical solutions for a range of block sizes and spacings for both the aligned and staggered arrangements.

(iii) Keller and Sacks (1964) tabulated numerical values of D_e/D for a regular lattice of impermeable, uniform cylinders of various radii. They confirmed Keller's asymptotic solution (12.22b), adjusted by (12.23a). Their results for cylinders of radius r agree with those of Bell and Crank (1974) for square blocks of side 2σ, when suitable scaled so that they have the same cross-sectional area, i.e. $r = 2\sigma/\sqrt{\pi}$. Thus, the geometric shape of the dispersed phase is not important when the areas are normalized, except when the blocks or cylinders are very close together. The weighted approximation (12.45) with (12.46) also represents satisfactorily the results of Keller and Sacks.

Baxley, Nicholas, and Couper (1968) tried a numerical-statistical approach to cope with randomly distributed particles of irregular shape. They considered that any two-phase system could be represented by a three-dimensional cubical system of small cubes of each phase. Any irregular shaped particle can be built to any degree of approximation by arranging a sufficient number of small cubes according to a predetermined plan. Also, elementary cubes of the discrete phase can be inserted into the continuum according to any specified statistical distribution. They simulated a two-phase system as a three-dimensional matrix in a computer. An IBM 7040 permitted $8 \times 8 \times 11 = 704$ elementary cubes and so the number of discrete phase cubes for a given volume fraction f was $704f$. The heterogeneous system was built up in the computer by placing small solid cubes in storage locations whose coordinates were determined by a constant-density random-number generator. Then a set of finite-difference equations for the composite was solved. The calculated results were compared with measured data for several liquid–solid and liquid–liquid systems. Unexpectedly, Jefferson's series–parallel approximation for spheres arranged on a regular lattice gave equally good results.

In the light of more recent work by Bell and Crank (1973) on rectangular blocks, it could be that the corners of Baxley's cubes are one serious source of error and that the finite-difference representation needs to be improved.

13

MOVING BOUNDARIES

13.1. Introduction

IN one general class of problem some of the diffusing molecules are immobilized and prevented from taking further part in the diffusion process. This may be due to a chemical reaction by which the diffusing molecules are either precipitated or form a new immobile chemical compound. Alternatively, we may have a less specific type of adsorption on fixed sites. We are concerned with immobilization processes which are irreversible, and so rapid compared with the rate of diffusion that they may be considered instantaneous. Only a limited number of molecules can be immobilized in a given volume of the medium. Examples are the diffusion of oxygen into muscle where oxygen combines with the lactic acid; the reaction of Cu^{2+} ions with CS_2 groups as they diffuse in cellulose xanthate, and the diffusion of periodate ions into cellulose fibres and their removal by combination with the glucose groups of the cellulose. An essential feature of the idealized problem of diffusion accompanied by the instantaneous and irreversible immobilization of a limited number of the diffusing molecules is that a sharp boundary surface moves through the medium. It separates a region in which all the sites are occupied from one in which none of them are. In front of the advancing boundary the concentration of freely diffusing molecules is zero, while behind it immobilization is complete. Fig. 13.1 illustrates the situation. The same mathematical problem is presented by heat flow in a medium which undergoes a phase change at some fixed temperature, accompanied by the absorption or liberation of latent heat. Such problems are often referred to as

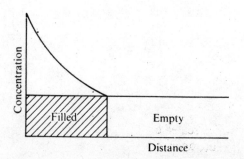

FIG. 13.1. Diffusion of particles which are trapped in immobile holes.

'Stefan problems' after J. Stefan (1890). An excellent survey of the literature dating from that time is given by Muehlbauer and Sunderland (1965). Bankoff (1964) has written a useful review article with numerous references. Rubenstein's book (1971) gives a systematic presentation of the mathematical aspects of Stefan problems.

The occurence of a sharp boundary does not necessarily imply a chemical reaction. When a liquid penetrant diffuses into a polymer sheet or filament, sharp boundaries are frequently observed under a microscope. They correspond to near-discontinuities in the gradient of the concentration–distance curve (Hartley 1946). They imply discontinuous changes in the diffusion coefficient–concentration relationship and present the same mathematical problem. We consider both analytical and numerical methods of obtaining mathematical solutions.

13.2. Discontinuous diffusion coefficients

We shall first consider diffusion coefficients which change discontinuously from one constant value to another at certain concentrations. They may be zero or infinite over parts of the concentration range. Such coefficients provide examples of extreme types of concentration-dependence and are useful in that they enable limits to be set to the extent to which concentration-dependence of the diffusion coefficient can modify the course of diffusion as indicated, for example, by the shape of the concentration–distance curve. Some special cases which do not require detailed calculation are considered first, and afterwards solutions are developed for more general cases.

13.2.1. *Special cases not requiring detailed calculation*

When the diffusion coefficient varies discontinuously with concentration in certain ways, the concentration–distance curves and the sorption– and desorption–time curves can be deduced by general reasoning from the known solutions for a constant diffusion coefficient. Some examples referring to uni-directional diffusion in a plane sheet are presented graphically in Figs. 13.2 and 13.3. These diagrams are largely self-explanatory and need only brief comments. The nomenclature is as follows: c is the concentration of diffusing substance at a distance x measured from the surface of the sheet in the direction of diffusion at time t, and D is the diffusion coefficient. Also M_t is the total amount of diffusing substance absorbed by or desorbed from unit area of a plane sheet of thickness l in time t. For sorption the sheet is initially free of diffusing substance and the surface is maintained at $c = C_1$ throughout; for desorption the initial concentration is C_1 throughout the sheet and the surface is maintained at zero concentration. In Fig. 13.2(a), over the concentration range $C_1 > c > C_x$ for which D is zero, the concentration gradient is infinite and there is no penetration into the sheet. The sorption

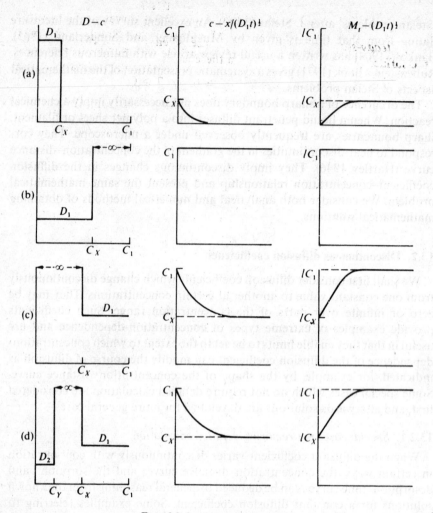

FIG. 13.2. Sorption behaviour.

behaviour is identical with that for a constant diffusion coefficient D_1 and surface concentration C_X, and the equilibrium total content of the sheet is lC_X. When D is infinite over any finite interval of concentration at the upper end of the concentration range as in Fig. 13.2(b), the existence of a finite concentration gradient anywhere in the range $C_1 > c > C_X$ would mean an infinite rate of transfer of diffusing substance. Consequently, the concentration must reach its final uniform value C_1 throughout the sheet infinitely rapidly, and this is true whatever the form of the D against c curve at low concentrations, even if D is zero. When D is infinite at low concentrations, but drops to a constant finite value D_1 at $c = C_X$ (Fig. 13.2(c)), the sheet attains a uniform concentration C_X throughout, infinitely rapidly, and the remainder of the sorption behaviour is as for a constant diffusion coefficient D_1. The sorption behaviour is precisely the same for a diffusion coefficient which is infinite for intermediate concentrations and has a constant finite value at high concentrations, as in Fig. 13.2(d), whatever the form of D at low concentrations, again even if D is zero. The curves of Fig. 13.3 for desorption follow by similar arguments and need no comment.

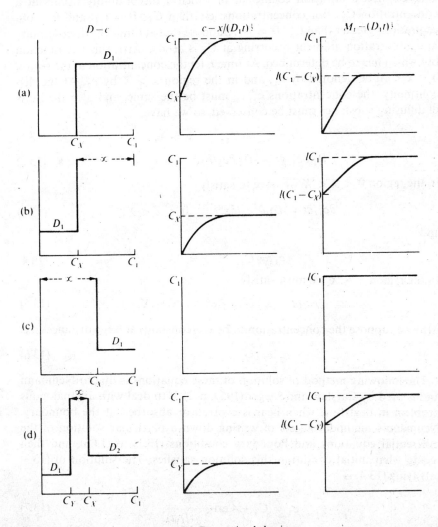

Fig. 13.3. Desorption behaviour.

13.2.2. *Diffusion coefficients having a discontinuity at one concentration*

The more general cases in which the diffusion coefficient changes discontinuously from one constant finite value to another, at one or more concentrations, require detailed calculation. At the concentration at which a discontinuous change in D occurs there is also a discontinuity in the concentration gradient, and the way in which this moves has to be determined. The problem can be stated mathematically as follows.

Suppose that diffusion takes place into a semi-infinite medium and that the surface $x = 0$ is maintained at a constant concentration C_1. We shall consider first a diffusion coefficient in which a discontinuity occurs at a concentration C_X. For concentrations less than C_X, $D = D_2$, and for concentrations greater than C_X, $D = D_1$. Suppose that at time t the discontinuity in concentration gradient occurring at C_X is at $x = X(t)$; this is a function of t which has to be determined. At time t, let the concentration in the region $0 < x < X$ be denoted by c_1, and in the region $x > X$ by c_2. At the discontinuity, the concentrations c_1, c_2 must be the same, and also the mass of diffusing substance must be conserved, so we have

$$c_1 = c_2 = C_X, \qquad x = X, \tag{13.1}$$

$$D_1 \, \partial c_1 / \partial x = D_2 \, \partial c_2 / \partial x, \qquad x = X. \tag{13.2}$$

In the region $0 < x < X$ we have to satisfy

$$\partial c_1 / \partial t = D_1 \, \partial^2 c_1 / \partial x^2, \qquad 0 < x < X, \tag{13.3}$$

and

$$c_1 = C_1, \qquad x = 0. \tag{13.4}$$

In the region $x > X$ we must satisfy

$$\partial c_2 / \partial t = D_2 \, \partial^2 c_2 / \partial x^2, \qquad x > X. \tag{13.5}$$

Also we suppose the concentration to be C_2 (constant) at large distances, i.e.

$$c_2 = C_2, \qquad x = \infty. \tag{13.6}$$

The following method of solution of these equations is due to Neumann and is used by Carslaw and Jaeger (1959, p. 283) to deal with an analogous problem in heat flow when heat is evolved or absorbed at the boundary. Neumann's method consists of writing down a particular solution of the differential equations and boundary conditions (13.1) to (13.6), and then seeing what initial condition this solution satisfies. The solution of (13.3) satisfying (13.4) is

$$c_1 = C_1 + A \, \text{erf} \frac{x}{2\sqrt{(D_1 t)}}, \tag{13.7}$$

where A is a constant. Also if B is a constant,

$$c_2 = C_2 + B \operatorname{erfc} \frac{x}{2\sqrt{(D_2 t)}}. \tag{13.8}$$

Then (13.1) requires

$$A \operatorname{erf} \frac{X}{2\sqrt{(D_1 t)}} = C_X - C_1, \tag{13.9}$$

and

$$B \operatorname{erfc} \frac{X}{2\sqrt{(D_2 t)}} = C_X - C_2, \tag{13.10}$$

Since (13.9) and (13.10) have to be satisfied for all values of t, X must be proportional to $t^{\frac{1}{2}}$, say,

$$X = k t^{\frac{1}{2}}, \tag{13.11}$$

where k is a constant to be determined. Using throughout (13.7), (13.8), (13.9), (13.10), and (13.11) in (13.2) we obtain

$$\frac{C_X - C_1}{g(k/2D_1^{\frac{1}{2}})} + \frac{C_X - C_2}{f(k/2D_2^{\frac{1}{2}})} = 0, \tag{13.12}$$

where g and f are functions given by

$$g\left(\frac{k}{2D_1^{\frac{1}{2}}}\right) = \pi^{\frac{1}{2}} \frac{k}{2D_1^{\frac{1}{2}}} \exp\left(k^2/4D_1\right) \operatorname{erf} \frac{k}{2D_1^{\frac{1}{2}}}. \tag{13.13}$$

$$f\left(\frac{k}{2D_2^{\frac{1}{2}}}\right) = \pi^{\frac{1}{2}} \frac{k}{2D_2^{\frac{1}{2}}} \exp\left(k^2/4D_2\right) \operatorname{erfc} \frac{k}{2D_2^{\frac{1}{2}}}. \tag{13.14}$$

When (13.12) is solved numerically, it gives k in terms of C_1, C_2, C_X, and the diffusion coefficients D_1 and D_2. The numerical solution is greatly facilitated by using Figs. 13.4, 13.5 and 13.6 which show graphs of f and g. The same graphs were shown in Danckwerts's paper (1950b).

Substituting $k = X/t^{\frac{1}{2}}$ in (13.9) and (13.10), we obtain

$$A \operatorname{erf} \frac{k}{2D_1^{\frac{1}{2}}} = C_X - C_1, \tag{13.15}$$

$$B \operatorname{erfc} \frac{k}{2D_2^{\frac{1}{2}}} = C_X - C_2, \tag{13.16}$$

from which A and B can be evaluated, knowing k, and hence the concentrations c_1 and c_2 follow from (13.7) and (13.8). When $t = 0$, $X = 0$ and $c_2 = C_2$.

This solution applies strictly only when the medium is semi-infinite, a condition which is effectively satisfied in the early stages of diffusion into a sheet of finite thickness. When the concentration at the centre of the sheet

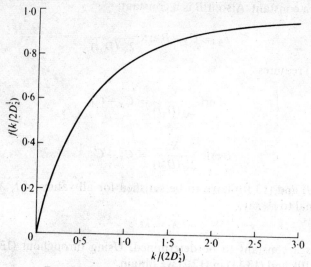

FIG. 13.4 $f(k/2D_2^{\frac{1}{2}})$ for positive values of $k/2D_2^{\frac{1}{2}}$.

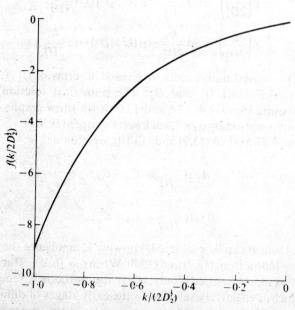

FIG. 13.5. $f(k/2D_2^{\frac{1}{2}})$ for negative values of $k/2D_2^{\frac{1}{2}}$.

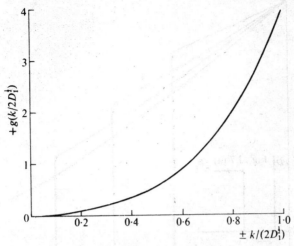

FIG. 13.6. g as a function $k/2D_1^{\frac{1}{2}}$.

becomes appreciable, this solution must be replaced by one in which the finiteness of the sheet is recognized. In general, it is necessary to continue the solution by numerical methods such as are discussed in Chapter 8. In at least one interesting case, however, a formal solution is possible after the sheet has ceased to be semi-infinite. This is when the diffusion coefficient is zero at low concentrations, i.e. $D_2 = 0$. Since this case can be solved completely without excessive labour, and as it is a limiting case of a diffusion coefficient which is small at low concentrations and increases with concentration increasing, it justifies detailed consideration.

13.2.3. Diffusion coefficient zero for concentrations less than C_X and constant and finite above C_X

When D_2 approaches zero, $k/2D_2^{\frac{1}{2}}$ becomes large, and by applying the asymptotic expansion

$$\exp(z^2)\,\mathrm{erfc}\,z = \frac{1}{\pi^{\frac{1}{2}}}\left(\frac{1}{z} - \frac{1}{2z^3} + \dots\right) \tag{13.17}$$

to (13.14) eqn (13.12) reduces to

$$\frac{C_X - C_1}{g(k/2D_1^{\frac{1}{2}})} + C_X - C_2 = 0. \tag{13.18}$$

Eqn (13.18) can be solved to give k and the solution follows from (13.7). From (13.2), $\partial c_2/\partial x \to \infty$ when $D_2 \to 0$, unless $\partial c_1/\partial x = 0$, which is the final steady state, and hence the concentration gradient is infinite at low concentrations. Examples of concentration curves, plotted against $x/2(D_1 t)^{\frac{1}{2}}$, are shown in Fig. 13.7 for $C_X = \frac{1}{3}C_1, \frac{1}{2}C_1, \frac{2}{3}C_1$, and $\frac{5}{6}C_1$ respectively, and

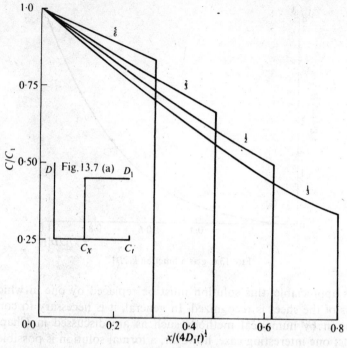

FIG. 13.7. Concentration–distance curves for D as in Fig. 13.7(a). Numbers on curves are values of C_X/C_1.

for $C_2 = 0$ in each case. They also apply to the problem of diffusion with precipitation discussed by Hermans (1947), who gives a graph effectively of $k/2D_1^{\frac{1}{2}}$ as a function of $(C_X - C_1)/(C_X - C_2)$, so that the root of (13.18) can be read off for any desired combination of values of C_1, C_2, and C_X. This is the graph shown in Fig. 13.6.

It is clear from the way in which the concentration–distance curves of Fig. 13.7 terminate abruptly that, from the time the vertical advancing front reaches the centre of a finite sheet, the diffusion coefficient is constant at D_1 over the whole of the remaining concentration range. The solution required is thus the well-known one for diffusion into a plane sheet, with a constant diffusion coefficient and a given initial concentration distribution through the sheet. In the present nomenclature, the solution (4.16) given in § 4.3.1 (p. 47) is

$$c_1 = C_1 + \frac{2}{l} \sum_{n=0}^{\infty} \exp\left\{-D_1(2n+1)^2\pi^2(t-t_0)/4l^2\right\} \sin\frac{(2n+1)\pi x}{2l}$$

$$\times \left\{\frac{2(-1)^{n+1}lC_1}{(2n+1)\pi} + \int_0^l f(x)\sin\frac{(2n+1)\pi x}{2l}\,dx\right\}, \qquad (13.19)$$

where the surface of the sheet $x = 0$ is maintained at C_1, the centre of the sheet is at $x = l$, and $f(x)$ is the concentration distribution through the sheet at time $t = t_0$. The total amount, M_t, of diffusing substance present in half the sheet at time t is obtained by integrating the right-hand side of (13.19) with respect to x between the limits 0 and l, and is given by

$$\frac{M_t}{lC_1} = 1 + 2 \sum_{n=0}^{\infty} \exp \{-D_1(2n+1)^2\pi^2(t-t_0)/4l^2\}$$

$$\times \left\{ \frac{2(-1)^n}{(2n+1)\pi} \int_0^l \frac{1}{lC_1} f(x) \sin \frac{(2n+1)\pi x}{2l} \, dx - \frac{4}{(2n+1)^2\pi^2} \right\}. \qquad (13.20)$$

In the present problem, $f(x)$ is given by (13.7) evaluated at time t_0, so that

$$f(x) = C_1 \left\{ 1 + \frac{A}{C_1} \operatorname{erf} \frac{x}{2(D_1 t_0)^{\frac{1}{2}}} \right\}, \qquad (13.21)$$

where A is given by (13.9). The time t_0 is defined as that at which $X = l$, so that it is related to the solution of (13.7) by the expression

$$\left(\frac{D_1 t_0}{l^2} \right)^{\frac{1}{2}} = \frac{D_1^{\frac{1}{2}}}{k}. \qquad (13.22)$$

By substituting in (13.20) from (13.21) and (13.22), M_t/lC_1 is expressed as a function of the single variable $D_1 t/l^2$. The integral in (13.20) is conveniently evaluated numerically. A family of curves showing M_t/lC_1 as a function of $(D_1 t/l^2)^{\frac{1}{2}}$ is shown in Fig. 13.8 for several values of C_x including $C_x = 0$, when the solution is simply that for a diffusion coefficient having the constant value D_1. The linear parts of these curves follow readily from the solution (13.7) since we have

$$\frac{\partial M_t}{\partial t} = -\left(D_1 \frac{\partial c_1}{\partial x} \right)_{x=0} = -\frac{A D_1^{\frac{1}{2}}}{(\pi t)^{\frac{1}{2}}},$$

and hence

$$\frac{M_t}{lC_1} = -\frac{2}{\pi^{\frac{1}{2}}} \frac{A}{C_1} \left(\frac{D_1 t}{l^2} \right)^{\frac{1}{2}}, \qquad (13.23)$$

where A is given by (13.9) and is negative for sorption.

We may note that these curves, which are universal for all values of D_1, are each linear at first, becoming concave downwards later though it might have been expected that they would show a different behaviour. Thus when the vertical front of a concentration–distance curve (Fig. 13.7) reaches the centre of the sheet, the region of zero diffusion coefficient is immediately removed and the subsequent sorption is governed by a constant and possibly high value of D, i.e. by D_1. On these grounds it is not at first unreasonable to

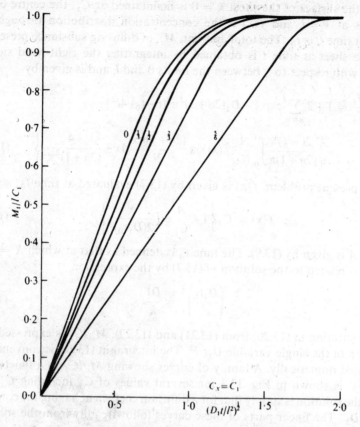

FIG. 13.8. Sorption curves for D as in Fig. 13.7(a). Numbers on curves are values of C_X/C_1.

expect the sorption to proceed more rapidly at this stage, i.e. for the gradient of a curve of Fig. 13.8 to be first constant and then to increase. Detailed calculation shows that this is not so, however, even for this limiting case of a discontinuous diffusion coefficient which is zero at low concentrations.

13.2.4. *Diffusion coefficients having discontinuities at two concentrations*

The above results can be extended to the case of a diffusion coefficient defined by

$$D = D_1, \qquad C_1 > c > C_X, \tag{13.24}$$

$$D = D_2, \qquad c < C_Y, \tag{13.25}$$

$$D = D_3, \qquad C_X > c > C_Y. \tag{13.26}$$

It is convenient to refer to this as a two-step diffusion coefficient. At the concentrations C_X and C_Y there are discontinuities in concentration gradient, and at each discontinuity conditions corresponding to (13.1) and (13.2) are to be satisfied. The method is so closely similar to that just described that the equations can be written down without explanation. Concentrations c_1, c_2, c_3 are associated with the ranges in which the diffusion coefficient has the values D_1, D_2, D_3 respectively. Then if A, B, and E are constants we have the following solutions:

$$c_1 = C_1 + A \operatorname{erf} \frac{x}{2(D_1 t)^{\frac{1}{2}}}, \qquad 0 < x < X, \tag{13.27}$$

$$c_2 = C_2 + B \operatorname{erfc} \frac{x}{2(D_2 t)^{\frac{1}{2}}}, \qquad x > Y, \tag{13.28}$$

$$c_3 = C_X + E \left\{ \operatorname{erf} \frac{x}{2(D_3 t)^{\frac{1}{2}}} - \operatorname{erf} \frac{X}{2(D_3 t)^{\frac{1}{2}}} \right\}, \qquad X < x < Y. \tag{13.29}$$

The conditions

$$c_1 = c_3 = C_X, \qquad x = X, \tag{13.30}$$

$$c_2 = c_3 = C_Y, \qquad x = Y, \tag{13.31}$$

require that

$$C_X - C_1 = A \operatorname{erf} \frac{X}{2(D_1 t)^{\frac{1}{2}}}, \tag{13.32}$$

$$C_Y - C_2 = B \operatorname{erfc} \frac{Y}{2(D_2 t)^{\frac{1}{2}}}, \tag{13.33}$$

$$C_Y - C_X = E \left\{ \operatorname{erf} \frac{Y}{2(D_3 t)^{\frac{1}{2}}} - \operatorname{erf} \frac{X}{2(D_3 t)^{\frac{1}{2}}} \right\}, \tag{13.34}$$

from which it follows that

$$X = k_1 t^{\frac{1}{2}}, \qquad Y = k_2 t^{\frac{1}{2}}, \tag{13.35}$$

where k_1 and k_2 are constants to be determined. By using the two conditions

$$D_1 \, \partial c_1 / \partial x = D_3 \, \partial c_3 / \partial x, \qquad x = X, \tag{13.36}$$

$$D_2 \, \partial c_2 / \partial x = D_3 \, \partial c_3 / \partial x, \qquad x = Y, \tag{13.37}$$

and eqns (13.27), (13.28), (13.29), (13.32), (13.33), (13.34), and (13.35) we derive the following two equations from which to evaluate k_1 and k_2.

$$\frac{D_1^{\frac{1}{2}}(C_X - C_1) \exp(-k_1^2/4D_1)}{\operatorname{erf}(k_1/2D_1^{\frac{1}{2}})} - \frac{D_3^{\frac{1}{2}}(C_Y - C_X) \exp(-k_1^2/4D_3)}{\operatorname{erf}(k_2/2D_3^{\frac{1}{2}}) - \operatorname{erf}(k_1/2D_3^{\frac{1}{2}})} = 0, \tag{13.38}$$

$$\frac{D_2^{\frac{1}{2}}(C_Y - C_2) \exp(-k_2^2/4D_2)}{\operatorname{erfc}(k_2/2D_2^{\frac{1}{2}})} + \frac{D_3^{\frac{1}{2}}(C_Y - C_X) \exp(-k_2^2/4D_3)}{\operatorname{erf}(k_2/2D_3^{\frac{1}{2}}) - \operatorname{erf}(k_1/2D_3^{\frac{1}{2}})} = 0. \tag{13.39}$$

Once k_1 and k_2 are determined the whole solution follows as before, and it is easy to see that the same initial condition holds, namely, that the region $x > 0$ is at a uniform concentration C_2. Eqns (13.38) and (13.39) can be solved numerically without excessive labour, but for any further extension to a three-or-more step diffusion coefficient the numerical work would probably be prohibitive. However, the two-step form includes several interesting types of diffusion coefficient and by suitable choice of D_1, D_2, D_3 and C_X, C_Y a reasonable approximation to many continuously varying diffusion coefficients can be obtained. Some results for two-step diffusion coefficients are discussed in Chapter 9.

13.3. Immobilizing reaction

The concentration plots in Fig. 13.7 have the form of Fig. 13.1. If we interpret C in §§ 13.2.2 and 13.2.3 to be the total concentration of molecules, i.e. both freely diffusing and immobilized, then C_X in (13.18) with $C_2 = 0$, can be identified as the concentration of the immobilized component. In order that the front can advance a distance δX we need to supply an amount $C_X \delta X$ of diffusant. The amount arriving at X in a time interval δt is $-D_1 \delta t \, \partial C/\partial x$. Conservation at the moving boundary therefore requires

$$-D_1 \partial C/\partial x = C_X \, dX/dt.$$

If this condition is substituted for (13.2) when $D_2 = 0$, we finally regain (13.18). In the case of heat flow C_X becomes $L\rho_1$ where L is the latent heat and ρ_1 the density of medium 1. Thus, a problem in diffusion with immobilization can be thought of formally as a concentration-dependent diffusion process in which D is zero when the total concentration C is in the range $0 \leqslant C < C_X$ and changes discontinuously to a finite value at $C = C_X$.

13.4. General problem of the moving boundary

We give now Danckwert's (1950b) treatment of the more general problem in which a moving interface is involved. In all the cases considered the two regions are separated by a plane surface and diffusion takes place only in the direction perpendicular to this plane. The concentration is initially uniform in each region. The process of diffusion may cause changes which bring about the disappearance or appearance of matter at the interface in one or both regions, and hence a resulting bodily movement of the matter in one or both regions relative to the interface. An example which is easy to visualize is afforded by the melting of ice in contact with water; as heat (here regarded as the diffusing substance) flows from water to ice, ice disappears and water appears at the interface, so that both ice and water are in bodily movement with respect to the interface. In all the cases considered the rates of bodily motion of the matter in the two regions with respect to the interface are directly proportional to each other, e.g. the volume of ice melted is proportional to the volume of water formed. Variations in specific volume or partial specific volume due to changes in concentration (or temperature) are ignored. The solutions derived below do not apply to systems in which convection currents are important or in which gradients of both concentration and temperature exist.

13.4.1. *The problem*

Consider two media which are free from convection currents but which may be in relative bodily motion along the x-axis, which is perpendicular to the interface. Position in medium 1 is specified by a coordinate in the x_1 system which is stationary with respect to medium 1; position in medium 2 by a coordinate in the x_2 system, stationary with respect to medium 2. The media are separated at time t by the plane $x_1 = X_1$, $x_2 = X_2$, which is initially at $x_1 = x_2 = 0$. Medium 1 occupies all or part of the space $X_1 < x_1 < \infty$, medium 2 all or part of the space $-\infty < x_2 < X_2$ (Fig. 13.9).

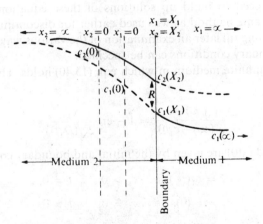

FIG. 13.9.

In both media there is a substance which moves by diffusion relative to the x_1 and x_2 coordinates and is transferred from one medium to the other. The concentration of the diffusing substance at time t is denoted by c_1 at x_1 and by c_2 at x_2. The following equations are obeyed in the two media,

$$\partial c_1/\partial t = D_1 \, \partial^2 c_1/\partial x_1^2, \qquad (13.40)$$

$$\partial c_2/\partial t = D_2 \, \partial^2 c_2/\partial x_2^2, \qquad (13.41)$$

where the diffusion coefficients D_1 and D_2 are assumed to be independent of c_1 and c_2.

At any time the concentrations $c_1(X_1)$, $c_2(X_2)$ at either side of the interface are assumed to be related by an equilibrium expression

$$c_2(X_2) = Qc_1(X_1) + R, \qquad (13.42)$$

where Q and R are constants, e.g. for the absorption of a gas obeying Henry's law, Q is the solubility of the gas and $R = 0$. The diffusing substance is conserved at the interface so that

$$D_1(\partial c_1/\partial x_1)_{x_1 = x_1} - D_2(\partial c_2/\partial x_2)_{x_2 = x_2} + c_1(X_1)\,dX_1/dt - c_2(X_2)\,dX_2/dt = 0.$$

$$(13.43)$$

We have already said that there is constant proportionality between the rates of movement of the two media relative to the interface and hence it follows that

$$X_2 = PX_1, \qquad (13.44)$$

where P is a constant determined by the conditions of the problem and may in some cases be zero.

We now proceed to build up solutions of these equations by what is essentially the same method as was used earlier for discontinuous diffusion coefficients, leaving till later an examination of the various ways in which the initial and boundary conditions can be specified.

Consider an infinite medium in which eqn (13.40) holds. Then a solution takes the form

$$\frac{c_1(\infty) - c_1}{c_1(\infty) - c_1(0)} = 1 - \operatorname{erf}\frac{x_1}{2\sqrt{(D_1 t)}}, \qquad (13.45)$$

where $c_1(\infty)$ and $c_1(0)$ are given by the initial and boundary conditions, i.e.

$$c_1 = c_1(\infty), \qquad x_1 > 0, \qquad t = 0, \qquad (13.46)$$

$$c_1 = c_1(0), \qquad x_1 = 0, \qquad t > 0. \qquad (13.47)$$

Similarly, the solution of (13.41) for corresponding conditions

$$c_2 = c_2(-\infty), \qquad x_2 < 0, \qquad t = 0, \qquad (13.48)$$

$$c_2 = c_2(0), \qquad x_2 = 0, \qquad t > 0, \qquad (13.49)$$

is

$$\frac{c_2(-\infty) - c_2}{c_2(-\infty) - c_2(0)} = 1 + \operatorname{erf}\frac{x_2}{2\sqrt{(D_2 t)}}. \qquad (13.50)$$

For the same conditions, the total amounts of diffusing substance, V_1 and

V_2, crossing the planes $x_1 = 0, x_2 = 0$ respectively in time t in the direction of decreasing x are

$$V_1 = 2\{c_1(\infty) - c_1(0)\}(D_1 t/\pi)^{\frac{1}{2}}, \tag{13.51}$$

$$V_2 = 2\{c_2(0) - c_2(-\infty)\}(D_2 t/\pi)^{\frac{1}{2}}. \tag{13.52}$$

The solutions (13.45) and (13.50) apply to an infinite medium. Each can be applied equally well to a region bounded by one or two x-planes either stationary or moving, provided that (i) the initial concentration at every point in this restricted region is the same as for the same value of x_1 or x_2 in the infinite medium; and (ii) the concentration at the boundary plane or planes is at all times the same as for the same value of x_1 or x_2 in the infinite medium. The problems to be discussed here concern media bounded by one or two planes, but they fulfil the above conditions and so the solutions for infinite media can be used. In other words we shall show that the solutions (13.45) and (13.50) are compatible with (i) eqns (13.42), (13.43), (13.44) and (ii) the conditions determined by the data of the problem which may be of one of two kinds described below as Class A and Class B. It should be noted that the values of c_1 outside medium 1 and of c_2 outside medium 2 have no physical significance.

13.4.2. Problems of Class A

Here the movement of one or both media relative to the boundary is caused by the transfer of diffusing substance across the interface. The conditions are that two of the quantities $c_1(\infty)$, $c_1(0)$, $c_1(X_1)$, $c_2(-\infty)$, $c_2(0)$ are specified and also that the magnitudes of X_1 and X_2 are at all times proportional to the amount of diffusing substance which has crossed the interface $(x_1 = X_1, x_2 = X_2)$. Hence we may write

$$\frac{dX_1}{dt} = S\left\{D_1\left(\frac{\partial c_1}{\partial x}\right)_{x_1 = X_1} + c_1(X_1)\frac{dX_1}{dt}\right\}, \tag{13.53}$$

where S is a constant of proportionality characteristic of the system and is the ratio of the magnitude of X_1 to the amount of diffusing substance which has crossed the interface in the direction of decreasing x. Combining this with (13.43) and (13.44) we have

$$\frac{dX_2}{dt} = PS\left\{D_2\left(\frac{\partial c_2}{\partial x_2}\right)_{x_2 = X_2} + c_2(X_2)\frac{dX_2}{dt}\right\}. \tag{13.54}$$

Substituting (13.42), (13.44), (13.45), and (13.50) in (13.53) and (13.54) we find

$$\frac{dX_1}{dt}\left\{\frac{1}{S} - c_1(X_1)\right\} = \{c_1(\infty) - c_1(0)\}\left(\frac{D_1}{\pi t}\right)^{\frac{1}{2}}\exp(-X_1^2/4D_1 t). \tag{13.55}$$

$$\frac{dX_1}{dt}\left\{\frac{1}{S}-PR-PQc_1(X_1)\right\} = \{c_2(0)-c_2(-\infty)\}\left(\frac{D_2}{\pi t}\right)^{\frac{1}{2}} \exp\left(-P^2X_1^2/4D_2t\right).$$

$$(13.56)$$

Also from (13.45), putting $x_1 = X_1, c_1 = c_1(X_1)$, we have

$$c_1(X_1) = c_1(0)+\{c_1(\infty)-c_1(0)\}\,\text{erf}\left\{\frac{X_1}{2\sqrt{(D_1t)}}\right\}. \qquad (13.57)$$

It is clear that (13.55), (13.56), and (13.57) can simultaneously be satisfied for all values of t, if and only if $X_1/t^{\frac{1}{2}}$ is constant. Put

$$X_1 = 2\alpha(D_1t)^{\frac{1}{2}}, \qquad\qquad (13.58)$$

so that (13.55), (13.56), and (13.57) become respectively

$$c_1(\infty)-c_1(0) = \pi^{\frac{1}{2}}\alpha\left\{\frac{1}{S}-c_1(X_1)\right\}\,\exp\,(\alpha^2), \qquad (13.59)$$

$$c_2(0)-c_2(\infty) = \pi^{\frac{1}{2}}\alpha\left(\frac{D_1}{D_2}\right)^{\frac{1}{2}}\left\{\frac{1}{S}-PR-PQc_1(X_1)\right\}\exp\left(\frac{\alpha^2P^2D_1}{D_2}\right). \qquad (13.60)$$

$$c_1(X_1) = c_1(0)+\{c_1(\infty)-c_1(0)\}\,\text{erf}\,\alpha, \qquad (13.61)$$

while substituting $x_2 = X_2$ in (13.50) and using (13.58) and (13.44) gives

$$\frac{Qc_1(X_1)+R-c_2(-\infty)}{c_2(0)-c_2(-\infty)} = 1+\text{erf}\left\{P\alpha\left(\frac{D_1}{D_2}\right)^{\frac{1}{2}}\right\}. \qquad (13.62)$$

Eqns (13.59) to (13.62) are independent and contain, besides physical constants and the parameters R and S, the six quantities $c_1(0)$, $c_1(\infty)$, $c_2(0)$, $c_2(-\infty)$, $c_1(X_1)$, and α. Hence if two of the concentrations are given, the four equations can be solved for the other three and α, and the concentrations c_1 and c_2 follow as functions of x and t from (13.45) and (13.50). The expressions so obtained for c_1 and c_2 satisfy both the initial and boundary conditions of the problem and the equation of diffusion, and therefore constitute the required solution. Substitution for α in (13.58) gives X_1, and hence X_2 from (13.44), in terms of t and known quantities. Finally we notice that the concentrations $c_1(X_1)$ and $c_2(X_2)$ at the interface are necessarily constant from (13.61) and (13.42).

13.4.3. Problems of Class B

Here the movements of the media on either side of the interface are not related to the amount of diffusing substance which has crossed the interface by eqns (13.53) and (13.54). Instead, three of the five concentrations $c_1(\infty)$, $c_1(0)$, $c_1(X_1)$, $c_2(0)$, $c_2(-\infty)$ are specified.

Substituting (13.42), (13.44), (13.45), and (13.50) into (13.43) gives

$$\{c_1(\infty) - c_1(0)\} \left(\frac{D_1}{\pi t}\right)^{\frac{1}{2}} \exp(X_1^2/4D_1 t)$$

$$+ \{c_2(0) - c_2(-\infty)\} \left(\frac{D_2}{\pi t}\right)^{\frac{1}{2}} \exp(-P^2 X_1^2/4D_2 t)$$

$$+ \frac{dX_1}{dt} \{(1 - PQ)c_1(X_1) - PR\} = 0. \tag{13.63}$$

Taken with (13.57) this can only be true for all values of t if $X_1/t^{\frac{1}{2}}$ is constant. On putting

$$X_1 = 2\beta t^{\frac{1}{2}}, \tag{13.64}$$

(13.63) becomes

$$\{c_1(\infty) - c_1(0)\} \left(\frac{D_1}{\pi}\right)^{\frac{1}{2}} \exp(-\beta^2/D_1)$$

$$+ \{c_2(0) - c_2(-\infty)\} \left(\frac{D_2}{\pi}\right)^{\frac{1}{2}} \exp(-P^2\beta^2/D_2)$$

$$+ \beta\{(\beta(1 - PQ)c_1(X_1) - PR\} = 0. \tag{13.65}$$

From (13.57)

$$c_1(X_1) = c_1(0) + \{c_1(\infty) - c_1(0)\} \operatorname{erf}(\beta/D_1^{\frac{1}{2}}), \tag{13.66}$$

while from (13.42) and (13.50)

$$c_2(X_2) = Qc_1(X_1) + R = c_2(0) + \{c_2(0) - c_2(-\infty)\} \operatorname{erf}(P\beta/D_2^{\frac{1}{2}}). \tag{13.67}$$

Eqns (13.65), (13.66), and (13.67) are independent and so, since three of the five concentrations $c_1(\infty)$, $c_1(X_1)$, $c_1(0)$, $c_2(0)$, $c_2(-\infty)$ are known, the values of the other two and of β can be determined. Hence X_1 can be found as a function of t from (13.64), and c_1 and c_2 as functions of x and t from (13.45) and (13.50).

13.4.4. Examples of Class A problems

(i) *Absorption by a liquid of a single component from a mixture of gases.* An ideal mixture of a soluble gas A and an insoluble gas is in contact with a liquid. Let the gas be medium 1 and the liquid medium 2. Initially the mole fraction of A in the gas is $c_1(\infty)$ and the concentration of dissolved A in the liquid, expressed as volume of gas per unit volume of liquid, is uniform and equal to $c_2(-\infty)$; $c_1(\infty)$ and $c_2(-\infty)$ are given. There is always equilibrium between the gas and the liquid at the interface, where Henry's law is assumed to be obeyed. It is further assumed that there is no appreciable change in the partial volume or temperature of the liquid when A is absorbed, and also that the diffusion of A in the liquid obeys the simple diffusion equation (13.41). If the origin of the x_1 system of coordinates moves so that there is no mass flow of the gaseous mixture across any plane of constant x_1, then eqn (13.40) is obeyed in the gas. For this to be so, X_1 must be equal at all times to the volume of component A absorbed by unit area of the liquid surface, i.e. $X_1 = V$. Thus the conditions of the problem correspond to those specified for Class A problems, with $S = 1$ in eqns (13.53) and (13.54).

The x_2 coordinate of the liquid surface does not change, and hence $X_2 = 0$, $c_2(X_2) = c_2(0)$ for all t; in equation (13.44), $P = 0$. Eqn (13.42) becomes

$$R = 0, \qquad c_2(X_2) = c_2(0) = Qc_1(X_1),$$

where Q is the solubility of A in the liquid expressed as volume of gaseous A in unit volume of liquid per unit mole fraction of A in the gas. Making the appropriate substitutions and eliminating the unknowns except α, eqns (13.59), (13.60), and (13.61) reduce to

$$\frac{\alpha(\pi D_1/D_2)^{\frac{1}{2}} - Qc_1(\infty) + c_2(-\infty)}{\alpha(\pi D_1/D_2)^{\frac{1}{2}} - Q + c_2(-\infty)} = \pi^{\frac{1}{2}}\alpha \exp(\alpha^2)\,\mathrm{erfc}\,\alpha = f(\alpha), \quad (13.68)$$

and putting $X_1 = V$ in (13.58) gives

$$V = 2\alpha(D_1 t)^{\frac{1}{2}}. \tag{13.69}$$

The function f has already been shown in Figs. 13.4 and 13.5 and, using these curves, α is readily evaluated from (13.68) by trial and error for known values of $D_1, D_2, Q, c_1(\infty)$, and $c_2(-\infty)$. The rate V at which A is absorbed by the liquid follows immediately from (13.69). It is also easy to show that

$$c_1(X_1) = \frac{c_1(\infty) - f}{1 - f} = \frac{c_2(-\infty) + \alpha(\pi D_1/D_2)^{\frac{1}{2}}}{Q}, \tag{13.70}$$

$$c_1 = \frac{c_1(\infty) + \{c_1(X_1) - c_1(\infty)\}}{\mathrm{erfc}\,\alpha}\,\mathrm{erfc}\,\frac{x_1}{2(D_1 t)^{\frac{1}{2}}}, \tag{13.71}$$

$$c_2 = Qc_1(X_1) + \{Qc_1(X_1) - c_2(-\infty)\}\,\mathrm{erf}\,\frac{x_2}{2(D_2 t)^{\frac{1}{2}}}. \tag{13.72}$$

The same equations may be used for the escape of dissolved gas A from solution. In this case α will be negative and the amount of A leaving the solution in time t is $-V$.

For further discussion of the physical conditions under which these solutions are applicable, Danckwerts's original paper (1950b) should be consulted.

A special case arises when $c_1(X_1)$, the concentration at the interface, is determined by some factor other than diffusion in a liquid. Examples quoted by Danckwerts are the isothermal evaporation of a liquid into still air, when

$c_1(X_1)$ is determined by vapour pressure, and the sorption of a gas by a liquid which is rapidly stirred without deforming the surface, so that $Qc_1(X_1) = c_2(-\infty)$. For such cases it is easy to show that (13.59) and (13.61) become

$$\frac{c_1(X_1)-c_1(\infty)}{c_1(X_1)-1} = f(\alpha), \tag{13.73}$$

where $f(\alpha)$ and α have the same significance as before.

(ii) *Tarnishing reactions.* A film of tarnish is formed on the surface of a metal by reaction with a gas. The reaction proceeds by diffusion of dissolved gas through the film to the surface of the metal, where its concentration $c_1(X_1)$ is assumed to be zero, i.e. the reaction is assumed to be so rapid that the rate of tarnishing is controlled entirely by the diffusion process. The outer surface of the film is constantly saturated with the gas.

Let the film be medium 1. Since $c_1(X_1)$ is not determined by diffusion in another medium, eqns (13.60) and (13.62) are not required. Let W be the mass fraction of the gaseous component in the compound which it forms with the metal, ρ the density of this compound (which is assumed independent of the concentration of the dissolved gas), and c_1 the concentration of the dissolved gas (expressed as mass of gas per unit volume of the film) at a distance x_1 beneath the surface of the film. The outer surface of the film is at $x_1 = 0$, the metal surface at $x_1 = X_1$, the film thickness being X_1. From (13.53)

$$S = -1(W\rho). \tag{13.74}$$

Furthermore, since $c_1(X_1) = 0$, $c_1(0)$, the saturated concentration of gas at the outer surface of the film, is given by

$$\frac{c_1(0)}{W\rho} = \pi^{\frac{1}{2}}\alpha \exp(\alpha^2)\operatorname{erf}\alpha = g(\alpha), \tag{13.75}$$

from (13.59) and (13.61), and

$$X_1 = 2\alpha(D_1 t)^{\frac{1}{2}}, \tag{13.76}$$

where D_1 is the diffusion coefficient of the dissolved gas in the film. Fig. 13.6 gives g as a function of α, and so values of α corresponding to given values of $c_1(0)/(W\rho)$ can be read directly from that graph and used to calculate film thickness as a function of time from (13.76). If $c_1(0) \ll W\rho$, expansion of $\exp(\alpha^2)$ and $\operatorname{erf}\alpha$ shows that $g = 2\alpha^2$ approximately and therefore in this case

$$X_1 = \left(\frac{2D_1 c_1(0)t}{W\rho}\right)^{\frac{1}{2}} \tag{13.77}$$

This means that, if the solubility of the gas in the film is sufficiently small, the film thickness is given without appreciable error by the formula obtained by assuming that the concentration gradient in the film is uniform. The problem of tarnishing was also solved by Booth (1948) who points out that the form (13.77) is a good approximation for systems usually encountered in practice.

13.4.5. *Examples of Class B problems*

(i) *Solution of gas in a liquid, followed by reaction with a solute.* The surface of the liquid is constantly saturated with the gas B, which is assumed to be undiluted with inert gases. The liquid contains a solute A which effectively reacts instantaneously and irreversibly with the dissolved gas. Diffusion of both A and B is assumed to obey the simple diffusion equation of the types (13.40) and (13.41). Any part of the liquid will contain A or B but not both. There will be a plane, $x_1 = X_1, x_2 = X_2$, at which the concentrations of both A and B will be zero. That part of the liquid containing A is taken to be medium 1, that containing B is medium 2. The surface of the liquid is permanently at $x_1 = x_2 = 0$, so that $X_1 = X_2$ and $P = 1$ from eqn (13.44). Also

$$c_1(X_1) = c_2(X_2) = 0, \tag{13.78}$$

and hence $R = 0$ in eqn (13.42). We assume n_1 moles of A to react with n_2 moles of B and so, in order to fulfil the conservation condition.(13.43), we put $c_1 = -m_1 n_2/n_1$, where m_1 is the concentration of A at x_1 in moles per unit volume and c_2 the concentration of B at x_2 in the same units. Further, $c_2(0)$ is the saturated concentration of B at the surface of the liquid, and $m_1(\infty)$ the initial concentration of A in the liquid. Eqns (13.65) and (13.67) then become

$$n_1 c_2(0) f(\beta/D_1^{\frac{1}{2}}) - n_2 m_1(\infty) g(\beta/D_2^{\frac{1}{2}}) = 0, \tag{13.79}$$

where f and g are the functions of $\beta/D_1^{\frac{1}{2}}$ and $\beta/D_2^{\frac{1}{2}}$ given respectively, as functions of α, in eqns (13.68) and (13.75) and plotted in Figs. 13.4, 13.5, and 13.6. Hence from (13.52) and (13.67) we find the volume of gaseous B absorbed in time t to be given by

$$V = \frac{2c_2(0)}{\operatorname{erf}(\beta/D_2^{\frac{1}{2}})} \left(\frac{D_2 t}{\pi}\right)^{\frac{1}{2}}. \tag{13.80}$$

The value of β can be found from (13.79) by trial and error, using Figs. 13.4, 13.5, and 13.6, and substituted in eqn (13.80) to obtain V. This problem is also treated by Hermans (1947).

If the gas is diluted by an insoluble component, the surface concentration $c_2(0)$ of B in the liquid is determined by diffusion in the gas. In this case, in order to obtain an exact solution, (13.69) and (13.73) must be used together with (13.79) and (13.80). A value of the concentration at the surface of the liquid must be found by trial and error such that the rate of absorption is the same when calculated from either set of equations. The quantities D_1, $c_1(\infty)$, and $c_1(X_1)$ in (13.69) and (13.73) refer respectively to the diffusion coefficient of B in the gas and to the mole fractions of B in the bulk of the gas and at the interface.

(ii) *Progressive freezing of a liquid.* The solution to this problem, given previously by Carslaw and Jaeger (1959), assumes that no change of volume occurs on freezing. The present equations take such a volume change into account. Suppose liquid initially occupies the region $x > 0$ and that freezing proceeds progressively, due to the removal of heat from the surface, $x = 0$, which is maintained at a constant temperature T_0. Subsequently, let the liquid be medium 1, and the solid, medium 2. Since no material crosses the surface maintained at T_0 this corresponds always to $x_2 = 0$. If there is a volume change on freezing there is relative movement of the planes $x_1 = 0$ and $x_2 = 0$. The diffuser in this case is heat. Take L to be the latent heat of fusion per unit mass, ρ_1, ρ_2 the densities, σ_1, σ_2 the specific heats, and D_1, D_2 the thermal diffusivities of liquid and solid respectively. The variation of these quantities with temperature is ignored. If K_1, K_2 are the thermal conductivities of the liquid and solid respectively, then

$$D_1 = K_1/(\rho_1\sigma_1); \qquad D_2 = K_2/(\rho_2\sigma_2). \tag{13.81}$$

The temperature of the liquid is initially T_∞ throughout and the solid–liquid interface $(x_1 = X_1, x_2 = X_2)$ is always at the melting-point T_X. The heat content of the solid at T_X is taken to be zero, and hence that of the liquid at T_X is L. In this problem, concentration signifies the heat content, and hence, since $c_2(X_2) = 0$, we have from (13.42)

$$c_1(X_1) = -R/Q = L/\rho_1. \tag{13.82}$$

Since $X_1/X_2 = \rho_2/\rho_1$, it follows from (13.44) that

$$P = \rho_1/\rho_2. \tag{13.83}$$

We also have

$$c_1 = \rho_1\sigma_1(T_1 - T_X + L/\sigma_1), \tag{13.84}$$

$$c_2 = \rho_2\sigma_2(T_2 - T_X), \tag{13.85}$$

where T_1 is the temperature at x_1 in the liquid and T_2 the temperature at x_2 in the solid.

Eliminating $c_1(0)$ and $c_2(-\infty)$ from (13.65) to (13.67) and substituting equations (13.81) to (13.85) we find

$$\frac{\sigma_1(T_\infty - T_X)}{f(\beta/D_1^{\frac{1}{2}})} + \frac{\sigma_2(T_0 - T_X)}{g\{\rho_1\beta/(\rho_2 D_2^{\frac{1}{2}})\}} + L = 0, \tag{13.86}$$

where f and g are the functions plotted with α as variable in Figs. 13.4, 13.5, and 13.6. The thickness of the solid formed up to time t is X_2, where

$$X_2 = X_1\rho_1/\rho_2 = 2\beta t^{\frac{1}{2}}\rho_1/\rho_2. \tag{13.87}$$

The temperature at any point in the solid is T_2, where

$$T_2 = T_X - (T_X - T_0)\left[1 - \frac{\text{erf}\,\{x_2/2(D_2 t)^{\frac{1}{2}}\}}{\text{erf}\,\{\rho_1 \beta/(\rho_2 D_2^{\frac{1}{2}})\}}\right], \tag{13.88}$$

while in the liquid it is T_1, where

$$T_1 = T_X - (T_X - T_\infty)\left[1 - \frac{\text{erfc}\,\{x_1/2(D_1 t)^{\frac{1}{2}}\}}{\text{erfc}\,\{\beta/D_1^{\frac{1}{2}}\}}\right]. \tag{13.89}$$

Hence if (13.86) is solved by trial and error using Figs. 13.4, 13.5, and 13.6, the resulting value of β can be used to calculate the thickness of the solid and the temperatures at any point as functions of time. The equations apply also, with suitable changes of nomenclature, to the melting of a solid which is at a uniform temperature.

It should be noted that a practical system only behaves in the way described by the above equations provided the density is uniform or increases steadily in a downward direction; otherwise convection currents arise.

13.5. Radially symmetric phase growth

We refer here to a particular problem with radial symmetry for which analytical solutions are easily found. A spherical or cylindrical new phase is growing from a negligible radius in an initially uniform medium and equilibrium conditions are maintained at the growing surface. Frank (1950) presents mathematical solutions earlier developed by Rieck (see Huber 1939). Other relevant papers by Ivantsov (1947), Zener (1949), and Horvay and Cahn (1961) include the growth of cylinders, paraboloids, and ellipsoids.

Following Frank we denote concentration (or temperature) by ϕ and write

$$\phi(r, t) = \phi(s) \tag{13.90}$$

where r is the radial coordinate and s is defined by

$$s = r/(Dt)^{\frac{1}{2}} \tag{13.91}$$

Then the usual diffusion equation for spherical symmetry becomes

$$\frac{\partial^2 \phi}{\partial s^2} = -\left(\frac{s}{2} + \frac{2}{s}\right)\frac{\partial \phi}{\partial s}. \tag{13.92}$$

Putting $\partial \phi/\partial s = p$ and integrating, we have

$$\partial \phi/\partial s = -As^{-2}\exp\left(-\tfrac{1}{4}s^2\right) \tag{13.93}$$

and hence

$$\phi = A\int_s^x z^{-2}\exp\left(-\tfrac{1}{4}z^2\right)\mathrm{d}z + \phi_r. \tag{13.94}$$

where ϕ_∞ is the value of ϕ at infinity and A is a constant. Integration by parts yields

$$\phi - \phi_\infty = A\{s^{-1} \exp\left(-\tfrac{1}{4}s^2\right) - \tfrac{1}{2}\pi^{\frac{1}{2}} \operatorname{erfc}\left(\tfrac{1}{2}s\right)\}$$

$$\equiv AF_3(s), \quad \text{say.} \tag{13.95}$$

Tables and graphs of $F_3(s)$ are given by Frank (1950).

The total flux of diffusant over a sphere of radius $R = S(Dt)^{\frac{1}{2}}$ is

$$-4\pi R^2 D(\partial\phi/\partial r)_{r=R} = 4\pi A D^{\frac{3}{2}} t^{\frac{1}{2}} \exp\left(-\frac{S^2}{4}\right) \tag{13.96}$$

using (13.95) and (13.91).

Frank takes q to be the amount of diffusant expelled per unit volume of new phase formed as growing proceeds and assumes the boundary condition.

$$4\pi A D^{\frac{3}{2}} t^{\frac{1}{2}} \exp\left(-\frac{S^2}{4}\right) = q\frac{d}{dt}(\tfrac{4}{3}\pi R^3)$$

$$= q2\pi S^3 D^{\frac{3}{2}} t^{\frac{1}{2}}, \tag{13.97}$$

so that

$$A = \tfrac{1}{2}qS^3 \exp\left(\frac{S^2}{4}\right). \tag{13.98}$$

Also, on the surface of the growing sphere, radius S, ϕ has the constant value

$$\phi_S = \tfrac{1}{2}qS^3 \exp\left(\frac{S^2}{4}\right) F_3(S) + \phi_\infty$$

$$= qf_3(S) + \phi_\infty, \quad \text{say.} \tag{13.99}$$

Frank (1950) tabulates the function $f_3(S)$, and shows graphs of S as a function of $(\phi_S - \phi_\infty)^3/q$ for the spherical, cylindrical, and linear cases.

For the two-dimensional, cylindrical growth Frank obtains the corresponding solutions

$$\partial\phi/\partial s = -As^{-1} \exp\left(-\tfrac{1}{4}s^2\right), \tag{13.100}$$

$$\phi - \phi_\infty = \tfrac{1}{2}A \int_{\frac{1}{4}s^2}^{\infty} z^{-1} e^{-z} \, dz$$

$$= -\tfrac{1}{2}A \operatorname{Ei}\left(-\tfrac{1}{4}s^2\right) \equiv AF_2(s), \tag{13.101}$$

where Ei is the exponential integral to be found in standard mathematical tables.

Frank tabulates $F_2(s)$. Using the same type of boundary condition as (13.97) we find

$$A = \tfrac{1}{2}qS^2 \exp\left(\tfrac{1}{4}S^2\right),$$

and so on the surface of the growing cylinder of radius S, ϕ has the constant value

$$\phi_S = -\tfrac{1}{4}qS^2 \exp(\tfrac{1}{4}S^2)\,\mathrm{Ei}(-\tfrac{1}{4}S^2) + \phi_\infty$$

$$= qf_2(S) + \phi_\infty. \tag{13.102}$$

We note that in the linear case

$$f_1(S) = \tfrac{1}{2}S \exp(\tfrac{1}{4}S^2)F_1(S) \tag{13.103}$$

and

$$F_1(s) = \pi^{\frac{1}{2}} \mathrm{erfc}(\tfrac{1}{2}s). \tag{13.104}$$

Readers should consult Frank's paper (1950) for related problems which include the cases in which both heat and moisture control phase growth, and where two dissolved substances diffuse independently to form the new phase.

Reiss and La Mer (1950) studied the growth of mono-dispersed aerosols and hydrosols. Their mathematical solutions assume that the flux of material through any surface in the diffusion field is much greater than the rate of change of concentration on that surface.

13.6. Approximate analytic solutions

Formal mathematical solutions have only been obtained for a limited number of problems, mostly in infinite or semi-infinite media. In other cases, for example when the surface concentration varies with time, or the medium is in the form of a cylinder or sphere, or the diffusion coefficient is concentration-dependent, some approximation must be made or a numerical procedure adopted.

13.6.1. *Steady-state approximation*

Stefan (1891) and Hill (1928) assumed that the concentration distribution behind the moving boundary approximated at any instant to the steady-state distribution which would be set up if the boundary were fixed in its position at that instant. We illustrate this approach with the following problem.

The surface of a semi-infinite sheet of uniform material is in contact with a well-stirred solution of extent l, and the solute diffuses into the sheet which is initially free of solute. The sheet contains S sites per unit volume, on each of which one diffusing molecule can be instantaneously and irreversibly immobilized. The concentration in the solution is always uniform and is initially C_B expressed as the number of molecules per unit volume of solution. The concentration at any time just within the surface of the sheet is taken to be that in the solution. We denote by $C(x, t)$ the concentration of freely diffusing molecules at any point x in the sheet at time t expressed in molecules per unit volume. The concentration of immobilized molecules is zero if C is zero and equal to S when C is non-zero. If we assume the diffusion coefficient D to be constant, the one-dimensional equation

$$\frac{\partial C}{\partial t} = D \frac{\partial^2 C}{\partial x^2}$$

(13. 05)

is to be solved subject to

$$C = 0, \qquad t = 0, \qquad x > 0, \tag{13.106}$$

$$l\, \partial C/\partial t = D\, \partial C/\partial x, \qquad x = 0, \qquad t \geqslant 0, \tag{13.107}$$

$$S\, dX/dt = -D\, \partial C/\partial x, \qquad x = X, \qquad t > 0. \tag{13.108}$$

Here $X(t)$ is the value of x at which $C = 0$, and denotes the position of the moving boundary. Condition (13.107) expresses the conservation of solute at the surface and (13.108) is derived in § 13.3 (p. 298).

Denote by $C_1(t)$ the value of C at the surface and throughout the well-stirred solution at time t. Thus $C_1(t) = C_B$ when $t = 0$ and decreases as diffusion proceeds. When the advancing boundary is at $x = X(t)$ the quasi-steady state is given by

$$C = C_1 \frac{X - x}{X}. \tag{13.109}$$

On differentiating (13.109) and substituting in (13.107) and (13.108) we find

$$\frac{l}{D} \frac{\partial C}{\partial t} = -\frac{C_1}{X} \tag{13.110}$$

and

$$\frac{S}{D} \frac{dX}{dt} = \frac{C_1}{X}. \tag{13.111}$$

Eliminating C_1 from (13.110) and (13.111), integrating the result and using the initial condition that

$$C_1 = C_B, \qquad X = 0, \tag{13.112}$$

we find

$$\frac{C_1}{C_B} = 1 - \frac{SX}{lC_B}. \tag{13.113}$$

Substitution for C_1 in (13.111), followed by integration, yields

$$\frac{Dt}{l^2} = -\frac{C_B}{S}\left\{\frac{SX}{lC_B} + \ln\left(1 - \frac{SX}{lC_B}\right)\right\}$$

$$= \frac{C_B}{S}\left\{\frac{C_1}{C_B} - \ln\left(\frac{C_1}{C_B}\right) - 1\right\}. \tag{13.114}$$

From (13.114) we find $X(t)$ and $C_1(t)$, and then $C(x, t)$ follows from (13.109). Crank (1957b) gave corresponding approximations for the cylinder and sphere. By comparison with accurate solutions obtained by the numerical procedures of § 13.7 he concluded that when S/C_B exceeds 10, the steady-state approximations are good enough for most practical purposes. Thus when $S/C_B = 10$, the approximated position of the moving boundary is in error by about 1 per cent and by about 3 per cent when $S/C_B = 5$.

Improved approximations starting from the steady state solution have been derived by Stefan (1891), Pekeris and Slichter (1939), and Kreith and Romie (1955).

13.6.2. *Goodman's integral method*

In an approximate method, Goodman (1958) also postulates a concentration profile behind the moving boundary, though not necessarily the steady-state one. It usually takes the form of a polynomial which is made to satisfy all the boundary conditions. It satisfies also an integrated form of the diffusion equation. The position of the moving boundary emerges as the solution of an ordinary differential equation with time as the independent variable. A review of integral methods and some applications are given by Goodman (1964).

We illustrate the procedure by applying it to the problem of the preceeding section, in the particular case of C_1 constant. We integrate (13.105) with respect to x from $x = 0$ to X and use (13.108) to give

$$\frac{\partial}{\partial t} \int_0^X C \, dx = -D\left\{\sigma \frac{dX}{dt} + \frac{\partial C}{\partial x}(0, t)\right\}, \tag{13.115}$$

where $\sigma = S/D$, and we have interchanged the order of integration and differentiation. The concentration profile is assumed to have the form

$$C = a(x - X) + b(x - X)^2 \tag{13.116}$$

where a, b are constants to be determined from the boundary conditions and $X = X(t)$. The condition

$$C = 0, \qquad x = X(t), \qquad t \geqslant 0, \tag{13.117}$$

is satisfied. We also need

$$C = C_1 = \text{constant}; \qquad x = 0, \qquad t \geqslant 0. \tag{13.118}$$

The condition (13.108) is not very convenient as it stands, and so we first differentiate (13.117) with respect to t and obtain

$$\frac{dC}{dt} = \frac{\partial C}{\partial x}\frac{dX}{dt} + \frac{\partial C}{\partial t} = 0, \qquad x = X(t). \tag{13.119}$$

Elimination of dX/dt between (13.108) and (13.119) with use of (13.105) gives

$$\left(\frac{\partial C}{\partial x}\right)^2 = \sigma \frac{\partial C}{\partial t} = \sigma D \frac{\partial^2 C}{\partial x^2}, \quad x = X(t). \tag{13.120}$$

By substituting (13.118) and (13.120) in the profile (13.116) we find, after some manipulation

$$a = \frac{\sigma D}{X}\{1-(1+\mu)^{\frac{1}{2}}\}, \qquad b = \frac{aX+C_1}{X^2}, \tag{13.121}$$

where $\mu = 2C_1/(\sigma D)$.

Finally, we use the profile (13.116), with a and b from (13.121), in (13.115) to obtain an ordinary differential equation for $X(t)$. Its solution which satisfies $X(0) = 0$ is

$$X = \alpha t^{\frac{1}{2}},$$

where

$$\frac{\alpha}{2D^{\frac{1}{2}}} = \sqrt{3}\left\{\frac{1-(1+\mu)^{\frac{1}{2}}+\mu}{5+\mu+(1+\mu)^{\frac{1}{2}}}\right\}^{\frac{1}{2}}. \tag{13.122}$$

In principle, the method can be applied to a variety of boundary conditions and geometric shapes. Goodman (1964) and Crank and Gupta (1972a) show how additional boundary conditions can be generated by successive differentiation in order to determine successive coefficients in higher-order polynomial profiles. Poots (1962b) has applied integral methods to cylinders and spheres.

Only isolated attempts have been made to extend the integral method to more than one space dimension. There are obvious difficulties in the choice of the profile of the moving boundary. Poots (1962a) obtained a solution of the problem of the solidification of a long bar of square cross-section which compared moderately well with the numerical solution of Allen and Severn (1962). Pleshanov (1962) used a quadratic polynomial for the approximate temperature profile in bodies symmetric in one, two or three dimensions. Budhia and Kreith (1973) studied melting in a wedge.

13.7. Finite-difference methods

Several numerical methods based on the finite-difference replacements of the diffusion equation described in Chapter 8 have been proposed. They differ in the way they treat the moving boundary and the grid on which numerical values are calculated. In general, the moving boundary will not coincide with a grid line in successive time steps δt if we take δt to be constant and predetermined. Douglas and Gallie (1955) chose each δt iteratively so that the boundary always moved from one grid line to the neighbouring one in an interval δt. This means that successive time intervals are of different durations.

Crank (1957a) proposed two methods: the first fixes the boundary by a change of the space variable; the second uses special finite-difference formulae based on Lagrangian interpolation for unequal intervals in the neighbourhood of the moving boundary, in order to track its progress between grid lines. Ehrlich (1958) employed Taylor's expansions in time and space near the boundary. Murray and Landis (1959) deformed the grid so that the number of space intervals between the moving boundary and an outer surface remained constant, with suitable transformation of the diffusion equation. Crank and Gupta (1972a, b) developed several procedures in which the grid is moved with the velocity of the boundary. A recent method (Chernous'ko 1969, 1970; Dix and Cizek 1970) of handling the diffusion equation, which offers a special advantage in Stefan problems, interchanges the concentration and space variables so that x becomes the dependent, and C and t the independent variables. If the boundary is known to occur at a fixed concentration its position comes naturally from the solution without special treatment.

Lazaridis (1970) has extended finite difference methods to two- and three-dimensional corners. An alternative approach is mentioned by Crank (1974).

13.7.1. *Fixing the boundary*

Crank (1957a) obtained numerical solutions for the problem posed in § 13.6.1 (p. 310) by using two transformations: (i) to remove the singularity at $x = 0, t = 0$ and (ii) to fix the advancing boundary. The conservation condition (13.107) was used in the integrated form

$$C = C_B - \frac{1}{l} \int_0^x (C+S)\,dx, \qquad x = 0, \qquad t \geqslant 0. \tag{13.123}$$

The singularity is handled by use of the new variables

$$c = C/C_B, \qquad s = S/C_B, \qquad \xi = x/(Dt)^{\frac{1}{2}}, \qquad \tau = (Dt/a^2)^{\frac{1}{2}}, \tag{13.124}$$

when the relevant equations are

$$2\frac{\partial^2 c}{\partial \xi^2} = \tau \frac{\partial c}{\partial \tau} - \xi \frac{\partial c}{\partial \xi}, \tag{13.125}$$

$$c = 0, \qquad \xi = \infty, \qquad \tau = 0, \tag{13.126}$$

$$-\frac{1}{2}\frac{d\xi_x}{d\tau} = \frac{1}{s\tau}\left(\frac{\partial c}{\partial \xi}\right)_x + \frac{\xi_x}{2\tau}, \qquad \tau > 0, \tag{13.127}$$

where $\xi_x = X/(Dt)^{\frac{1}{2}}$ is the position of the moving boundary in the transformed variables. Also

$$c = 0, \qquad \xi \geqslant \xi_x, \qquad \tau \geqslant 0, \tag{13.128}$$

while (13.123) becomes

$$c = 1 - \frac{a}{l}\int_0^{\xi_x} (c+s)\tau\,d\xi, \qquad \xi = 0, \qquad \tau \geqslant 0. \tag{13.129}$$

When $\tau = 0$, these equations have an analytical solution for $0 \leqslant \xi \leqslant \xi_X$,

$$c = 1 + B \operatorname{erf}(\tfrac{1}{2}\xi),$$

which satisfies $c = 1$, $\xi = 0$, $\tau = 0$. The constants B and ξ_X are determined from (13.127) and (13.128) with $\tau = 0$, since

$$1 + B \operatorname{erf}(\tfrac{1}{2}\xi_X) = 0 \tag{13.130}$$

and

$$\pi^{\frac{1}{2}}\xi_X \operatorname{erf}(\tfrac{1}{2}\xi_X) \exp(\tfrac{1}{2}\xi_X^2) = 2/s. \tag{13.131}$$

The function on the left-side of (13.131) is plotted in Fig. 13.6.

We now fix the moving boundary by writing

$$\eta = \xi/\xi_X, \qquad \tau = \tau. \tag{13.132}$$

Essentially the same transformation was used previously by Landau (1950) in a consideration of ablating slabs in which the melt is continuously and immediately removed from the surface. Citron (1960) extended the treatment to include temperature-dependent thermal properties. Sanders (1960) obtained an analytical solution for a particular specified motion of the free boundary.†

Using (13.132) in the present problem we obtain

$$2\frac{\partial^2 c}{\partial \eta^2} = \xi_X^2 \tau \frac{\partial c}{\partial \tau} + \frac{2}{s}\left(\frac{\partial c}{\partial \eta}\right)_1 \frac{\partial c}{\partial \eta}, \qquad 0 < \eta < 1, \tag{13.133}$$

$$-\frac{1}{2}\frac{d\xi_X}{d\tau} = \frac{1}{s\tau\xi_X}\left(\frac{\partial c}{\partial \eta}\right)_1 + \frac{\xi_X}{2\tau}, \qquad \eta = 1, \tag{13.134}$$

$$c = 1 - \frac{a}{l}\int_0^1 (c+s)\tau\xi_X \, d\eta, \qquad \eta = 0, \tag{13.135}$$

when we have used $(\partial c/\partial \eta)_1$ to mean $(\partial c/\partial \eta)_{\eta=1}$. The problem is now that of finding values of ξ_X and $(\partial c/\partial \eta)_1$ which are mutually consistent with (13.134) and for which the solution of (13.133) satisfies (13.135). Crank (1957a, b) used a convenient iterative finite-difference scheme to obtain numerical solutions for a plane sheet, cylinder and sphere in which the fraction $s = S/C_B$, i.e. the ratio of the number of sites per unit volume to the number of molecules in unit volume of solution, takes the values 1, 2, 5, 10. For each value of s, solutions were obtained for an unlimited amount of solute, and for two restricted amounts chosen such that in the final equilibrium state respectively 50 per cent and 90 per cent of the solute has entered the medium. For each s, $A/\pi a^2$ is calculated from the expression for the fraction of the total amount of solute

† Spalding and Gibson (1971) coupled this transformation with the idea of a 'penetration distance' (see § 7.3, p. 129) in an economized finite-difference procedure for solving concentration-dependent and other non-linear diffusion problems.

finally in the cylinder, for example, which is

$$\frac{1+s}{1+A/(\pi a^2)},$$

where a is the radius of the cylinder and A the cross-sectional area of the solution in which the cylinder is immersed. Table 13.1 show values of $(Dt/a^2)^{\frac{1}{2}}$ at which the total uptake is $\frac{1}{4}, \frac{1}{2}, \frac{3}{4}$, expressed as a fraction of its equilibrium value. Tables 13.2 and 13.3 show values of $(Dt/a^2)^{\frac{1}{2}}$ at which the advancing boundary is $\frac{1}{4}, \frac{1}{2}$, and $\frac{3}{4}$ of the way towards the centre. The time at which the centre itself is reached are shown only for the plane sheet as they are difficult to determine accurately for the cylinder and sphere.

Meadley (1971) considered the back diffusion of solute in a layer of solution when the free surface recedes due to evaporation of the solvent. He fixed the receding surface by a transformation essentially the same as (13.132).

13.7.2. *Diffusion on both sides of a moving boundary*

For the problem considered in § 13.7.1, the total concentration falls to zero at the advancing boundary, c.f. Fig. 13.7. A more general problem arises when diffusion occurs on both sides of the boundary, as in the example of § 13.2.2 (p. 290). We can still use the transformations of § 13.7.1 to fix the boundary, as long as the sheet is behaving semi-infinitely, i.e. the concentration at the centre of the sheet has not changed to within the accuracy of the calculations. After that, Crank (1957a) proposed either to use two transformations to fix both the moving boundary and the central plane of the sheet or to employ Lagrangian interpolation formulae to track the boundary. The second approach is described in the next section.

13.7.3. *Lagrange interpolation*

We return to the x, t plane and develop finite-difference approximations to derivatives based on functional values which are not necessarily equally spaced in the argument. They are generalized forms of the formulae developed in Chapter 8.

The Lagrangian interpolation formula is

$$f(x) = \sum_{j=0}^{n} l_j(x) f(a_j), \qquad (13.136)$$

where

$$l_j(x) = \frac{p_n(x)}{(x - a_j) p_n'(a_j)}, \qquad (13.137)$$

$$p_n(x) = (x - a_0)(x - a_1) \dots (x - a_{n-1})(x - a_n) \qquad (13.138)$$

and $p_n'(a_j)$ is its derivative with respect to x, at $x = a_j$. We restrict attention to three-point formulae, i.e. $n = 2$, and find that

$$\frac{1}{2}\frac{d^2 f(x)}{dx^2} = \frac{f(a_0)}{(a_0 - a_1)(a_0 - a_2)} + \frac{f(a_1)}{(a_1 - a_0)(a_1 - a_2)} + \frac{f(a_2)}{(a_2 - a_0)(a_2 - a_1)} \qquad (13.139)$$

and

$$\frac{df(x)}{dx} = l'_0(x)f(a_0) + l'_1(x)f(a_1) + l'_2(x)f(a_2), \qquad (13.140)$$

where

$$l'_0(x) = \frac{(x - a_1) + (x - a_2)}{(a_0 - a_1)(a_0 - a_2)}, \qquad l'_1(x) = \frac{(x - a_2) + (x - a_0)}{(a_1 - a_0)(a_1 - a_2)},$$

and

$$l'_2(x) = \frac{(x - a_0) + (x - a_1)}{(a_2 - a_0)(a_2 - a_1)}. \qquad (13.141)$$

We apply these formulae in the neighbourhood of the moving boundary. Taking the plane sheet as an example, we consider it to comprise M layers each of thickness δx, and let the boundary $x = X$ at time t be somewhere in the $(m+1)$th layer (Fig. 13.10). Thus $X = (m+p)\delta x$ where p is fractional and $0 < p < 1$. If for $x < X$ we identify

$$f(a_0) = c_{m-1}, \qquad f(a_1) = c_m, \qquad f(a_2) = c_X, \qquad (13.142)$$

then (13.139) and (13.140) become

$$\frac{\partial^2 c}{\partial x^2} = \frac{2}{(\delta x)^2}\left\{\frac{c_{m-1}}{p+1} - \frac{c_m}{p} + \frac{c_X}{p(p+1)}\right\}, \qquad x = m\delta x \qquad (13.143)$$

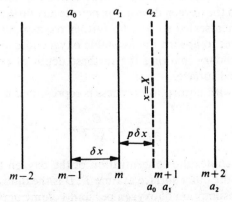

FIG. 13.10.

and

$$\frac{\partial c}{\partial x} = \frac{1}{\delta x} \left\{ \frac{pc_{m-1}}{p+1} - \frac{p+1}{p} c_m + \frac{2p+1}{p(p+1)} c_x \right\}, \qquad x = X. \qquad (13.144)$$

Similarly for $x > X$ we have

$$\frac{\partial^2 c}{\partial x^2} = \frac{2}{(\delta x)^2} \left\{ \frac{c_x}{(1-p)(2-p)} - \frac{c_{m+1}}{1-p} + \frac{c_{m+2}}{2-p} \right\}, \qquad x = (m+1)\delta x, \qquad (13.145)$$

and

$$\frac{\partial c}{\partial x} = \frac{1}{\delta x} \left\{ \frac{2p-3}{(1-p)(2-p)} c_x + \frac{2-p}{1-p} c_{m+1} - \frac{1-p}{2-p} c_{m+2} \right\}, \qquad x = X. \qquad (13.146)$$

We use these formulae for the space derivatives together with the usual explicit or implicit replacements of time derivatives (see Chapter 8) in the diffusion equation itself, and in the conditions on the moving boundary $x = X$. For points other than $m\delta x$, X and $(m+1)\delta x$, we use the usual finite-difference formula developed in Chapter 8 for equal intervals, including any condition on $x = 0$.

13.7.4. *Moving grid system*

Crank and Gupta (1972a) dealt with a moving-boundary problem arising from the diffusion of oxygen in absorbing tissue by using Lagrangian-type formulae and a Taylor series near the boundary. Subsequently (1972b) they developed an alternative method based on a moving grid.

In the physical problem, oxygen is allowed to diffuse into a medium, and some of the oxygen is absorbed, thereby being removed from the diffusion process. The oxygen concentration at the surface of the medium is maintained constant. This first phase of the problem continues until a steady state is reached, in which the oxygen does not penetrate any further into the medium. The surface is then sealed so that no further oxygen passes in or out. The medium continues to absorb the available oxygen already in it, and consequently the boundary marking the furthest depth of penetration recedes towards the sealed surface.

The diffusion-with-absorption process is represented by the equation

$$\frac{\partial C}{\partial T} = D \frac{\partial^2 C}{\partial X^2} - m, \qquad (13.147)$$

where $C(X, T)$ denotes the concentration of the oxygen free to diffuse at a distance X from the outer surface at time T; D is the diffusion constant; and m, the rate of consumption of oxygen per unit volume of the medium, is also assumed constant. The problem has two parts:

(i) *Steady-state.* During the initial phase, when oxygen is entering through the surface, the boundary condition there is

$$C = C_0, \qquad X = 0, \qquad T \geqslant 0, \tag{13.148}$$

where C_0 is constant. A steady-state is ultimately achieved, in which $\partial C / \partial T = 0$ everywhere, when both the concentration and its space derivative are zero at a point $X = X_0$. No oxygen diffuses beyond this point, and we have

$$C = \partial C / \partial X = 0, \qquad X \geqslant X_0. \tag{13.149}$$

The required solution in the steady state is easily found to be

$$C = \frac{m}{2D}(X - X_0)^2, \tag{13.150}$$

where $X_0 = \sqrt{(2DC_0/m)}$.

(ii) *Moving-boundary problem.* After the surface has been sealed, the point of zero concentration, originally at $X = X_0$, recedes towards $X = 0$. This second phase of the problem can be expressed in terms of the variables

$$x = \frac{X}{X_0}, \qquad t = \frac{DT}{X_0^2}, \qquad c = \frac{D}{mX_0^2} = \frac{C}{2C_0} \tag{13.151}$$

by the equations

$$\frac{\partial c}{\partial t} = \frac{\partial^2 c}{\partial x^2} - 1, \qquad 0 \leqslant x \leqslant x_B(t), \tag{13.152}$$

$$\partial c / \partial x = 0, \qquad x = 0, \qquad t \geqslant 0, \tag{13.153}$$

$$c = \partial c / \partial x = 0, \qquad x = x_B(t), \qquad t \geqslant 0, \tag{13.154}$$

$$c = \tfrac{1}{2}(1 - x)^2, \qquad 0 \leqslant x \leqslant 1, \qquad t = 0. \tag{13.155}$$

The surface is sealed at $t = 0$ and $x_B(t)$ denotes the position of the moving boundary in the reduced space variable.

An approximate analytical solution can be used to get away from the singularity at $x = 0, t = 0$ due to the instantaneous sealing of the surface. In the range $0 \leqslant t \leqslant 0 \cdot 020$, the expression

$$c(x, t) = \tfrac{1}{2}(1 - x)^2 - 2\sqrt{\left(\frac{t}{\pi}\right)} \exp\left\{-\left(\frac{x}{2\sqrt{t}}\right)^2\right\} + x \operatorname{erfc}\left(\frac{x}{2\sqrt{t}}\right), \qquad 0 \leqslant x \leqslant 1, \tag{13.156}$$

is sufficiently accurate for most purposes. During this time the boundary has not moved to within the accuracy of the calculations. Once the boundary begins to move, a grid system can be used, moving with the velocity of the moving boundary (Crank and Gupta 1972b). This has the effect of transferring the unequal space interval from the neighbourhood of the moving boundary to the sealed surface. An improved degree of smoothness in the calculated motion of the boundary was obtained compared with the results of using Lagrange interpolation.

We divide the region $0 \leqslant x \leqslant 1$ into n intervals each of width Δx such that $x_i = i\,\Delta x$; $i = 0, 1, \dots n$ and $n\,\Delta x = 1$. We denote by c_i^j the values of c at $(i\,\Delta x, j\,\Delta t)$, $j = 0, 1, 2, \dots$. We start at $t = 0$ with the steady-state solution (13.155). In the first interval Δt we evaluate c_{n-1}^1, and also the new position of the boundary, which has moved from $x = 1$ to $x = 1 - \varepsilon$, say, as in Fig. 13.11.

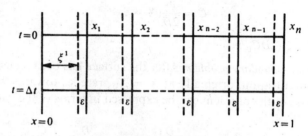

FIG. 13.11. Moving grid. (From Crank and Gupta 1972b).

The whole grid is now moved a distance ε to the left, as indicated by the broken lines. We wish to evaluate values of c^0 and the second space derivatives at each of the points $x_1 - \varepsilon, x_2 - \varepsilon, \dots, x_{n-1} - \varepsilon, 1 - \varepsilon$, at $t = 0$. Crank and Gupta (1972b) describe two methods of doing this. In the first method, interpolations are based on cubic splines, and a tridiagonal set of algebraic equations needs to be solved for each step in time. The second method, which we shall describe, is based on polynomial approximations and is simpler to apply. In return, we lose some of the continuity which the splines secure.

We represent $c(x)$ at any time between the two points x_i, x_{i+1} by

$$c_{i,i+1} = \alpha + \beta x + \gamma x^2 + \mu x^3, \tag{13.157}$$

where $\alpha = \alpha(i, i+1)$, etc. We employ the usual expressions

$$\frac{\partial^2 c_i}{\partial x^2} = \frac{c_{i-1} - 2c_i + c_{i+1}}{(\Delta x)^2}, \qquad i = 1, 2, \dots, n-1, \tag{13.158}$$

and at the surface $x = 0$,

$$\frac{\partial^2 c_0}{\partial x^2} = \frac{2}{\xi^2}(c_1 - c_0), \tag{13.159}$$

where $\xi = x_1 - x_0$. At $x = x_1$, the Lagrangian type formula is used to allow for the unequal interval ξ, namely

$$\frac{\partial^2 c_1}{\partial x^2} = 2\left\{\frac{c_0}{\xi(\xi+\Delta x)} - \frac{c_1}{\xi\,\Delta x} + \frac{c_2}{\Delta x(\xi+\Delta x)}\right\}. \tag{13.160}$$

From (13.157) we obtain

$$c''_{i,i+1} = 6\mu x + 2\gamma, \tag{13.161}$$

where $c'' = \partial^2 c/\partial x^2$. Assuming the function values to be known at any time $j\,\Delta t$, when the distance of the moving boundary from the surface $x = 0$ is $\xi^j + r\,\Delta x$, the method proceeds as follows. Obtain the second derivatives at each point of the grid at time $j\,\Delta t$, using the relations (13.158) and (13.159). By inserting $c_i, c_{i+1}, c''_i, c''_{i+1}$ into (13.157) and (13.161), we derive the coefficients α, β, γ, μ, and hence determine the interpolating polynomial (13.157) for each interval x_i to x_{i+1}.

The value of c_r^{j+1}, i.e. at the point neighbouring the moving boundary and time $(j+1)\,\Delta t$, follows from the simple explicit relationship:

$$(c_r^{j+1} - c_r^j)/\Delta t = (c'')_r^j - 1, \tag{13.162}$$

where $(c'')_r^j$ denotes the value of c'' at x_r at $j\,\Delta t$. The Taylor's series for c_r obtained by expanding about the moving point can be written as

$$c_r = c(x_0) - l\left(\frac{\partial c}{\partial x}\right)_{x=x_B} + \tfrac{1}{2}l^2\left(\frac{\partial^2 c}{\partial x^2}\right)_{x=x_B} - \tfrac{1}{6}l^3\left(\frac{\partial^3 c}{\partial x^3}\right)_{x=x_B} + \dots, \tag{13.163}$$

where $l(0 \leqslant l \leqslant \Delta x)$ is the distance of the moving boundary from x_r. We can derive the extra derivatives by differentiating (13.154) with respect to t, i.e.

$$\frac{dc}{dt} = \left(\frac{\partial c}{\partial x}\right)_{x=x_B}\frac{dx_B}{dt} + \left(\frac{\partial c}{\partial t}\right)_{x=x_B} = 0. \tag{13.164}$$

Using (13.152) and (13.154) in (13.164) we obtain

$$\partial^2 c/\partial x^2 = 1, \qquad x = x_0. \tag{13.165}$$

Similarly, by further differentiation we find

$$\partial^3 c/\partial x^3 = -dx_B/dt, \qquad x = x_B. \tag{13.166}$$

Provided the boundary is not moving too quickly we can use the Taylor series to obtain approximately

$$l = \sqrt{(2c_r)}. \tag{13.167}$$

Therefore, once c_r^{j+1} is known from (13.162) we find the position of the moving boundary from (13.167). Hence its movement ε^{j+1} in the interval from $j\,\Delta t$ to $(j+1)\,\Delta t$ is given by

$$\varepsilon_r^{j+1} = \Delta x - l^{j+1} \tag{13.168}$$

Having got ε we then interpolate the values of $c(x)$ and $c''(x)$ at $t = j\,\Delta t$ at the points $x_1 - \varepsilon,\, x_2 - \varepsilon, \ldots,\, x_n - \varepsilon$ using (13.157) and (13.161). The values of $c(x)$ at the new points $x_1,\, x_2 \ldots x_r$ at time $(j+1)\,\Delta t$ follow at once from

$$\frac{c^{j+1}(x_i^{j+1}) - c^j(x_i^{j+1})}{\Delta t} = c''(x_i^{j+1}) - 1, \qquad (13.169)$$

remembering that

$$x_i^{j+1} = x_i^j - \varepsilon^{j+1}, \qquad i = 1, 2 \ldots, r,$$

together with

$$\frac{c_0^{j+1} - c_0^j}{\Delta t} = c''(x_0) - 1, \qquad x = 0. \qquad (13.170)$$

We have seen that the space interval $x_1 - x_0$ varies from one time step to the next. As we proceed in steps Δt we eventually find that the points x_0 and x_1 come so close together that the values of c there are not significantly different, to the accuracy of calculating. We then replace ξ by $\xi + \Delta x$ to get values at the next time step and proceed as before.

Crank and Gupta showed that their solutions obtained by the different numerical procedures were in reasonable agreement. Furthermore, a slightly modified form of Goodman's integral method shows that quite good approximations are provided by the following analytic expressions:

$$c(0, t) = \tfrac{1}{2} - 2\sqrt{(t/\pi)}, \qquad (13.171)$$

$$x_B(t) = 1 - \exp\left\{-2\left(\frac{t_1 - t}{t - t_0}\right)^{\frac{1}{4}}\right\}, \qquad (13.172)$$

$$c(x, x_0) = \left(1 - \frac{x}{x_0}\right)^2 \left\{\tfrac{1}{2}x^2 + 4c_0\left(1 - \frac{x}{x_0}\right) - 3c_0\left(1 - \frac{x}{x_0}\right)^2\right\}. \qquad (13.173).$$

These solutions are applicable only for the time interval $t_0 \leqslant t \leqslant t_1$ where $t_0 = 4/25\pi$ and $t_1 = \pi/16$. For $t \leqslant t_0$, the approximate solution for small times (13.156) holds and from (13.171) t_1 is seen to be the total time for the concentration everywhere to become zero.

An alternative form of moving grid was used by Murray and Landis (1959). They kept the number of space intervals between $x = 0$ and $x = X$ constant and equal to n, say. Thus $\delta x = X(t)/n$ is different in each time step. They differentiated partially with respect to time t, following a given grid point instead of at constant x. We have for the point $i\delta x$

$$\left(\frac{\partial c}{\partial t}\right)_i = \left(\frac{\partial c}{\partial x}\right)_t \left(\frac{\mathrm{d}x}{\mathrm{d}t}\right)_i + \left(\frac{\partial c}{\partial t}\right)_x. \qquad (13.174)$$

Murray and Landis assumed a general grid point at x to move with velocity

dx/dt, where

$$\frac{1}{x}\frac{dx}{dt} = \frac{1}{X}\frac{dX}{dt}.$$ (13.175)

Then eqn (13.152) become

$$\left(\frac{\partial c}{\partial t}\right)_i = \frac{x_i}{X}\frac{dX}{dt}\frac{\partial c}{\partial x} + \frac{\partial^2 c}{\partial x^2} - 1.$$ (13.176)

Finite-difference solutions can be obtained when dX/dt is determined by an appropriate boundary condition. The idea implicit in (13.174) can be applied to any moving grid. If it moves bodily, as assumed by Crank and Gupta, all grid points have the same velocity, dX/dt, the relation (13.175) is not needed and the x_i/X factor in (13.176) becomes unity.

13.7.5. *Isotherm migration method*

The aim of all the methods described so far has been to determine how the concentration at a given point changes with time. An alternative method was proposed by Chernous'ko (1969, 1970) and independently by Dix and Cizek (1970). It has a particular advantage where phase changes are involved, and so we describe it in this chapter, though its use is not confined to moving boundary problems. In this method, the way in which a fixed temperature moves through the medium is calculated. Hence it is referred to as the 'isotherm migration method'. In the diffusion analogy we trace the motion of contours of constant concentration.

When a phase change or other boundary occurs at a prescribed temperature, its motion emerges naturally from the solution without any special treatment. Also the need to evaluate temperature or concentration-dependent parameters at each time step is avoided. The appropriate values are simply carried along with the isotherm. These advantages are partly offset by some increase in complexity of the finite-difference equations and difficulties with some boundary conditions.

Instead of expressing concentration C as a function of x and t, we now re-write the equations so that x becomes a function of C and t. The dependent variable becomes $x(C, t)$ instead of $C(x, t)$. We note that with reference to the diffusion equation in one dimension we can write

$$\left(\frac{\partial C}{\partial x}\right)_t = \left(\frac{\partial x}{\partial C}\right)_t^{-1}; \qquad \left(\frac{\partial x}{\partial t}\right)_c = -\left(\frac{\partial C}{\partial t}\right)_x\left(\frac{\partial x}{\partial C}\right)_t,$$

since for C held constant

$$dC = \left(\frac{\partial C}{\partial x}\right)_t dx + \left(\frac{\partial C}{\partial t}\right)_x dt = 0.$$

Starting from the usual diffusion equation

$$\frac{\partial C}{\partial t} = D \frac{\partial^2 C}{\partial x^2},$$ (13.177)

and using the above relations we find

$$\frac{\partial C}{\partial t} = -\left(\frac{\partial x}{\partial t}\right)\left(\frac{\partial x}{\partial C}\right)^{-1} = D\frac{\partial}{\partial x}\left(\frac{\partial x}{\partial C}\right)^{-1} = -D\left(\frac{\partial x}{\partial C}\right)^{-3}\frac{\partial^2 x}{\partial C^2},$$

i.e.

$$\frac{\partial x}{\partial t} = D\left(\frac{\partial x}{\partial C}\right)^{-2}\frac{\partial^2 x}{\partial C^2}.$$ (13.178)

If we consider, as an example, the simplest form of the problem in § 13.6.1. (p. 310) in which the condition (13.107) on $x = 0$ is replaced by

$$C = C_0 = \text{constant}, \qquad x = 0, \qquad t \geqslant 0,$$ (13.179)

then we can rewrite the condition (13.108) on the moving boundary

$$S\frac{dX}{dt} = -D\left(\frac{\partial x}{\partial C}\right)^{-1}, \qquad C = 0, \qquad t \geqslant 0,$$ (13.180)

remembering that we took $C = 0$ on $x = X(t)$. The condition (13.179) becomes

$$x = 0, \qquad C = C_0, \qquad t \geqslant 0.$$ (13.181)

In order to start the solution we need to use an analytic or approximate solution suitable for small times. This provides values of x for equally spaced values of C at a given small time t from which we can proceed in steps δt to obtain values of x on a grid in the C, t-plane by applying the usual finite-difference formulae to the transformed equations.

If we let x_i^n be the value of x at $C = i\delta C, t = n\delta t$, we find the diffusion equation (13.178) becomes approximately

$$x_i^{n+1} = x_i^n + 4\,\delta t\left\{\frac{x_{i+1}^n - 2x_i^n + x_{i-1}^n}{(x_{i-1}^n - x_{i+1}^n)^2}\right\},$$ (13.182)

and from (13.180) we have

$$X_0^{n+1} = X_0^n - \frac{D\,\delta t\,\delta C}{S(X_0^n - X_1^n)}$$ (13.183)

Crank and Phahle (1973) obtained results by this method, for the problem of melting a block of ice, that agreed well with the exact analytical solution and with solutions obtained by Goodman's integral method and Lagrangian interpolation.

13.8. Other methods

Methods based on applications of Green's functions lead to integro-differential equations for the position of the phase interface. Short descriptions of these and other methods including Boley's (1961) fictitious boundary conditions or 'imbedding' techniques, and the variational methods of Biot (1970) and Biot and Daughaday (1962) are given by Bankoff (1964) and Goodman (1964), together with references to other papers.

Fox (1974) has written an extensive survey in a book which also contains accounts of new research and of practical problems.

14

DIFFUSION AND CHEMICAL·REACTION

14.1. Introduction

THE problem discussed in this chapter is that of the absorption of one substance by another through which it can diffuse and with which it can also react chemically. This can be regarded as a problem in diffusion in which some of the diffusing substance becomes immobilized as diffusion proceeds, or as a problem in chemical kinetics in which the rate of reaction depends on the rate of supply of one of the reactants by diffusion. There are numerous practical examples of processes involving simultaneous diffusion and chemical reaction of one sort or another. Thus diffusion may take place within the pores of a solid body which can absorb some of the diffusing substance, or we may have diffusion occurring through a gel and an immobile product resulting from the attraction of the diffusing molecules to fixed sites within the medium. Examples involving diffusion into living cells and micro-organisms can be cited from biology and biochemistry. In contrast with the sharp boundary problems of Chapter 13, here we assume that some un-occupied reacting sites are always available. Chemical reactions in high polymer substances are often considerably dependent on the mobility of the reactants as well as on the kinetics of the reaction itself.

14.2. Instantaneous reaction

If the reaction by which the immobilized reactant is formed proceeds very rapidly compared with the diffusion process, local equilibrium can be assumed to exist between the free and immobilized components of the diffusing substance. In the simplest case, the concentration S of immobilized substance is directly proportional to the concentration C of substance free to diffuse, i.e.

$$S = RC. \tag{14.1}$$

In the particular case of diffusion with adsorption on to internal surfaces or sites (14.1) is referred to as a linear adsorption isotherm.

When diffusion is accompanied by absorption, the usual equation for diffusion in one dimension has to be modified to allow for this, and becomes

$$\frac{\partial C}{\partial t} = D\frac{\partial^2 C}{\partial x^2} - \frac{\partial S}{\partial t}, \tag{14.2}$$

if the diffusion coefficient D is constant. On substituting for S from (14.1) we have

$$\frac{\partial C}{\partial t} = \frac{D}{R+1} \frac{\partial^2 C}{\partial x^2}, \tag{14.3}$$

which is seen to be the usual form of equation for diffusion governed by a diffusion coefficient given by $D/(R+1)$. Clearly the effect of the instantaneous reaction is to slow down the diffusion process. Thus if $R+1 = 100$, the overall process of diffusion with reaction is slower than the simple diffusion process alone by a hundredfold. In fact, if the linear relationship (14.1) holds, solutions of the diffusion-with-reaction problem for given initial and boundary conditions are the same as for the corresponding problem in simple diffusion, except that the modified diffusion coefficient $D/(R+1)$ is to be used. This is true irrespective of whether the diffusion-with-reaction occurs in a plane sheet, cylinder, or sphere, or any other geometric shape, and whether diffusion occurs in one dimension or more.

14.2.1. Non-linear isotherm

If the relationship between S and C is not linear but is of the form

$$S = RC^n, \tag{14.4}$$

for example, where R and n are constants, then (14.2) still holds, but (14.3) becomes non-linear and solutions of the diffusion-with-reaction problem in this case can only be obtained by numerical methods of integration such as those described in Chapter 8. Some numerical solutions have been obtained (Crank 1948b, Crank and Godson 1947) describing the uptake of a restricted amount of solute by a cylinder, when diffusion within the cylinder is accompanied by adsorption and where the concentrations of free and adsorbed solute are related by an equation such as (14.4). In this example, 90 per cent of the total amount of solute is taken up by the cylinder in the final equilibrium state. Figs. 14.1 and 14.2 show the effect of the exponent n on the over-all rate of uptake of solute and on the way in which it is distributed through the cylinder at a given time.

If R is large so that $\partial C/\partial t$ may be neglected compared with $\partial S/\partial t$, (14.2) may be written

$$\frac{\partial S}{\partial t} = \frac{\partial}{\partial x}\left(D\frac{\partial C}{\partial x}\right) = \frac{\partial}{\partial x}\left\{\frac{D}{n}\left(\frac{1}{R}\right)^{1/n} S^{(1-n)/n} \frac{\partial S}{\partial x}\right\}, \tag{14.5}$$

on substituting for C from (14.4). Thus we see that diffusion accompanied by a non-linear reaction described by (14.4) is formally the same as diffusion governed by a diffusion coefficient which is not constant, but which depends on the total concentration of diffusing substance, free and immobile. This statement holds whatever the relationship between the concentration of

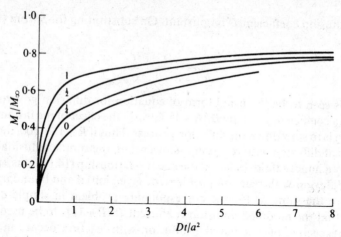

FIG. 14.1. Diffusion with adsorption into a cylinder of radius a. Numbers on curves are values of exponent n in (14.4).

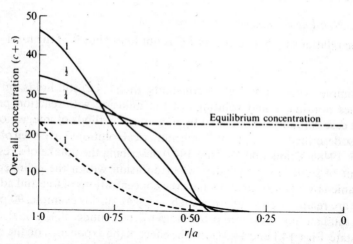

FIG. 14.2. Concentration distributions in a cylinder of radius a when $Dt/a^2 = 5$. Numbers on curves are values of exponent n in (14.4). ———— 90 per cent of solute enters cylinder. — — — — concentration at surface of cylinder is constant.

free and immobile components, provided it is non-linear. It need not necessarily be of the form of (14.4). If in (14.4) n is fractional, e.g. as in the well-known Freundlich type of adsorption isotherm, then the effective diffusion coefficient in (14.5), which is given by

$$(D/n)(1/R)^{1/n}S^{(1-n)/n},$$

increases as the concentration S is increased. If on the other hand $n > 1$, the effective diffusion coefficient decreases as the concentration is increased.

Standing, Warwicker, and Willis (1948) have considered an extension of this argument to the case in which the product of the reaction is not immobile but can itself diffuse at a rate different from that of the free component.

Vieth and Sladek (1965) considered an isotherm which is a combination of a linear component and a non-linear one of the Langmuir type. The linear part corresponds to gas dissolved in the amorphous regions of a glassy polymer; the non-linear, to gas trapped in small holes or microvoids in the polymer. The trapped gaseous molecules are immobile. The diffusion equation takes the form

$$D\frac{\partial^2 C_D}{\partial x^2} = \frac{\partial C_D}{\partial t}\left[1 + \frac{C_H'(b/k_D)}{\{1+(b/k_D)C_D\}^2}\right],$$

where C_D is the concentration of gas free to diffuse with diffusion coefficient D, C_H' is the saturation capacity of the microvoids, and b and k_D are parameters in the Langmuir and linear isotherms respectively. Vieth and Sladek obtained finite-difference solutions pertaining to diffusion into a plane sheet from a limited volume of gas. They found that their mathematical model described experimental data satisfactorily, and offered a way of estimating diffusion rates from sorption data for polymers and also for porous catalysts.

14.3. Irreversible reaction

If the diffusing substance is immobilized by an irreversible first-order reaction so that the rate of removal of diffusing substance is kC, where k is a constant, then the equation for diffusion in one dimension becomes

$$\frac{\partial C}{\partial t} = D\frac{\partial^2 C}{\partial x^2} - kC, \tag{14.6}$$

provided the diffusion coefficient D is assumed to be constant. This is also the equation representing the conduction of heat along a wire which loses heat from its surface at a rate proportional to its temperature. Some solutions relating to this latter problem are given by Carslaw and Jaeger (1959, p. 134), where use is made of the transformation $C = C' \exp(-kt)$ which reduces (14.6) to the usual equation in one dimension with C' as dependent variable. Odian and Kruse (1969) used solutions of (14.6) and its steady-state equivalent in discussing the effects of diffusion on radiation-induced graft polymerization.

Danckwerts (1951) has shown how solutions of (14.6) and of the general equation

$$\frac{\partial C}{\partial t} = D\nabla^2 C - kC \tag{14.7}$$

can be deduced by simple transformation of the solutions of the corresponding problems in diffusion without reaction. He applies his method to two types of surface boundary condition. In the first case the surface is in equilibrium at all times with the surrounding atmosphere or solution, and the surface concentration has the constant value C_0. In the second case the rate of evaporation or absorption at any time is given by

$$D\, \partial C/\partial N = \alpha(C_0 - C_s), \tag{14.8}$$

where C_s is the actual concentration on the surface at that time, C_0 the equilibrium surface concentration attained after infinite time, and $\partial C/\partial N$ the concentration gradient measured in the outward direction along the normal to the surface. Here α is a constant. For convenience in writing we put $h = \alpha/D$ in what follows.

In the problems to be considered, therefore, the initial and boundary conditions are

$$C = 0, \quad t = 0, \qquad \text{at all points in the medium}, \tag{14.9}$$

and either

$$C = C_0, \quad t > 0, \qquad \text{at all points on the surface}, \tag{14.10}$$

or

$$\partial C/\partial N = h(C_0 - C), \quad t > 0, \qquad \text{at all points on the surface.} \tag{14.11}$$

14.3.1. *Danckwert's method*

Let C_1 be the solution of the equation for diffusion in one dimension in the absence of reaction, i.e. of

$$\frac{\partial C_1}{\partial t} = D\frac{\partial^2 C_1}{\partial x^2}, \tag{14.12}$$

for the same boundary conditions as are imposed on C in (14.9), (14.10), and (14.11). We shall give the argument for diffusion in one dimension and it will be clear, as Danckwerts shows, that it can be applied equally well to the general equation (14.7). If C_1 is a solution of (14.12), the solution of (14.6) for the same boundary conditions of the above type is

$$C = k\int_0^t C_1\, e^{-kt'}\, dt' + C_1\, e^{-kt}. \tag{14.13}$$

This is easily shown as follows. By differentiating (14.13) with respect to t we obtain

$$\frac{\partial C}{\partial t} = kC_1\, e^{-kt} - kC_1\, e^{-kt} + \frac{\partial C_1}{\partial t}e^{-kt} = \frac{\partial C_1}{\partial t}e^{-kt}, \tag{14.14}$$

and on differentiating twice with respect to x we have

$$\frac{\partial^2 C}{\partial x^2} = k\int_0^t \frac{\partial^2 C_1}{\partial x^2}e^{-kt'}\, dt' + \frac{\partial^2 C_1}{\partial x^2}e^{-kt} \tag{14.15}$$

Substituting from (14.12) in (14.15) gives

$$D\frac{\partial^2 C}{\partial x^2} = k \int_0^t \frac{\partial C_1}{\partial t} e^{-kt'} dt' + \frac{\partial C_1}{\partial t} e^{-kt} \tag{14.16}$$

and finally using (14.14) we find

$$D\frac{\partial^2 C}{\partial x^2} = kC + \frac{\partial C}{\partial t}, \tag{14.17}$$

so that (14.13) is a solution of (14.6).

Furthermore, when $t = 0$, we see from (14.13) that $C = C_1$ and so C obeys the required initial conditions. For points at which $C_1 = C_0$, for all t,

$$C = kC_0 \int_0^t e^{-kt'} dt' + C_0 e^{-kt} = C_0. \tag{14.18}$$

For points at which

$$\partial C_1 / \partial N = h(C_0 - C_1), \tag{14.19}$$

we find

$$\frac{\partial C}{\partial N} = k \int_0^t \frac{\partial C_1}{\partial N} e^{-kt'} dt' + \frac{\partial C_1}{\partial N} e^{-kt}$$

$$= hk \int_0^t (C_0 - C_1) e^{-kt'} dt' + h(C_0 - C_1) e^{-kt}$$

$$= h(C_0 - C). \tag{14.20}$$

The solution (14.13) therefore satisfies the required initial and boundary conditions if C_1 does.

If the quantity, M_t, of diffusing substance absorbed in time t is required, we use in general

$$F = \frac{\partial M_t}{\partial t} = \iint D\frac{\partial C}{\partial N} dS = k \int_0^t F_1 e^{-kt'} dt' + F_1 e^{-kt}. \tag{14.21}$$

The area integral is taken over the whole surface of the absorbing medium and F_1 is the corresponding rate of absorption when no chemical reaction takes place, i.e.

$$F_1 = \frac{dM_1}{dt} = \iint D\frac{\partial C_1}{\partial N} dS = \frac{d}{dt} \iiint C_1 \, dx \, dy \, dz. \tag{14.22}$$

The volume integral is taken over the whole volume of the absorbing medium.

Many of the solutions for diffusion without reaction are available in the form of infinite series which can be written

$$C_1/C_0 = 1 - \sum f(x, y, z) e^{-vt}. \tag{14.23}$$

Numerous examples have been given in Chapters 3–6. Here f and v are different for each term in the series, but are not functions of t; v is not a function of x, y, z. Applying (14.13) to (14.23) we find

$$\frac{C}{C_0} = 1 - \sum f(x, y, z) \left[\frac{k + v \exp\left[-t(k+v)\right]}{k+v} \right] \qquad (14.24)$$

Similarly, since in such cases F_1 is of the form

$$F_1/C_0 = \sum g\, e^{-vt}, \qquad (14.25)$$

where g is different for each term in the series but is not a function of x, y, z, t, we obtain from (14.21)

$$\frac{F}{C_0} = \sum g \left[\frac{k + v \exp\left\{-t(k+v)\right\}}{k+v} \right], \qquad (14.26)$$

and

$$M_t = \int_0^t F\, \mathrm{d}t' = C_0 \sum g \left[\frac{kt(k+v) - v \exp\left\{-t(k+v)\right\} + v}{(k+v)^2} \right] \qquad (14.27)$$

14.3.2. *Examples*

The method can be applied generally for boundary conditions of the type (14.10) or (14.11). It will be illustrated by the examples quoted by Danckwerts.

(i) *Sphere with surface evaporation condition.* We take the sphere to be of radius a. Then the solution for diffusion without reaction is (Chapter 6, eqn (6.40))

$$\frac{C_1}{C_0} = 1 - \frac{2ha^2}{r} \sum_{n=1}^{\infty} \frac{\exp\left(-D\alpha_n^2 t\right)}{\{a^2\alpha_n^2 + ah(ah-1)\}} \frac{\sin r\alpha_n}{\sin a\alpha_n}, \qquad (14.28)$$

where the α_ns are the roots of

$$a\alpha \cot a\alpha + ah - 1 = 0, \qquad (14.29)$$

which are given in Table 6.2. Hence

$$F_1 = 4\pi a^2 D \left(\frac{\partial C_1}{\partial r}\right)_{r=a} = 8\pi h^2 C_0 Da^2 \sum_{n=1}^{\infty} \frac{\exp\left(-D\alpha_n^2 t\right)}{a\alpha_n^2 + h(ah-1)}, \qquad (14.30)$$

and from (14.26)

$$F = 8\pi h^2 C_0 Da^2 \sum_{n=1}^{\infty} \frac{k + D\alpha_n^2 \exp\left\{-t(k+D\alpha_n^2)\right\}}{(k+D\alpha_n^2)\{a\alpha_n^2 + h(ah-1)\}}, \qquad (14.31)$$

and from (14.27)

$$M_t = 8\pi h^2 C_0 Da^2 \sum_{n=1}^{\infty} \frac{kt(k+D\alpha_n^2) - D\alpha_n^2[\exp\left\{-t(k+D\alpha_n^2)\right\} - 1]}{(k+D\alpha_n^2)^2\{a\alpha_n^2 + h(ah-1)\}}. \qquad (14.32)$$

(ii) *Rectangular block or parallelepiped with constant surface concentration.*
We take the block to occupy the space $-a < x < a$, $-b < y < b$, $-c < z < ç$, so that the edges are of lengths $2a$, $2b$, $2c$. The solution for diffusion without reaction is

$$\frac{C_1}{C_0} = 1 - \frac{64}{\pi^3} \sum_{l=0}^{\infty} \sum_{m=0}^{\infty} \sum_{n=0}^{\infty} \frac{(-1)^{l+m+n}}{(2l+1)(2m+1)(2n+1)} \cos \frac{(2l+1)\pi x}{2a}$$

$$\times \cos \frac{(2m+1)\pi y}{2b} \cos \frac{(2n+1)\pi z}{2c} \exp(-t\alpha_{l,m,n}) \tag{14.33}$$

where

$$\alpha_{l,m,n} = \frac{\pi^2 D}{4} \left\{ \left(\frac{2l+1}{a} \right)^2 + \left(\frac{2m+1}{b} \right)^2 + \left(\frac{2n+1}{c} \right)^2 \right\}. \tag{14.34}$$

Hence

$$F_1 = \frac{d}{dt} \int_{-a}^{a} \int_{-b}^{b} \int_{-c}^{c} C_1 \, dx \, dy \, dz$$

$$= \left(\frac{64}{\pi^3} \right)^2 C_0 abc \sum_{l=0}^{\infty} \sum_{m=0}^{\infty} \sum_{n=0}^{\infty} \frac{\alpha \, e^{-t\alpha}}{(2l+1)^2(2m+1)^2(2n+1)^2}, \tag{14.35}$$

and from (14.26)

$$F = \left(\frac{64}{\pi^3} \right)^2 C_0 abc \sum_{l=0}^{\infty} \sum_{m=0}^{\infty} \sum_{n=0}^{\infty} \frac{\alpha[k + \alpha \exp\{-t(k+\alpha)\}]}{(k+\alpha)(2l+1)^2(2m+1)^2(2n+1)^2}, \tag{14.36}$$

where by α is understood $\alpha_{l,m,n}$ given by (14.34). Similarly M_t follows readily from (14.27).

(ii) *Infinite cylinder with surface evaporation condition.* We take the cylinder to have radius a and consider diffusion to be entirely radial. The expression for diffusion without reaction is (Chapter 5, eqn (5.46))

$$\frac{C_1}{C_0} = 1 - \frac{2h}{a} \sum_{n=1}^{\infty} \frac{J_0(r\alpha_n) \exp(-D\alpha_n^2 t)}{(h^2 + \alpha_n^2) J_0(a\alpha_n)}, \tag{14.37}$$

where the α_ns are the roots of

$$\alpha J_1(a\alpha) = h J_0(a\alpha), \tag{14.38}$$

and J_0, J_1 are Bessel functions, of the first kind, of order zero and one respectively. Values of α_n are to be found in Table 5.2. Hence

$$F_1 = \frac{d}{dt} \int_0^a 2\pi r C_1 \, dr = 4\pi h^2 C_0 D \sum_{n=1}^{\infty} \frac{\exp(-D\alpha_n^2 t)}{h^2 + \alpha_n^2}, \tag{14.39}$$

and from (14.26)

$$F = 4\pi h^2 C_0 D \sum_{n=1}^{\infty} \frac{k + D\alpha_n^2 \exp\{-t(k + D\alpha_n^2)\}}{(h^2 + \alpha_n^2)(k + D\alpha_n^2)}. \tag{14.40}$$

Also M_t is easily found from (14.27). Here M_t and F refer to unit length of cylinder.

Other results may be derived in this way or directly from (14.13) or (14.21). Some examples of expressions for the rate of uptake F are given below:

(iv) *Sphere (radius a) with constant surface concentration*

$$F = 8\pi a D C_0 \sum_{n=1}^{\infty} \frac{ka^2 + Dn^2\pi^2 \exp\{-t(k + Dn^2\pi^2/a^2)\}}{ka^2 + Dn^2\pi^2}. \qquad (14.41)$$

(v) *Infinite cylinder (radius a) with constant surface concentration.* Here F is the rate of uptake through unit length of the curved surface of the cylinder:

$$F = 4\pi D C_0 \sum_{n=1}^{\infty} \frac{k + D\alpha_n^2 \exp\{-t(k + D\alpha_n^2)\}}{k + D\alpha_n^2}, \qquad (14.42)$$

where the α_ns are the roots of $J_0(a\alpha) = 0$.

(vi) *Finite cylinder (radius a, length 2l) with constant surface concentration.* Here F is the rate of total uptake through the ends and the curved surface:

$$F = \frac{64DlC_0}{\pi} \sum_{n=0}^{\infty} \sum_{m=1}^{\infty} \left\{ \frac{\beta_{m,n}}{\alpha_m^2(2n+1)^2} \right\} \left[\frac{k + D\beta_{m,n} \exp\{-t(k + D\beta_{m,n})\}}{k + D\beta_{m,n}} \right], \qquad (14.43)$$

where the α_ns are the roots of $J_0(a\alpha) = 0$, and

$$\beta_{m,n} = \alpha_m^2 + (2n+1)^2\pi^2/(4l^2). \qquad (14.44)$$

(vii) *Semi-infinite solid with surface evaporation condition.* The solution is obtained by application of (14.21) to the corresponding solution without reaction (Chapter 3, eqn (3.35)). Here F is the rate of uptake per unit area of plane surface and

$$F = \frac{hDC_0}{h^2D - k}[h(Dk)^{\frac{1}{2}} \operatorname{erf}(kt)^{\frac{1}{2}} + h^2D \operatorname{erfc} h(Dt)^{\frac{1}{2}} \exp\{t(h^2D - k)\} - k]. \qquad (14.45)$$

(viii) *Semi-infinite medium with constant surface concentration.* The mathematical problem and its solution are identical with those for the conduction of heat along a thin rod which loses heat from its surface at a rate proportional to its temperature (Danckwerts 1950a; Carslaw and Jaeger 1959). The expression for the concentration is

$$\frac{C}{C_0} = \tfrac{1}{2} \exp\{-x\sqrt{(k/D)}\} \operatorname{erfc}\left\{ \frac{x}{2\sqrt{(Dt)}} - \sqrt{(kt)} \right\}$$

$$+ \tfrac{1}{2} \exp\{x\sqrt{(k/D)}\} \operatorname{erfc}\left\{ \frac{x}{2\sqrt{(Dt)}} + \sqrt{(kt)} \right\} \qquad (14.46)$$

Also

$$F = C_0\sqrt{(Dk)} \left\{ \operatorname{erf}\sqrt{(kt)} + \frac{e^{-kt}}{\sqrt{(\pi kt)}} \right\}, \tag{14.47}$$

and

$$M_t = C_0\sqrt{(D/k)}\{(kt + \tfrac{1}{2})\operatorname{erf}\sqrt{(kt)} + \sqrt{(kt/\pi)}\,e^{-kt}\}. \tag{14.48}$$

The dimensionless quantity $M_t k^{\frac{1}{2}}/(C_0 D^{\frac{1}{2}})$ is plotted as a function of kt in Fig. 14.3. From this graph, M_t can be obtained at any time t for any combination of the variables C_0, D, and k.

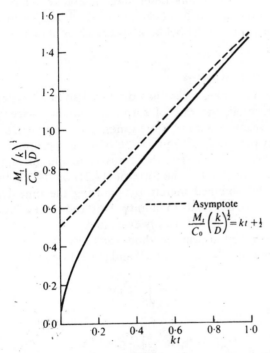

FIG. 14.3. Diffusion with an irreversible reaction into a semi-infinite medium with constant surface concentration.

When kt is large so that $\operatorname{erf}\sqrt{(kt)}$ approaches unity, (14.46), (14.47), and (14.48) become

$$C/C_0 = \exp(-xk^{\frac{1}{2}}/D^{\frac{1}{2}}), \tag{14.49}$$

$$F = C_0(Dk)^{\frac{1}{2}}, \tag{14.50}$$

$$M_t = C_0(Dk)^{\frac{1}{2}}\left(t + \frac{1}{2k}\right). \tag{14.51}$$

Eqns (14.49) and (14.50) show that the concentration at any point, and the rate of sorption each tend to a steady value and the total amount taken up increases linearly with time as in Fig. 14.3.

When kt is very small we find by expanding erf $\sqrt{(kt)}$ and $\exp(-kt)$ and neglecting powers of kt higher than the first, that (14.47) and (14.48) become

$$F = C_0(1+kt)\{D/(\pi t)\}^{\frac{1}{2}} \tag{14.52}$$

and

$$M_t = 2C_0(1+\tfrac{1}{2}kt)(Dt/\pi)^{\frac{1}{2}}, \tag{14.53}$$

which reduce to the well-known solutions for diffusion without reaction when $k = 0$. Further solutions describing the uptake of a restricted amount of diffusing substance which is simultaneously destroyed by an irreversible reaction are given in § 14.4.4 below as special cases of more general solutions for a reversible reaction.

14.3.3. Steady-state solutions

As with the semi-infinite case just discussed so in all systems of the type under consideration, the rate of sorption and the concentration at any point tend to steady values at large times. The steady-state solutions may be obtained by putting $\partial C/\partial t$ equal to zero in the appropriate form of (14.7), which may then be solved to give the steady-state solution directly. When the expression for C or F takes the form of (14.24) or (14.26), the steady-state solution may be obtained merely by omitting the time-dependent term, which tends to zero as t tends to infinity. This leads to a solution in the form of a series which is often not readily evaluated.

In some cases the following method leads to a more convenient form of solution. On putting $t = \infty$ in (14.13) and (14.21) we find

$$C_\infty/k = \int_0^\infty C_1 e^{-kt}\, dt, \tag{14.54}$$

and

$$F_\infty/k = \int_0^\infty F_1 e^{-kt}\, dt. \tag{14.55}$$

The expressions on the right-hand sides of these equations are the Laplace transforms of C_1 and F_1 respectively, the parameter (usually written as p or s in transform notation) taking the value k. Thus, as an example, for a semi-infinite solid with surface evaporation we have, in the absence of reaction,

$$F_1 = hDC_0 \exp(h^2 Dt)\, \text{erfc}\, \{h(Dt)^{\frac{1}{2}}\}. \tag{14.56}$$

The Laplace transform of the right-hand side of (14.56) is

$$hDC_0/\{k+h\sqrt{(Dk)}\},$$

and hence from (14.55)

$$F_\infty = \frac{hDC_0}{1 + h\sqrt{(D/k)}}. \tag{14.57}$$

Some other steady-state solutions are given below, preceded by references to the corresponding expressions describing the approach to the steady state.

(i) *Sphere with surface evaporation condition* (eqn (14.31)).

$$\frac{F_\infty}{C_0} = 4\pi a^2 hD \left\{ \frac{(ka^2/D)^{\frac{1}{2}} \coth (ka^2/D)^{\frac{1}{2}} - 1}{ah + (ka^2/D)^{\frac{1}{2}} \coth (ka^2/D)^{\frac{1}{2}} - 1} \right\}. \tag{14.58}$$

(ii) *Infinite cylinder with surface evaporation condition* (eqn (14.40)).

$$\frac{F_\infty}{C_0} = 2\pi ahD \left\{ \frac{(k/D)^{\frac{1}{2}} I_1(ka^2/D)^{\frac{1}{2}}}{(k/D)^{\frac{1}{2}} I_1(ka^2/D)^{\frac{1}{2}} + hI_0(ka^2/D)^{\frac{1}{2}}} \right\}. \tag{14.59}$$

Here I_0 and I_1 are modified Bessel functions of the first kind of zero and first order respectively.

(iii) *Sphere with constant surface concentration* (eqn (14.41)).

$$\frac{F_\infty}{C_0} = 4\pi aD \{(ka^2/D)^{\frac{1}{2}} \coth (ka^2/D)^{\frac{1}{2}} - 1\}. \tag{14.60}$$

(iv) *Infinite cylinder with constant surface concentration* (eqn (14.42)).

$$\frac{F_\infty}{C_0} = 2\pi a(kD)^{\frac{1}{2}} \frac{I_1(ka^2/D)^{\frac{1}{2}}}{I_0(ka^2/D)^{\frac{1}{2}}}. \tag{14.61}$$

A further application of the use of eqn (14.13) is to the problem of the extraction of a dissolved substance from a drop of liquid which is rising or falling through another liquid. It is described in Danckwert's paper (1951).

14.4. Reversible reaction

The most general case for which formal mathematical solutions have so far been obtained is that in which the reaction is first-order and reversible. The behaviour to be expected when the reaction is reversible depends on the relative rates of diffusion and reaction. Thus we have seen in § 14.2 (p. 326) that when the reaction is very rapid we can assume that the immobilized component is always in equilibrium with the component free to diffuse, and diffusion is the rate-controlling process. At the other extreme is the case of diffusion being so rapid compared with the reaction that the concentrations of diffusing substance and immobilized product are effectively uniform throughout the medium and the behaviour is controlled solely by the reversible reaction. Solutions of the more general case in which the rates of diffusion and reaction are comparable have been obtained by Wilson (1948) and Crank (1948c).

The general problem can conveniently be stated in terms of a solute diffusing from a solution into a plane sheet of material. The modifications necessary for corresponding, alternative problems, such as those of a sphere or cylinder suspended in a vapour, will be obvious. Suppose an infinite sheet of uniform material of thickness $2a$ is placed in a solution and that the solute is allowed to diffuse into the sheet. As diffusion proceeds, a first-order, reversible reaction occurs and a product, which is non-diffusing, is formed. The sheet occupies the space $-a \leqslant x \leqslant a$, and there is a restricted amount of solution which occupies the space $-l-a \leqslant x \leqslant -a$, $a \leqslant x \leqslant l+a$. The concentration of solute in the solution is always uniform and is initially C_0, while initially the sheet is free from solute. Let C be the concentration of solute free to diffuse within the sheet and S that of the immobilized solute, each being expressed as amount per unit volume of sheet.

The diffusion is governed by the equation

$$\frac{\partial C}{\partial t} = D\frac{\partial^2 C}{\partial x^2} - \frac{\partial S}{\partial t}, \tag{14.62}$$

and we consider the simultaneous reaction to be of the type

$$\partial S/\partial t = \lambda C - \mu S. \tag{14.63}$$

Here D is the diffusion coefficient and λ and μ are the rate constants of the forward and backward reactions respectively. Thus the immobilized solute is formed at a rate proportional to the concentration of solute free to diffuse, and disappears at a rate proportional to its own concentration. We require solutions of (14.62) and (14.63) with the initial condition

$$S = C = 0, \qquad -a < x < a, \qquad t = 0, \tag{14.64}$$

and with a boundary condition expressing the fact that the rate at which solute leaves the solution is equal to that at which it enters the sheet over the surfaces $x = \pm a$. This condition is

$$l\,\partial C/\partial t = \mp D\,\partial C/\partial x, \qquad x = \pm a, \qquad t > 0. \tag{14.65}$$

We here assume that the concentration of solute free to diffuse just within the surface of the sheet is the same as that in the solution. This may not be so and there may be a distribution factor K, which is not unity, such that the concentration just within the sheet is K times that in the solution. This can clearly be allowed for by using a modified length of solution l/K in place of l in (14.65) and elsewhere. Mathematical solutions follow for these equations and for corresponding equations for the cylinder and sphere.

14.4.1. *Mathematical solutions*

(i) *Plane sheet*. Solutions of the equations of § 14.4 can be obtained by the method of Laplace transforms. Writing $\bar{C}$ and $\bar{S}$ for the Laplace transforms of C and S respectively, so that

$$\bar{C} = \int_0^\infty C e^{-pt}\,dt, \qquad \bar{S} = \int_0^\infty S e^{-pt}\,dt, \qquad (14.66)$$

we have the following equations for $\bar{C}$ and $\bar{S}$

$$p\bar{C} = -p\bar{S} + D\,\partial^2\bar{C}/\partial x^2, \qquad (14.67)$$

$$p\bar{S} = \lambda\bar{C} - \mu\bar{S}, \qquad (14.68)$$

$$-lC_0 + pl\bar{C} = -D\,\partial\bar{C}/\partial x, \qquad x = a. \qquad (14.69)$$

On eliminating $\bar{S}$ from (14.67) and (14.68), and replacing the partial derivative by an ordinary derivative since t does not appear, we find

$$\frac{d^2\bar{C}}{dx^2} + k^2\bar{C} = 0, \qquad k^2 = -\frac{p}{D}\frac{p+\lambda+\mu}{p+\mu}, \qquad (14.70)$$

of which the solution that gives $\bar{C}$ an even function of x is

$$\bar{C} = F(p)\cos kx. \qquad (14.71)$$

The function $F(p)$ is determined by the boundary condition (14.69) and it follows immediately that

$$\bar{C} = \frac{lC_0 \cos kx}{pl \cos ka - kD \sin ka}. \qquad (14.72)$$

The derivation of C is straightforward as in Chapter 2, § 2.4.3 (ii), and after some reduction gives

$$C = \frac{lC_0}{1+(R+1)a} + \sum_{n=1}^{\infty} \frac{C_0 \exp(p_n t)}{1 + \left\{1 + \dfrac{\lambda\mu}{(p_n+\mu)^2}\right\}\left\{\dfrac{a}{2l} + \dfrac{p_n}{2Dk_n^2} + \dfrac{p_n^2 la}{2D^2 k_n^2}\right\}} \frac{\cos k_n x}{\cos k_n a}, \qquad (14.73)$$

where the p_ns are the non-zero roots of

$$\frac{lp_n}{D} = k_n \tan k_n a, \qquad k_n^2 = -\frac{p_n}{D}\frac{p_n+\lambda+\mu}{p_n+\mu}, \qquad (14.74)$$

and $R = \lambda/\mu$ is the partition factor between immobilized and free solute. The expression for S differs from (14.73) only by having an extra factor $\lambda/(p_n+\mu)$ multiplying the nth term, including the term $p_n = 0$. Writing M_t for the total amount of solute, both free to diffuse and immobilized, in half the sheet at time t, and M_∞ for the corresponding quantity in the final equilibrium state attained theoretically after infinite time, we have

$$\frac{M_t}{M_\infty} = 1 - \sum_{n=1}^{\infty} \frac{(1+\alpha)\exp(p_n t)}{1 + \left\{1 + \dfrac{\lambda\mu}{(p_n+\mu)^2}\right\}\left\{\dfrac{a}{2l} + \dfrac{p_n}{2Dk_n^2} + \dfrac{p_n^2 la}{2D^2 k_n^2}\right\}}, \tag{14.75}$$

where the p_ns are given by (14.74) and where

$$\alpha = l/(R+1)a, \qquad M_\infty = lC_0/(1+\alpha). \tag{14.76}$$

(ii) *Cylinder.* The case of the cylinder was considered by Wilson (1948), using a slightly different method. The final result for a cylinder of radius a, in a solution occupying a region of cross-sectional area A, is

$$\frac{M_t}{M_\infty} = 1 - \sum_{n=1}^{\infty} \frac{(1+\alpha)\exp(p_n t)}{1 + \left\{1 + \dfrac{\lambda\mu}{(p_n+\mu)^2}\right\}\left\{\dfrac{\pi a^2}{A} + \dfrac{Ap_n^2}{4\pi D^2 k_n^2}\right\}}, \tag{14.77}$$

where

$$\alpha = A/\{\pi a^2(R+1)\}, \qquad M_\infty = AC_0/(1+\alpha), \tag{14.78}$$

and the p_ns and k_ns are given by

$$\frac{Ap_n}{2\pi Da} = k_n \frac{J_1(k_n a)}{J_0(k_n a)}, \qquad k_n^2 = -\frac{p_n}{D}\frac{p_n+\lambda+\mu}{p_n+\mu}. \tag{14.79}$$

The expression for the concentration of free solute is

$$C = \frac{AC_0}{A+(R+1)\pi a^2} + \sum_{n=1}^{\infty} \frac{C_0 \exp(p_n t)}{1 + \left\{1 + \dfrac{\lambda\mu}{(p_n+\mu)^2}\right\}\left\{\dfrac{\pi a^2}{A} + \dfrac{Ap_n^2}{4\pi D^2 k_n^2}\right\}}\frac{J_0(k_n r)}{J_0(k_n a)}. \tag{14.80}$$

The expression for S differs from (14.80) only by having an extra factor $\lambda/(p_n+\mu)$ multiplying the nth term, including the term $p_n = 0$.

(iii) *Sphere.* On introducing new variables rC and rS the equations for the spherical case take essentially the same form as those for the plane case, and the same method of solution leads to

$$\frac{M_t}{M_\infty} = 1 - \sum_{n=1}^{\infty} \frac{(1+\alpha)\exp(p_n t)}{1 + \left\{1 + \dfrac{\lambda\mu}{(p_n+\mu)^2}\right\}\left\{\dfrac{2\pi a^3}{V} - \dfrac{p_n}{2Dk_n^2} + \dfrac{Vp_n^2}{8\pi aD^2 k_n^2}\right\}}, \tag{14.81}$$

where the p_ns and k_ns are given by

$$\frac{V}{4\pi Da}p_n - 1 = -k_n a \cot k_n a, \qquad k_n^2 = -\frac{p_n}{D}\frac{p_n+\lambda+\mu}{p_n+\mu}, \tag{14.82}$$

and

$$\alpha = \frac{3V}{4\pi a^3(R+1)}, \qquad M_\infty = \frac{VC_0}{1+\alpha}. \qquad (14.83)$$

The expression for C is

$$C = \frac{3VC_0}{3V+4\pi a^3(R+1)}$$

$$+ \sum_{n=1}^{\infty} \frac{C_0 \exp(p_n t)}{1+\left\{1+\dfrac{\lambda\mu}{(p_n+\mu)^2}\right\}\left\{\dfrac{2\pi a^3}{V}-\dfrac{p_n}{2Dk_n^2}+\dfrac{Vp_n^2}{8\pi aD^2k_n^2}\right\}} \; \frac{a}{r}\frac{\sin k_n r}{\sin k_n a} \qquad (14.84)$$

and that for S differs from (14.84) only by having an extra factor $\lambda/(p_n+\mu)$ multiplying the nth term, including the term $p_n = 0$.

14.4.2. *Physical significance of the mathematical solutions*

When mathematical solutions are as complicated in form as those in § 14.4.1 their physical significance is not immediately obvious. Consider eqns (14.74) and put

$$x = k^2 a^2, \qquad y = pa^2/D, \qquad \xi = a^2(\lambda+\mu)/D, \qquad \eta = a^2\mu/D. \quad (14.85)$$

Then

$$ly/a = \sqrt{x}\,\tan\sqrt{x}, \qquad (14.86)$$

$$x = -\frac{y(y+\xi)}{y+\eta} \qquad (14.87)$$

are the equations to be solved for the roots. Graphs of eqns (14.86) and (14.87) are sketched in Fig. 14.4 to show the general location of the roots. The graph of (14.86) is the same for all ξ and η, and from the figure it is easy to see qualitatively how the roots vary with ξ and η. When corresponding transformations to those of (14.85) are applied to eqns (14.79) for the cylinder and (14.82) for the sphere, the resulting equations are of the same form as (14.86) and (14.87), so that Fig. 14.4 can be taken as showing qualitatively the location of the roots for all three cases. There is a root for which $k_n a$ is imaginary given by

$$ly/a = \sqrt{(-x)}\tanh\sqrt{(-x)}. \qquad (14.88)$$

The general expression for M_t/M_∞ therefore comprises a unit term from $p_n = 0$, a term for which $k_n a$ is imaginary, and two infinite series of terms corresponding to the intersections of the two branches of (14.87) with successive branches of (14.86).

The relative importance of the various terms depends on the parameter η. It is interesting to see quantitatively what happens to the general solution

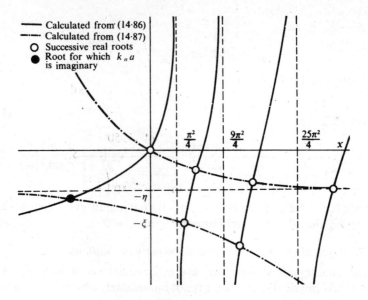

Fig. 14.4. Location of roots of (14.86) and (14.87).

for the extreme values of this parameter which correspond to very fast and very slow reactions. The roots of (14.87) are given by

$$2y = -(x+\xi) \pm \sqrt{\{(x+\xi)^2 - 4x\eta\}}, \tag{14.89}$$

the two infinite series arising from the alternative sign. For extreme values of η the roots are readily obtained by using the appropriate binomial expansion of the term under the square root sign in (14.89). Proceeding in this way with the aid of Fig. 14.4, it is not difficult to show that if η is very large, that is the reaction is very rapid compared with diffusion, the terms in the general solution for M_t/M_∞ which arise by taking the negative sign in (14.89) vanish, as does also the term from the imaginary root. The terms from the positive sign lead to

$$\frac{M_t}{M_\infty} = 1 - \sum_{n=1}^{\infty} \frac{2\alpha(1+\alpha)\exp(-\beta k_n^2 a^2 t)}{1+\alpha+\alpha^2 k_n^2 a^2}, \tag{14.90}$$

where the $k_n a$s approach the roots of

$$\tan k_n a = -\alpha k_n a, \tag{14.91}$$

and

$$\beta = D/(R+1)a^2. \tag{14.92}$$

This is the solution obtained by Wilson (1948) for the case of an immobilizing reaction which is rapid compared with diffusion.

If, however, η is small because the reaction is infinitely slow ($\mu = 0$), we find that the terms arising from the positive sign in (14.89) vanish provided a/l is not zero, which case is treated separately in § 14.4.3. The terms from the negative sign combine with that from the imaginary root to give

$$\frac{M_t}{M_\infty} = \left(1 - \frac{R}{R+1}\frac{l}{l+a}\right)\left\{1 - \sum_{n=1}^{\infty} \frac{2(l/a)(1+l/a)\exp(-Dk_n^2 t)}{1 + l/a + (l/a)^2 k_n^2 a^2}\right\}, \qquad (14.93)$$

where the k_ns are given by

$$\tan k_n a = -(l/a)k_n a. \qquad (14.94)$$

The whole term in the second bracket in (14.93) is to be recognized as the expression for simple diffusion from a finite bath, i.e. diffusion in the absence of any immobilized component. Eqn (14.93) also describes a simple diffusion process, therefore, and M_t/M_∞ changes from zero at $t = 0$ to $1 - Rl/\{(R+1)\times(l+a)\}$ at $t = \infty$, which is easily shown to be the fractional uptake of solute to be expected in the absence of immobilized solute. Thus (14.93) indicates the behaviour to be expected on general argument, namely that for an infinitely slow reaction the sheet takes up, by simple diffusion, only the fraction of solute which it can accommodate in the freely diffusing state and none in the immobilized state.

If on the other hand $\mu a^2/D$ is small because D is very large, all terms in the general solution vanish except the one associated with the imaginary root and we are left with

$$\frac{M_t}{M_\infty} = 1 - \frac{R}{R+1}\frac{l}{l+a}\exp\left\{-\left(1 + \frac{Ra}{l+a}\right)\mu t\right\}. \qquad (14.95)$$

This expression is readily deduced from elementary considerations when diffusion is so rapid that the concentration of solute is effectively uniform through the sheet at all times.

The type of behaviour observed in a practical system for which $\mu a^2/D$ is very small, depends on the time scale of the experiment. If this is such that the reaction occurs very slowly compared with the duration of the experiment the simple diffusion behaviour of eqn (14.93) is observed. If on the other hand diffusion is very rapid compared with the time scale of the experiment, the simple first-order reaction of eqn (14.95) is observed.

14.4.3 Numerical evaluation

When eqns (14.74) and (14.75) are written in terms of p_n/μ and $k_n a$, we see that M_t/M_∞ can be calculated as a function of Dt/a^2 if three parameters are known. The parameters are l/a, that is the ratio of the volumes occupied by solution and sheet respectively; the partition factor R; and the modified rate constant for the reaction, $\mu a^2/D$. Alternatively, since $R = \lambda/\mu$, a solution is defined by l/a and the two rate constants $\mu a^2/D$ and $\lambda a^2/D$. In some cases it is more useful to relate R to the fraction of the total amount of solute which is in the sheet finally, i.e. to M_∞/lC_0, by the relations (14.76). For the cylinder the corresponding parameters are $\pi a^2/A$, R, and $\mu a^2/D$, and for the sphere $4\pi a^3/3V$, R, and $\mu a^2/D$.

Once the roots p_n, $k_n a$ are obtained, the evaluation of each of the expressions for M_t/M_∞ for the plane sheet, cylinder, and sphere is straightforward provided l, A, and V are finite. The cases of l, A, and V infinite, however, need further consideration because the convergence of terms for which p_n approaches $-\mu$ can be very slow and numerical evaluation becomes awkward and laborious, particularly for small $\mu a^2/D$. For the plane sheet when $\alpha = \infty$, eqn (14.75) reduces to

$$\frac{M_t}{M_\infty} = 1 - \sum_{n=1}^{\infty} \frac{2D^2 k_n^2 (1+p_n/\mu)^2 \exp(p_n t)}{(R+1)p_n^2 a^2 \{(1+p_n/\mu)^2 + R\}}, \qquad (14.96)$$

where now $k_n a = (n+\tfrac{1}{2})\pi$. As we saw in § 14.4.2, there are two infinite series in the general expression for M_t/M_∞. We shall confine attention for the moment to the series associated with the positive square root in eqn (14.89) since these are the terms for which p_n approaches $-\mu$ when Dk_n^2/μ is large. Substituting for $1+p_n/\mu$ from the second of eqns (14.74) we find

$$\frac{M_t}{M_\infty} = 1 - \sum_{n=1}^{\infty} \frac{2(R+1+p_n/\mu)^2 \exp(p_n t)}{(R+1)\{(1+p_n/\mu)^2 + R\}k_n^2 a^2}. \qquad (14.97)$$

If $p_n = -\mu$ to the order of accuracy required, after the first r non-zero roots, we have

$$\frac{M_t}{M_\infty} = 1 - \sum_{n=1}^{r} \frac{2(R+1+p_n/\mu)^2 \exp(p_n t)}{(R+1)\{(1+p_n/\mu)^2 + R\}k_n^2 a^2}$$
$$- \frac{R}{R+1}\left(1 - \sum_{n=1}^{r} \frac{2}{k_n^2 a^2}\right)\exp(-\mu t), \qquad (14.98)$$

approximately, since

$$\sum_{n=1}^{\infty} \frac{2}{k_n^2 a^2} = 1. \qquad (14.99)$$

The relationship (14.99) follows, for example, from (14.90) when $\alpha = \infty$, since $M_t/M_\infty = 0$ when $t = 0$. The error involved in use of the approximate form (14.98) is less than

$$\left\{\frac{(R+1+p_{r+1}/\mu)^2 \exp(p_{r+1}t)}{(R+1)\{(1+p_{r+1}/\mu)^2 + R\}} - \frac{R}{R+1}\exp(-\mu t)\right\}\left(1 - \sum_{n=1}^{r} \frac{2}{k_n^2 a^2}\right), \qquad (14.100)$$

and can be made as small as desired by choice of r. We may note in passing that since in (14.97) p_n/μ and $1+p_n/\mu$ occur only with R, this a more convenient form of expression for computation than (14.96) when p_n/μ is small or near -1, particularly for large R, since it is less sensitive to the accuracy of the roots p_n/μ.

When $l/a = \infty$, it follows from (14.88) that there is no root for which $k_n a$ is imaginary. The complete expression for M_t/M_∞ is therefore that of (14.98) together with terms arising from the negative sign in the roots of (14.89). On using (14.98) numerical evaluation of M_t/M_∞ for l/a infinite is straightforward. The corresponding formulae for the plane sheet and the sphere are easily derived.

14.4.4. *Irreversible reaction*

A special case of the above solutions of particular interest is that of an irreversible reaction, when the rate of formation of immobilized solute is directly proportional to the concentration of free solute. In this case $\mu = 0$, but λ is non-zero so that $R = \infty$, $\alpha = 0$, $M_\infty = lC_0$. The solution for the plane sheet, for example, for these values follows immediately from (14.74) and (14.75) provided l is finite.

The solution for the case of $l = \infty$ is less obvious. When $\mu = 0$, the imaginary root (p_i, k_i) is given by

$$lp_i = -Dk_i' \tanh k_i'a, \qquad k_i'^2 = (p_i + \lambda)/D, \qquad (14.101)$$

where $k_i = ik_i'$, and so when $l = \infty$,

$$p_i = 0, \qquad k_i'^2 = \lambda/D. \qquad (14.102)$$

When $\alpha = 0$, $\mu = 0$, and p_i is small, we can expand $\exp(p_i t)$ in powers of $p_i t$ and write (14.75) as

$$M_t = lC_0 - \frac{lC_0(1 + p_i t)}{1 + \dfrac{a}{2l} - \dfrac{p_i}{2Dk_i'^2} - \dfrac{p_i^2 la}{2D^2 k_i'^2}} - \sum_{n=1}^{\infty} \frac{lC_0 \exp(p_n t)}{1 + \dfrac{a}{2l} + \dfrac{p_n}{2Dk_n^2} + \dfrac{p_n^2 la}{2D^2 k_n^2}}, \qquad (14.103)$$

from which, when $l = \infty$, so that lp_i is given by (14.101) and (14.102), we have finally

$$\frac{M_t}{aC_0} = \frac{Dt}{a^2} q \tanh q + \tfrac{1}{2} \operatorname{sech}^2 q + \frac{1}{2q} \tanh q - \sum_{n=1}^{\infty} \frac{2D^2 k_n^2 \exp(p_n t)}{a^2 p_n^2}, \qquad (14.104)$$

where

$$k_n a = (n + \tfrac{1}{2})\pi, \qquad k_n^2 = -(p_n + \lambda)/D, \qquad q = \sqrt{(\lambda a^2/D)}. \qquad (14.105)$$

The first term on the right-hand side of (14.104) gives the rate of uptake of solute due to the chemical reaction in the final steady state.

The forms of (14.77) and (14.79) for the cylinder, and of (14.81) and (14.82) for the sphere when $\mu = 0$, $\alpha = 0$, are obvious. When $A = \infty$ we have for the cylinder

$$\frac{M_t}{\pi a^2 C_0} = 1 + \frac{2Dt}{a^2} q \frac{I_1(q)}{I_0(q)} - \frac{I_1^2(q)}{I_0^2(q)} - \sum_{n=1}^{\infty} \frac{4D^2 k_n^2 \exp(p_n t)}{p_n^2}, \quad (14.106)$$

where

$$J_0(k_n a) = 0, \qquad k_n^2 = -(p_n + \lambda)/D, \qquad q = \sqrt{(\lambda a^2/D)}. \quad (14.107)$$

When $V = \infty$ we have for the sphere

$$\frac{3M_t}{4\pi a^3 C_0} = \frac{3Dt}{a^2} (q \coth q - 1) - \tfrac{3}{2} \operatorname{cosech}^2 q + \tfrac{3}{2} \coth q - \sum_{n=1}^{\infty} \frac{6D^2 k_n^2 \exp(p_n t)}{a^2 p_n^2},$$

$$(14.108)$$

where

$$k_n a = n\pi, \qquad k_n^2 = -(p_n + \lambda)/D, \qquad q = \sqrt{(\lambda a^2/D)}. \quad (14.109)$$

The solutions (14.104), (14.106), (14.108) can of course be obtained directly by use of the Laplace transform or otherwise. When $A = \infty$ the concentration at the surface of the cylinder is constant and (14.106) is the integrated form of (14.42). Similarly (14.108) is the integrated form of (14.41).

14.4.5. *Desorption*

We have considered diffusion into a plane sheet initially free of solute. There is the complementary problem in which all the solute is at first uniformly distributed through the sheet and subsequently diffuses out into the solution. If free and immobilized solute are considered to be initially in equilibrium everywhere in the sheet, it is easily seen that the mathematical solutions presented above for sorption also describe desorption, provided M_t is taken to mean the amount of solute leaving the sheet up to time t, and M_∞ the corresponding amount after infinite time. For desorption from a plane sheet, for example, we want solutions of eqns (14.62) and (14.63) satisfying the condition (14.65), but with (14.64) replaced by

$$C = C_0, \qquad S = S_0, \qquad -a < x < a, \qquad t = 0. \quad (14.110)$$

Writing

$$C_1 = C_0 - C, \qquad S_1 = S_0 - S, \quad (14.111)$$

it is easy to see that (14.110) and the other equations for desorption are identical with (14.62), (14.63), (14.64), and (14.65) with C_1, S_1 written for C, S respectively, remembering only that $\lambda C_0 - \mu S_0 = 0$, if we have equilibrium throughout the sheet at $t = 0$. Hence the solutions for C_1 for desorption are readily deduced from those for C for sorption, as are also the expressions for M_t/M_∞ relating to desorption.

14.4.6. *Calculated results*

Evaluation of the expressions of § 14.4.4 for the irreversible reaction and any particular set of parameters is comparatively simple and straightforward. On the other hand a considerable amount of painstaking labour is involved in the evaluation of the formulae for the reversible reaction even when there is an infinite amount of solute, and it is therefore worth while to present some numerical values. In Tables 14.1, 14.2, 14.3, 14.4 calculated values of M_t/M_∞ are tabulated for the plane sheet, cylinder, and sphere. The arrangement of the tables was decided by ease of presentation. Three values of the partition factor R are included and for each R three rates of reaction. All tabulated solution refer to an infinite amount of diffusing substance. The values of M_t/M_∞ are believed to be correct to within ± 1 in the third decimal place.

Solutions for the plane sheet are illustrated in Figs. 14.5, 14.6, and 14.7. Those for the cylinder and sphere show the same general features, differing only in detail. Fig. 14.5 shows how the sorption curves change in shape and position as $\mu a^2/D$ is varied between the two extremes given by $\mu = \infty$ (infinitely rapid reaction) and $\mu = 0$ (simple diffusion with no reaction). By plotting against $(\mu t)^{\frac{1}{2}}$ as in Fig. 14.6 we can show the approach to the curve for $D = \infty$ (purely reaction controlled) as $\mu a^2/D$ tends to zero because of D becoming large. We see also in Fig. 14.6 that the discontinuity in the gradient of the curve for D infinite appears as a 'shoulder' when $\mu a^2/D = 0.01$. As $\mu a^2/D$ is increased further the shoulder disappears leaving a sorption curve

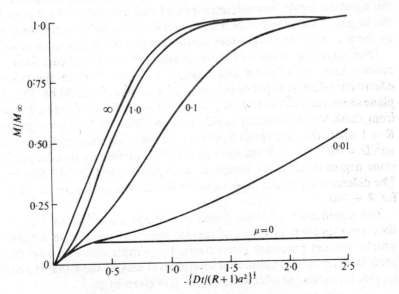

FIG. 14.5. Plane sheet, $R = 10$. Numbers on curves are values of $\mu a^2/D$.

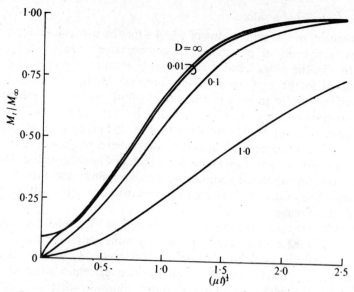

FIG. 14.6. Plane sheet, $R = 10$. Numbers on curves are values of $\mu a^2/D$.

with a point of inflexion. At still higher $\mu a^2/D$ the inflexion becomes less noticeable as in Fig. 14.5 and the curves have the simple shape commonly associated with diffusion. Fig. 14.7 shows the influence of the parameter R, the partition factor between immobilized and free solute. As R is increased the height of the shoulder decreases, and if curves for $R = 100$ were plotted on the present scale no shoulder would be detected for any value of $\mu a^2/D$.

One interesting feature of these results is that they indicate limits to the relative rates of diffusion and reaction outside of which the reaction is effectively infinitely rapid or infinitely slow as the case might be. Thus for the plane sheet, the values of M_t/M_∞ for $\mu a^2/D = 10$ differ by only a few per cent from those for an infinitely rapid reaction. The differences are greatest for $R = 1$ where they are about 5 per cent. At the other extreme the solution for $\mu a^2/D = 0.01$, $R = 1$, is the same as that for an infinitely slow reaction to the same degree of accuracy except at small times (see Fig. 14.7 for example). The differences increase in this case as R increases, being about 20 per cent for $R = 100$.

The significance of these limits is perhaps easier to appreciate when they are expressed in terms of the half-times of the simple diffusion and simple reaction processes respectively. For simple diffusion into the plane sheet from an infinite amount of solution it is easy to show that the half-time t_d, that is the time at which $M_t/M_\infty = \frac{1}{2}$, is given by

$$Dt_d/a^2 = -(4/\pi^2)\ln \pi^2/16 = 0.2, \qquad (14.112)$$

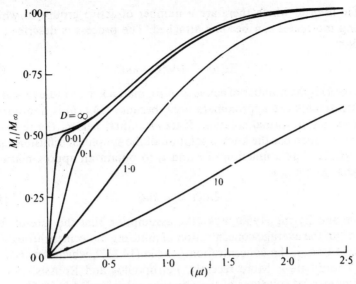

FIG. 14.7. Plane sheet, $R = 1$. Numbers on curves are values of $\mu a^2/D$.

approximately. On the other hand if immobilized solute is formed from a uniform, constant concentration C_0 of free solute according to the equation

$$\partial S/\partial t = \lambda C_0 - \mu S, \tag{14.113}$$

this reaction has a half-time t_r given by

$$\mu t_r = \ln 2 = 0.7, \tag{14.114}$$

approximately, and combining (14.112) and (14.113) we have for the ratio of the half-times,

$$t_r/t_d = 3.5D/(\mu a^2). \tag{14.115}$$

Thus we can say that if the diffusion is more than 1000 times faster than the reaction, the behaviour of the diffusion–reaction process is roughly the same as it would be if diffusion were infinitely rapid. On the other hand if the half-times for diffusion and reaction are comparable, the behaviour approximates to that for an infinitely rapid reaction. These statements indicate orders of magnitude only. They apply also to the cylinder and sphere.

14.5. A bimolecular reaction

Katz, Kubu, and Wakelin (1950) and also Reese and Eyring (1950) have considered the problem of diffusion accompanied by an immobilizing reaction which is bimolecular. They consider diffusion in a cylinder, representing a textile fibre in which there are a number of active groups to which the diffusing molecules can become attached. The process is described by the equation

$$\partial C/\partial t = D\nabla^2 C - knC, \qquad (14.116)$$

where $n(r, t)$ is the number of active groups unattacked at radius r and time t. Both treatments are approximate only, because (14.116) is non-linear and not amenable to formal solution. Katz et al. first neglect the term knC in (14.116) and then use the known solution of the simplified diffusion equation which yields C as a function of r and t, to obtain an approximation to n through the equation

$$\partial n/\partial t = -knC. \qquad (14.117)$$

Reese and Eyring (1950) make the assumption that the rate of reaction depends on the average concentration of diffusing molecules throughout the fibre, and not on the concentration C as in (14.117), which depends on the radial coordinate r. More recently, Petropoulos and Roussis (1969b) discussed numerical solutions of the complete equation (14.116). They referred also to solutions obtained by Lebedev (1966). Nicolson and Roughton (1951) applied the Crank–Nicolson numerical method to the diffusion of CO and O_2 through the outer membrane of the red blood corpuscle, accompanied by diffusion and chemical reaction in a core of concentrated haemoglobin. The equations take the form

$$D_1 \frac{\partial^2 p}{\partial x^2} = a_1 \frac{\partial p}{\partial t} + k'py - k(y_0 - y),$$

$$\frac{\partial y}{\partial t} = -k'py + k(y_0 - y),$$

in the haemoglobin core, and

$$D_2 \frac{\partial^2 p'}{\partial x^2} = a_2 \frac{\partial p'}{\partial t}$$

in the surface membrane. Further applications of the mathematical treatment have been made by Klug, Kreuzer, and Roughton (1956). The subject has been reviewed by Roughton (1959).

14.6. Reduced sorption curves

Weisz (1967) suggested that when fractional sorption curves are plotted against a modified time variable there is relatively little difference between their behaviour even for the extreme cases of a 'weak' linear isotherm and a 'strong' irreversible sorption characterized by the type of sharp advancing boundary discussed in Chapter 13. Weisz used a time variable T given by

$$T = \frac{Dt}{R^2} \frac{P}{b} \frac{C_0}{C_f},$$

where t is real time, D the diffusion coefficient of the mobile component, R the radius of the cylinder or sphere, P the fractional volume in which diffusion occurs, b a tortuosity factor, and C_0 and C_f are the surface and the total final concentrations throughout the medium. A second paper by Weisz and Hicks (1967) gives numerical results.

Further papers by Weisz and co-workers are referred to by Ott and Rys (1973) who present a general model of diffusion with immobilization including cases of competitive sorption of more than one diffusing species.

14.7. Constant reaction rate

Solutions of the equation

$$\frac{\partial C}{\partial t} = \nabla^2 C \pm A,$$

where A represents a constant rate of generation or removal of diffusing substance, have been adequately covered for the standard geometrical shapes by Carslaw and Jaeger (1959). Goldenberg (1951, 1952) solved the heat equation for the case of a constant rate of heat generation in a sphere embedded in an infinite medium having identical (1951) and different (1953) diffusion properties.

15

SIMULTANEOUS DIFFUSION OF HEAT AND MOISTURE

15.1. Introduction

THE problem to be discussed in this chapter is that of the diffusion of one substance through the pores of a solid body which may absorb and immobilize some of the diffusing substance with the evolution or absorption of heat. This heat will itself diffuse through the medium and will affect the extent to which the solid can absorb the diffusing substance. We thus have two processes, the transfer of moisture and the transfer of heat, which are coupled together, and we cannot in general consider one process without considering the other simultaneously. For convenience we shall refer to diffusion through pores, but the theory will apply to alternative systems provided only that some of the diffusing substance is immobilized and that heat is given out in the process. Thus the case of a dissolved substance diffusing through a gel would be included. Equations of the same form would also be obtained, neglecting thermal effects, for the diffusion of two substances through a porous solid, each capable of replacing the other in absorption by the solid.

15.2. Uptake of water by a textile fibre

A simple illustration of the coupling between the transfer of heat and that of moisture is afforded by the uptake of moisture by a single textile fibre. Water penetrates the fibre by diffusion and for a wool fibre a diffusion coefficient of about 10^{-7} cm^2 s^{-1} can be taken as representative. The time for a single fibre to reach say 80 per cent of its final uptake of moisture depends not only on the diffusion coefficient but also on the diameter of the fibre— the thinner the fibre the more quickly it absorbs. Now the average diameter of a fibre is so small (rather less than 10^{-3} cm) that for a diffusion coefficient of 10^{-7} cm^2 s^{-1} we expect 80 per cent absorption to be reached in about 2 seconds. King and Cassie (1940) attempted to demonstrate this conclusion experimentally by measuring the rate of sorption of water vapour by a small mass of wool (0.25 g) suspended by a sensitive spring-balance in an evacuated chamber into which water vapour was introduced at 23.5 mm pressure. By suspending the wool in an evacuated chamber King and Cassie eliminated diffusion through any surrounding atmosphere, and they expected to measure directly a rate of uptake governed solely by diffusion within the fibres. In view of the small diameter of the fibre we expect this uptake to be a matter of only a few seconds. This does not appear to be at first sight supported by the uptake curve they observed, which reached 80 per cent only after an hour or so.

We must remember, however, that when water vapour is absorbed by wool a large amount of heat is evolved which produces a considerable increase in temperature. King and Cassie showed that when this temperature rise is taken into account the relatively low rate of sorption can be reconciled with the statement that an individual fibre reaches equilibrium with its surroundings effectively instantaneously. Thus the uptake of moisture depends on the vapour pressure and also on temperature; the uptake is greater the higher the vapour pressure but is decreased by a rise in temperature. In the experiment we have described, the uptake would have been more than 30 per cent if the temperature had remained constant. Because the temperature rose to over 65 °C, however, the uptake immediately acquired was much less than 30 per cent. A temporary equilibrium is reached, in which the uptake is in equilibrium with the external vapour pressure at the modified temperature produced by the heat evolved as the vapour is absorbed. King and Cassie (1940) calculated this temporary equilibrium uptake to be 2·3 per cent and the temperature 80 °C. Subsequently heat is lost and as cooling proceeds the uptake of moisture increases. The uptake curve observed by King and Cassie was thus essentially a cooling curve. The uptake and temperature curves calculated in this way agree well with the corresponding experimental curves.

This experiment in a vacuum may seem artificial, but it illustrates that the immediate reaction of a mass of fibres when presented with a new atmosphere is to modify that atmosphere to be in equilibrium with itself at the expense of only a slight change in its own moisture content.

15.3. Two possible equilibrium conditions

We consider now an example quoted by Cassie (1940a) having a more direct bearing on the problem in which we are interested. Suppose a sheet of wool fibres conditioned to 45 per cent R.H. at 20 °C, so that its moisture uptake is 10 per cent, is suddenly placed in a stream of air at 65 per cent R.H. and the same temperature. The wool and air can come to equilibrium in two very different ways:

1. By an increase in the uptake of the wool until it is in equilibrium with 65 per cent R.H. The regain needed is 14 per cent.
2. By an increase in the temperature of the wool and air till the new vapour pressure represents only 45 per cent R.H., the uptake of the wool remaining essentially unchanged at 10 per cent. It is easy to calculate from vapour pressure tables that this will be so if the temperature rises to 25 °C.

Of these two possible equilibrium conditions the first, involving a considerable change in moisture uptake, can only be achieved after a large volume of air has passed through the wool. The second, involving a temperature rise, is easily attained almost immediately because of the large heat of sorption. Enough heat is produced by a relatively small increase in moisture uptake to raise the temperature of the wool to 26 °C ($\frac{1}{4}$ per cent will do it if the heat capacity of the air is neglected). For this reason the first equilibrium set up is the one in which the temperature rises but the regain is essentially unchanged. This is only a pseudo-equilibrium, however, because if we continue to blow air at 20 °C over the wool the final temperature must be 20 °C and the final regain 14 per cent. Here we have the essential feature of the propagation of humidity and temperature changes in textiles or similar materials, namely the existence of two equilibrium states—a temporary one set up quickly and involving no change in uptake, and a permanent one set up relatively slowly and involving a change in moisture content.

15.4. Propagation of two disturbances

Clearly, therefore, when the air stream first passes through, a front, separating the original and pseudo-equilibrium conditions, moves through the textile with the speed of the air flow if we neglect the heat capacity of the textile. Thus a fast disturbance representing change of temperature and moisture in the air without change of regain, is followed by a much slower disturbance bringing a change of regain. Furthermore, the same general behaviour is to be expected when the transfer of heat and moisture occurs by diffusion rather than as a result of aerodynamic flow of air.

Mathematical equations describing the diffusion phenomena in detail have been set up and from them it is possible to calculate how concentration, temperature, and total moisture uptake vary with time and distance through the medium till final equilibrium is attained. Henry (1939, 1948) first gave the

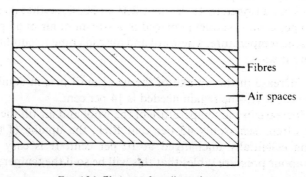

Fibres

Air spaces

FIG. 15.1. Element of textile package.

theory for diffusion of humid air into a textile package and Cassie (1940b) later gave the corresponding theory for air forced through the textile, neglecting diffusion effects entirely. Since then Daniels (1941) has taken into account diffusion of heat and moisture in an air stream forced through the textile. Armstrong and Stannett (1966a) extended the work of Cass, and others to include the effects of the diffusion of moisture in cylindrical fibres. In plane sheets which are thinner, movement of both heat and moisture were allowed for (Armstrong et al. 1966b). We now give Henry's treatment of the diffusion problem.

15.5. Equations for diffusion of heat and moisture

15.5.1. Equilibrium equation

Fig. 15.1 shows diagrammatically an element of a textile package occupied partly by fibres and partly by air spaces. This is much oversimplified but serves to fix ideas. We have said that a fibre can always be considered as in equilibrium with its immediate surroundings. We shall further assume linear dependence on both temperature and moisture content and write

$$M = \text{constant} + \sigma C - \omega T, \qquad (15.1)$$

where C is the concentration of water vapour in the air spaces expressed in g cm^{-3}, M is the amount of moisture absorbed by unit mass of fibre, σ and ω are constants. We shall consider the equilibrium uptake of moisture by a fibre to be related to water vapour concentration and temperature T by the linear relation (15.1). This is a necessary assumption if the theory is to proceed; clearly in practice it is only an approximation which is reasonable over small ranges of humidity and temperature.

15.5.2. Vapour diffusion equation

Consider an element of a textile package. We can derive two equations, one expressing the rate of change of concentration and the other the rate of change of temperature.

The rate of change of concentration is governed by the following:

(i) Diffusion of vapour through the air spaces and through the fibres, both these being proportional to the concentration gradient in the usual way. Diffusion through the pores will in many cases be greater than through the fibres, but even if this is not so we can represent both processes by one term if we make the assumption that the vapour in the fibre is always in equilibrium with that in the air in the immediate neighbourhood and the absorption isotherm is linear as in (15.1).

(ii) The absorption or desorption of moisture by the fibres from the air spaces.

Thus we can say,

Net amount of vapour entering element by diffusion
= increase in moisture in air + increase in moisture in fibres.   (15.2)

If a fraction v of the total volume of the package is occupied by air and $1 - v$ by fibre of density ρ_s then, expressing (15.2) mathematically, the equation governing the movement of vapour can be written

$$vgD_A\frac{\partial^2 C}{\partial x^2} = v\frac{\partial C}{\partial t} + (1-v)\rho_s\frac{\partial M}{\partial t},$$ (15.3)

where C and M have been defined and D_A is the diffusion coefficient for moisture in air. The factor g allows for the fact that the diffusion is not along straight air channels but through a matrix of intertwined fibres and any diffusion along the fibres themselves can also be allowed for in this factor. It is a factor which can be measured by permeability measurements under steady-state conditions.

15.5.3. Heat diffusion equation

The rate at which the temperature of the element changes is determined by

(i) conduction of heat through air and fibres;
(ii) the heat evolved when moisture is absorbed by fibres.

Thus

Increase in heat content of fibres
= net amount of heat entering by conduction + heat evolved
as fibres absorb moisture,   (15.4)

and this is expressed mathematically by the equation

$$\alpha\rho\frac{\partial T}{\partial t} = K\frac{\partial^2 T}{\partial x^2} + q\rho\frac{\partial M}{\partial t},$$ (15.5)

where α is the specific heat of the fibres, K the over-all heat conductivity of the package, ρ the density of the package, expressed as mass of fibre per unit overall volume, and q is the heat evolved when 1 g of water vapour is absorbed by the fibres. In writing (15.5) the reasonable assumption has been made that the heat content of the air is negligible compared with that of the fibres.

One vital point to be noticed is that both eqn (15.3) and (15.5), that is the vapour equation and the temperature equation, involve M, the amount of moisture in the fibres. It is at once obvious that the two processes, the transfer of moisture and the transfer of heat, are coupled together in this way and that we cannot in general consider one process without considering the other simultaneously.

15.5.4. *Assumptions underlying the mathematical theory*

It is worth while to enumerate some of the assumptions on which the theoretical treatment is based. The main ones are as follows:

(i) The linear dependence of M on C and T to which reference has already been made.

(ii) The quantities D_A, K, α, ρ are assumed constant and independent of moisture concentration and temperature.

(iii) The heat of sorption q is assumed independent of regain though in practice it is not so.

(iv) Hysteresis of sorption is neglected, that is the equilibrium equation (15.1) is assumed to hold whether the fibre is gaining or losing moisture.

(v) The relative volumes occupied by fibre and air are assumed not to change as diffusion proceeds, i.e. v is assumed constant. In actual fact, as the fibres sorb moisture they swell and occupy progressively more space, and the air correspondingly less. This is thought to be unimportant except for very dense packages.

(vi) No account has been taken of the influence of capillarity in the air spaces. This will be appreciable only at very high humidities or in water.

(vii) The fibres have been assumed to reach equilibrium with their immediate surroundings instantaneously. There is some evidence that relatively slow changes of fibre structure may occur as the moisture is taken up, and that while most of the uptake is effectively instantaneous there may be a slow drift of moisture content persisting for some time. The information on this at the moment is too sparse for it to be taken into account even if the mathematics permitted.

15.6. Solution of the equations

Using Henry's nomenclature the equations to be solved can be written

$$D'' \frac{\partial^2 C}{\partial x^2} - \frac{\partial C}{\partial t} = \gamma \frac{\partial M}{\partial t}, \tag{15.6}$$

$$\mathscr{D}'' \frac{\partial^2 T}{\partial x^2} - \frac{\partial T}{\partial t} = -\varepsilon \frac{\partial M}{\partial t}, \tag{15.7}$$

$$\partial M/\partial C = \sigma; \qquad \partial M/\partial T = \omega, \tag{15.8}$$

where

$$D'' = gD_A \qquad \mathscr{D}'' = K/(\alpha\rho); \qquad \gamma = (1-v)/(v\rho_s); \qquad \varepsilon = q/\alpha. \tag{15.9}$$

Eliminating M, we obtain

$$D'' \frac{\partial^2 C}{\partial x^2} - (1 + \gamma\sigma) \frac{\partial C}{\partial t} + \gamma\omega \frac{\partial T}{\partial t} = 0, \tag{15.10}$$

$$\mathscr{D}'' \frac{\partial^2 T}{\partial x^2} - (1 + \varepsilon\omega) \frac{\partial T}{\partial t} + \varepsilon\sigma \frac{\partial C}{\partial t} = 0, \tag{15.11}$$

which can be written more simply in the forms

$$D \frac{\partial^2 C}{\partial x^2} - \frac{\partial}{\partial t}(C - \lambda t) = 0, \tag{15.12}$$

$$\mathscr{D} \frac{\partial^2 T}{\partial x^2} - \frac{\partial}{\partial t}(T - \nu C) = 0, \tag{15.13}$$

where the significance of $D, \mathscr{D}, \lambda$, and ν is immediately obvious. These equations are now in a form analogous in some respect to those representing two coupled vibrations, and they may be treated in the usual manner, provided λ and ν are assumed constant, by the introduction of 'normal coordinates', which are linear combinations of C and T, of the form $rC + sT$. With proper choice of the ratio r/s, eqns (15.12) and (15.13), expressed in terms of the new coordinates, give rise to two simple diffusion equations, each containing only one of the normal coordinates. The quantities represented by the latter consequently diffuse according to the usual law, and independently of each other, just as the normal modes of vibration of coupled mechanical systems, once started, continue without mutual interference. In order to find the normal coordinates, we multiply (15.12) by r/D and (15.13) by $s/\mathscr{D}$, and add, obtaining

$$\frac{\partial^2}{\partial x^2}(rC + sT) - \frac{\partial}{\partial t}\left\{\left(\frac{r}{D} - \frac{s\nu}{\mathscr{D}}\right)C + \left(\frac{s}{\mathscr{D}} - \frac{r\lambda}{D}\right)T\right\} = 0. \tag{15.14}$$

If this is to be expressible as a simple diffusion equation for $rC + sT$ we must have

$$\frac{r/D - s\nu/\mathscr{D}}{r} = \frac{s/\mathscr{D} - r\lambda/D}{s},$$

i.e.

$$\frac{1}{D} - \frac{s}{r}\frac{\nu}{\mathscr{D}} = \frac{1}{\mathscr{D}} - \frac{r}{s}\frac{\lambda}{D} = \mu^2 \quad \text{say.} \tag{15.15}$$

This is a quadratic in r/s and gives the two values of r/s required to form the two normal coordinates. Also, μ^2 will have two values, corresponding to the roots of r/s. Eqn (15.14) can now be written

$$\frac{\partial^2}{\partial x^2}(rC + sT) - \frac{\partial}{\partial t}\mu^2(rC + sT) = 0, \tag{15.16}$$

which is a simple diffusion equation for $rC + sT$ with diffusion coefficient $1/\mu^2$.

We can now determine the normal coordinates and their diffusion coefficients. Elimination of r/s between the eqns (15.15) gives

$$\left(\mu^2 - \frac{1}{D}\right)\left(\mu^2 - \frac{1}{\mathscr{D}}\right) = \frac{\lambda v}{D\mathscr{D}}, \tag{15.17}$$

the roots of which are

$$\mu^2 = \frac{D + \mathscr{D} \pm \sqrt{\{(\mathscr{D} - D)^2 + 4\lambda v D \mathscr{D}\}}}{2D\mathscr{D}}. \tag{15.18}$$

15.6.1. Diffusion rates

If we write $1/\mu_1^2 = D_1$ and $1/\mu_2^2 = D_2$ in order to preserve the representation of a diffusion coefficient by the letter D, and if we put $D/\mathscr{D} = u$, then we have from (15.18)

$$\frac{D_1}{D} = \frac{2}{u + 1 + \sqrt{\{(1 - u)^2 + 4u\lambda v\}}}, \tag{15.19}$$

$$\frac{D_2}{D} = \frac{2}{u + 1 - \sqrt{\{(1 - u)^2 + 4u\lambda v\}}}. \tag{15.20}$$

Fig. 15.2 shows a nomogram, reproduced from Henry's paper (1948) for solving these equations. If a straight line be placed so as to cut the two straight scales at the appropriate values for u and λv, then the points at which it cuts the circular scale give D_1/D and D_2/D respectively. To find $D_1/\mathscr{D}$ and $D_2/\mathscr{D}$, $1/u$ must be used in place of u. Usually it is convenient to express D_1 in terms of D and D_2 in terms of $\mathscr{D}$.

If λv is small compared with $(\mathscr{D} - D)^2/(4D\mathscr{D})$, i.e. if either the coupling between the two diffusion processes is weak or D and $\mathscr{D}$ are of widely different magnitudes, the roots become

$$\left. \begin{aligned} \mu_1^2 &= \frac{1}{D} + \frac{\lambda v}{\mathscr{D} - D} \\ \mu_2^2 &= \frac{1}{\mathscr{D}} - \frac{\lambda v}{\mathscr{D} - D} \end{aligned} \right\} \tag{15.21}$$

approximately. Thus as the coupling becomes weaker μ_1^2 tends towards $1/D$, and μ_2^2 tends towards $1/\mathscr{D}$, so that we may speak of $\pm \mu_1$ as the vapour roots, and $\pm \mu_2$ as the thermal roots, though when coupling exists all roots are concerned with both diffusion processes to a greater or lesser extent.

15.6.2. Expressions for concentration and temperature changes

Proceeding from eqn (15.17) or (15.18) we see that

$$\mu_1^2 + \mu_2^2 = 1/D + 1/\mathscr{D},$$

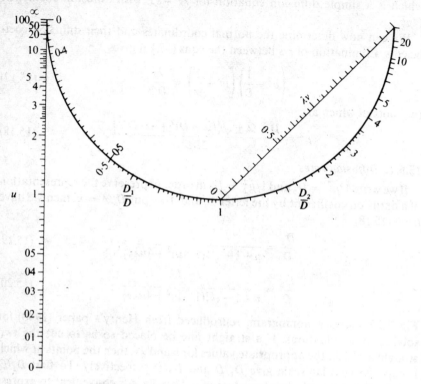

Fig. 15.2. Nomogram for D_1/D and D_2/D in terms of u and λv.

i.e.

$$\mathscr{D}(1 - \mu_1^2 D) = -D(1 - \mu_2^2 \mathscr{D}). \tag{15.22}$$

Using (15.15) and (15.22) we find that

$$\left.\begin{aligned}
\frac{s_1}{r_1} &= \frac{1 - \mu_1^2 D}{\bar{v}} \frac{\mathscr{D}}{D} = -\frac{1 - \mu_2^2 \mathscr{L}}{v} \\
\frac{r_2}{s_2} &= \frac{1 - \mu_2^2 \mathscr{D}}{\lambda} \frac{D}{\mathscr{L}} = -\frac{1 - \mu_1^2 D}{\lambda}
\end{aligned}\right\} \tag{15.23}$$

Hence if we choose r_1 and s_2 to be equal to unity, the normal coordinates are determined, and the solution of the differential equations (15.6), (15.7), (15.8) can be written

$$\Delta C - \frac{1 - \mu_2^2 \mathscr{D}}{v} \Delta T = \psi_1(x, y, z, t), \tag{15.24}$$

$$\Delta T - \frac{1-\mu_1^2 D}{\lambda} \Delta C = \psi_2(x, y, z, t), \tag{15.25}$$

wherè ΔC and ΔT are the deviations of C and T from some given values (e.g. the initial), and the ψs are solutions of ordinary diffusion equations with diffusion coefficients equal to $1/\mu_1^2$ and $1/\mu_2^2$ respectively, and subject to appropriate boundary and initial conditions. Solving for C and T we get

$$\Delta C = \phi_1(x, y, z, t) + \frac{1-\mu_2^2 \mathscr{D}}{v} \phi_2(x, y, z, t), \tag{15.26}$$

$$\Delta T = \phi_2(x, y, z, t) + \frac{1-\mu_1^2 D}{\lambda} \phi_1(x, y, z, t), \tag{15.27}$$

where

$$\phi_1(x, y, z, t) = \frac{1-\mu_2^2 D}{D(\mu_1^2 - \mu_2^2)} \psi_1(x, y, z, t), \tag{15.28}$$

and

$$\phi_2(x, y, z, t) = \frac{1-\mu_1^2 \mathscr{D}}{\mathscr{D}(\mu_2^2 - \mu_1^2)} \psi_2(x, y, z, t). \tag{15.29}$$

The ϕs, being proportional to the ψs, are also solutions of the ordinary diffusion equation with diffusion coefficients equal to $1/\mu_1^2$ and $1/\mu_2^2$ respectively, and with boundary and initial conditions which give the required conditions for C and T.

The physical interpretation of these equations derived by Henry is that each diffusion 'wave' of vapour is accompanied by a temperature 'wave' proceeding at the same rate, whose magnitude is proportional to that of the vapour diffusion wave, the relation between the two depending only upon the properties of the materials. Similarly, the main temperature 'wave' is accompanied by a subsidiary vapour diffusion wave. Even if only one of the external conditions, say the vapour concentration, alters, there will nevertheless be the complete set of two vapour 'waves' and two temperature 'waves', though the latter may be small if the coupling is weak.

Henry points out that similar reasoning to that just given still holds if eqns (15.12) and (15.13) have on their right-hand sides, instead of zero, any functions of the independent variables x, y, z, and t. Such equations would correspond to the case where either the diffusing substance or heat, or both, are being set free or absorbed in a manner additional to and independent of the diffusion processes, and independent of C and T. There might, for example, be a chemical reaction causing both the diffusing substance and heat to be evolved throughout the medium. In such cases the solution to the problem of the simultaneous transfer of heat and vapour reduces to the sum of solutions for the ordinary diffusion of the 'normal coordinates' with appropriate rates of evolution throughout the medium of the quantities they represent.

15.6.3. *Solutions for sudden changes of external conditions*

The above equations enable the solution of the coupled diffusion problem to be obtained in terms of the solution for the ordinary diffusion problem with the same boundary conditions. Suppose a specimen of any shape is in equilibrium with its surroundings and that at time $t = 0$ the concentration C and temperature T of the diffusing substance are suddenly altered to $C + \Delta_0 C$ and $T + \Delta_0 T$ respectively at the boundary, and maintained constant. Let the solution to the *simple* diffusion problem, assuming diffusion constants $D_1 = 1/\mu_1^2$ and $D_2 = 1/\mu_2^2$, be

$$\Delta C = \Delta_0 C f_1(x, y, z, t), \tag{15.30}$$

$$\Delta T = \Delta_0 T f_2(x, y, z, t). \tag{15.31}$$

The specimen eventually reaches equilibrium with the new conditions so that f_1 and f_2 must increase from 0 to 1 as t increases from 0 to infinity. The form of these functions depends on the shape of the specimen and typical solutions have been given in earlier chapters for plane sheets, cylinders, and spheres, for example. Clearly f_1 and f_2, multiplied by suitable constants depending upon the initial conditions at the boundary, are the functions ϕ_1 and ϕ_2 in eqns (15.26) and (15.27). Thus, remembering that the system must eventually come to equilibrium with $\Delta C = \Delta_0 C$ and $\Delta T = \Delta_0 T$, and that f_1 and f_2 are then both unity, we find

$$\Delta C = \frac{1}{D(\mu_1^2 - \mu_2^2)} [\{(1 - \mu_2^2 D) \Delta_0 C - \lambda \Delta_0 T\} f_1 - \{(1 - \mu_1^2 D) \Delta_0 C - \lambda \Delta_0 T\} f_2],$$

$$\tag{15.32}$$

$$\Delta T = \frac{1}{\mathscr{D}(\mu_1^2 - \mu_2^2)} [\{(1 - \mu_2^2 \mathscr{D}) \Delta_0 T - v \Delta_0 C\} f_1 - \{(1 - \mu_1^2 \mathscr{D}) \Delta_0 T - v \Delta_0 C\} f_2].$$

$$\tag{15.33}$$

An alternative form of these equations, put forward by Henry (1939) is

$$\Delta C = \Delta_0 C f_1 - \frac{(1 - \mu_1^2 D) \Delta_0 C - \lambda \Delta_0 T}{D(\mu_1^2 - \mu_2^2)} (f_2 - f_1). \tag{15.34}$$

$$\Delta T = \Delta_0 T f_2 - \frac{(1 - \mu_2^2 \mathscr{D}) \Delta_0 T - v \Delta_0 C}{\mathscr{D}(\mu_1^2 - \mu_2^2)} (f_2 - f_1). \tag{15.35}$$

The first term on the right-hand side of each equation represents the 'wave which would result if there were no coupling between the diffusion processes and which may be referred to as the 'permanent wave'. The second term represents the result of the coupling and may be called the 'temporary wave' since it starts at zero, increases to a maximum, and then diminishes to zero again.

In many cases it is required to know the rate at which the amount of diffusing substance absorbed M varies. To determine this we substitute

$$\Delta M = \sigma \Delta C - \omega \Delta T \tag{15.36}$$

in (15.34) and (15.35), and after some rearrangement we obtain

$$\Delta M = \Delta_0 M f_1 + \left\{ \left(\mu_1^2 - \frac{1}{D} - \frac{\varepsilon\omega}{\mathscr{L}''} \right) \Delta_0 M + \left(\frac{1}{\mathscr{L}''} - \frac{1}{D''} \right) \omega \Delta_0 T \right\} \left(\frac{f_2 - f_1}{\mu_1^2 - \mu_2^2} \right), \tag{15.37}$$

where from (15.10) and (15.12)

$$D'' = D(1 + \gamma\sigma), \tag{15.38}$$

and from (15.11) and (15.13)

$$\mathscr{L}'' = \mathscr{L}(1 + \varepsilon\omega). \tag{15.39}$$

If it is desired to include both the amount absorbed and the amount in the pores, and M' is the total amount of diffusing substance contained in unit mass of solid, then from (15.36) and the definition of γ in (15.9) we obtain

$$\Delta M' = \left(1 + \frac{1}{\gamma\sigma} \right) \Delta M + \frac{\omega}{\gamma\sigma} \Delta T, \tag{15.40}$$

which may be evaluated using (15.35) and (15.37). If $\gamma\sigma$ is large, i.e. if much more diffusing substance is absorbed by the solid than is contained in the pores, M' is nearly equal to M and (15.40) is not needed. Henry points out that this is so for baled cotton, for example, for which $\gamma\sigma$ may be as much as 7000. In this case also, since the terms in the expression $(\mu_1^2 - 1/D - \varepsilon\omega/\mathscr{L}'')$ occurring in (15.37) nearly cancel if $\varepsilon\omega$ is fairly large compared with unity, e.g. for cotton it is about 10 or 30, (15.37) may be written approximately

$$\Delta M = \Delta_0 M f_1 + \left\{ -\frac{u^2}{(1 + u^2)(1 + \varepsilon\omega) + u} \Delta_0 M \right.$$
$$\left. + \frac{1/\mathscr{L}'' - 1/D''}{\mu_1^2 - \mu_2^2} \omega \Delta_0 T \right\} (f_2 - f_1), \tag{15.41}$$

where u is written for $D/\mathscr{L}$. Here $\omega \Delta_0 T$ is the total change in the equilibrium value of M which would be produced by the change in temperature $\Delta_0 T$, if C remained constant. Henry (1939) has given numerical values of the various constants involved in the above equations for the case of baled cotton at densities 0·2, 0·4, 0·6 g/cm^{-3} and has written (15.40) in numerical form for each density.

15.6.4. *Change in external humidity only*

Henry (1948) has explored further the case in which no change is made in the external temperature, only the external humidity being changed. Then we can put $\Delta_0 T = 0$ in eqn (15.32) and by substituting $1/D_1$ and $1/D_2$ for μ_1^2 and μ_2^2 respectively, and then rearranging we obtain for the ratio of the change in C at any instant to the equilibrium value of the change,

$$\frac{\Delta C}{\Delta_0 C} = \frac{D_1(D_2 - D)f_1 - D_2(D_1 - D)f_2}{D(D_2 - D_1)}. \qquad (15.42)$$

This can be written very simply as

$$\Delta C / \Delta_0 C = (1 - p)f_1 + pf_2, \qquad (15.43)$$

where

$$p = \frac{D_2(D - D_1)}{D(D_2 - D_1)}. \qquad (15.44)$$

Thus p is a dimensionless quantity that can be expressed in terms of the ratios D_1/D and D_2/D. We have already seen in eqns (15.19) and (15.20) that these ratios can be calculated directly from $u = D/\mathscr{D}$ and λv, and so p can also be obtained from u and λv. The necessary relationship is

$$(2p - 1)^2 = \frac{(u - 1)^2}{(u - 1)^2 + 4uv\lambda}. \qquad (15.45)$$

Henry's nomogram (1948) for calculating p in terms of u and λv is reproduced in Fig. 15.3. Since the right-hand side of (15.45) is unaltered if $1/u$ is substituted for u, the scale for u from 1 to ∞ enables all possible values of u to be dealt with. It also follows from this symmetry that the same numerical value for p can be used to handle the problem in which the external temperature is changed instead of the humidity.

Putting $\Delta_0 T = 0$ in (15.41), we get

$$\frac{\Delta M}{\Delta_0 M} = f_1 - \frac{u^2 (f_2 - f_1)}{(1 + u)^2(1 + \varepsilon\omega) + u} = (1 + n)f_1 - nf_2, \qquad (15.46)$$

where

$$n = \frac{u^2}{(1 + u)^2(1 + \varepsilon\omega) + u}. \qquad (15.47)$$

Henry's nomogram for calculating n in terms of u and $1/(1 + \varepsilon\omega)$ is reproduced in Fig. 15.4. Under the conditions for which (15.41) and (15.46) apply we have

$$1 + \varepsilon\omega = 1/(1 - \lambda v), \qquad (15.48)$$

approximately.

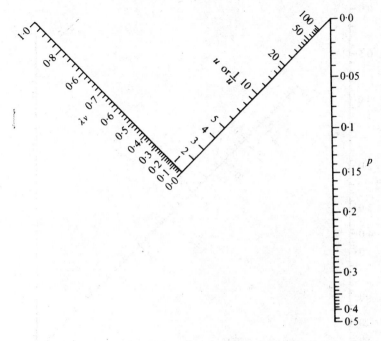

FIG. 15.3. Nomogram for p in terms of u and $\lambda \chi$.

Henry has given a table of values of p and n and also of the diffusion coefficients D, $\mathscr{D}$, D_1, D_2 for cotton packages of two densities and three relative humidities and temperatures. For the experimental data on which his calculations are based his original paper (1948) should be consulted. The two densities are $0.2\,\mathrm{g\,cm^{-3}}$ and $0.5\,\mathrm{g\,cm^{-3}}$ at temperatures $20\,°\mathrm{C}$, $50\,°\mathrm{C}$, and $80\,°\mathrm{C}$, and the relative humidities (R.H.) are 20 per cent, 65 per cent, and 90 per cent. The results are shown in Table 15.1, reproduced from Henry's paper (1948), the significance of the figures being clear from the labelled cells. Figures for $80\,°\mathrm{C}$ and 90 per cent relative humidity are omitted because under these conditions the vapour pressure would be a considerable fraction of atmospheric pressure, and fluid flow would make an appreciable contribution to the process. From this table and eqns (15.43) and (15.46), it is a simple matter to calculate how concentration and total moisture content vary for a package of known size and shape. When dealing with total moisture content, f_1 and f_2 are, of course, functions only of time. As an example, suppose the package is in the form of a large flat sheet then we have (§ 4.3.2, p. 48)

$$f_1 = 1 - \frac{8}{\pi^2}(e^{-m} + \tfrac{1}{9}e^{-9m} + \tfrac{1}{25}e^{-25m} + \dots),\qquad(15.49)$$

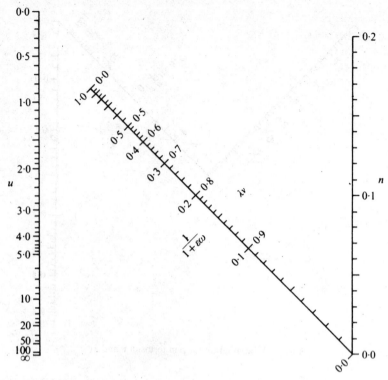

Fig. 15.4. Nomogram for n in terms of u and λv or $1/(1 + \varepsilon \omega)$.

where $m = \pi^2 D_1 t/l^2$, l being the thickness of the sheet. The function f_2 is the same with D_1 replaced by D_2.

The following points about the solutions are of particular interest.

(i) In Table 15.1 it is seen that, as the temperature rises, the moisture diffusion coefficient D, which is at first much smaller than the thermal diffusion coefficient $\mathscr{D}$, eventually becomes much larger. Furthermore, the diffusion coefficient D_1 for the slower normal function is always less than either D or $\mathscr{D}$ but never by a factor of less than $\frac{1}{2}$. On the other hand, D_2 is always much greater than D or $\mathscr{D}$. The figures for n show that the secondary wave is of very small amplitude, as judged by the total moisture content, at low temperatures, but becomes appreciable at high temperatures. The figures for p show that in the case of an external thermal disturbance only, the secondary wave may be important even at room temperatures.

(ii) Since both f_1 and f_2 always change from 0 to 1 we see from (15.35) that the change in temperature associated with an external change of humidity only ($\Delta_0 T = 0$), is a transient one, which increases from zero to a maximum and disappears again. This transient temperature has been observed by Cassie and Baxter (1940).

(iii) If the initial disturbance is one of temperature only there is a corresponding transient change in moisture content.

(iv) The effect of size or shape is all included in the function f and so conversion from one package to another is relatively simple. In particular t and l always appear as t/l^2 in (15.49), for example, and we have the usual dependence of the time scale on the square of the linear dimensions, familiar in simple diffusion problems.

(v) For comparisons of calculated and experimental values the papers of Cassie and Baxter (1940) and of Daniels (1941) should be consulted. The former describes a pure diffusion experiment, the latter a flow-under-pressure experiment with diffusion as a complicating feature. It is not to be expected that the theoretical treatment of a subject as complicated as this can reproduce all the features accurately and quantitatively. Its value lies in clarifying the mechanism by which heat and moisture are transferred, in making possible rough estimates of time scales under given conditions, and in particular, showing how the times for a given package can be estimated from measurements on a standard package.

15.7. Surface temperature changes accompanying the sorption of vapours

A much simpler problem, but one of practical interest, is to calculate the temperature change which accompanies the sorption of vapour by a solid, in cases where the temperature rises due to the heat of condensation given up at the surface of the solid and the heat of mixing can be neglected.

We consider a plane sheet suspended in a vapour maintained at constant pressure. The sheet is taken to be so thin that effectively all the vapour enters through the plane faces and a negligible amount through the edges. Uptake of vapour therefore occurs by uni-directional diffusion through the sheet and the amount taken up is assumed proportional to the square root of time. The heat of mixing is taken to be a negligible fraction of the heat of sorption which thus becomes simply the heat of vaporization. This assumption carries the implication that during sorption heat is evolved only at the two surfaces of the sheet. Values of heat conductivity, specific heat, and density of the sheet are taken to be constant, independent of temperature and vapour content. The calculated results are based in the first place on the assumption that heat is lost from the surfaces by radiation only, according to Stefan's law. For small temperature differences, however, both radiation and convection losses proceed at rates directly proportional to the temperature difference between the surface and its surroundings, and so convection losses can be allowed for by adjusting the radiation constant.

15.7.1. *Mathematical equations*

Taking θ to be the temperature difference between an element of the sheet and the vapour in which the sheet is suspended, θ_0 the temperature of this vapour assumed constant, x the space coordinate perpendicular to the surface of the sheet (the surfaces being at $x = 0$ and $x = l$), and t the time during which uptake of vapour has occurred, the equation governing the conduction of heat through the sheet is

$$\frac{\partial \theta}{\partial t} = \frac{K}{\rho \alpha} \frac{\partial^2 \theta}{\partial x^2}. \tag{15.50}$$

Here K is heat conductivity, ρ density, and α specific heat of the substance of the sheet, all expressed in c.g.s. units. At each surface, heat is continuously evolved, as vapour is taken up, at a rate given by $\frac{1}{2}\rho l L \, dR/dt$, where R is the regain at time t, i.e. mass of vapour taken up per unit initial mass of sheet, and L the heat of vaporization. Some of this heat is lost by radiation from the surface at a rate which for small temperature differences is given by $k\theta$, where $k = 4\sigma_0 \theta_0^3$ and σ_0 is Stefan's constant. The remainder of the heat enters the sheet by conduction at a rate given by $-K \, \partial\theta/\partial x$. The equation expressing conservation of heat at the surface $x = 0$ is therefore

$$\frac{1}{2}\rho l L \frac{dR}{dt} - k\theta + K \frac{\partial \theta}{\partial x} = 0, \qquad x = 0. \tag{15.51}$$

There is a corresponding condition for the surface $x = l$, but for purposes of calculation it is preferable to consider only half the sheet and to use the condition

$$\partial\theta/\partial x = 0, \qquad x = \tfrac{1}{2}l, \tag{15.52}$$

because the uptake of vapour and the heat flow are symmetrical about the central plane of the sheet. We need therefore to evaluate solutions of (15.50) with the boundary conditions (15.51) and (15.52) and the initial condition

$$\theta = 0, \qquad 0 < x < \tfrac{1}{2}l, \qquad t = 0. \tag{15.53}$$

We consider the regain R at time t to be given by

$$R = B(t/l^2)^{\frac{1}{2}}, \tag{15.54}$$

where B is a constant for a given experiment. It follows from eqn (4.20), for example, that

$$B = 4C_0 (D/\pi)^{\frac{1}{2}}, \tag{15.55}$$

where C_0 is the equilibrium regain attained theoretically after infinite time, and D an average diffusion coefficient for the concentration range 0 to C_0 of vapour. The constant B is thus dependent on the experiment considered, i.e. on the system, the temperature and the vapour pressure, but is independent of the thickness of the sheet.

On introducing the non-dimensional variables

$$\phi = \left(\frac{K\alpha}{\rho}\right)^{\frac{1}{2}} \frac{\theta}{BL}, \qquad \tau = \frac{Kt}{\rho\alpha l^2}, \qquad X = \frac{x}{l}, \qquad (15.56)$$

and substituting for R in (15.51) from (15.54), eqns (15.50), (15.51), (15.52), and (15.53) become

$$\frac{\partial\phi}{\partial\tau} = \frac{\partial^2\phi}{\partial X^2}, \qquad (15.57)$$

$$\frac{1}{4\tau^{\frac{1}{2}}} - \frac{kl}{K}\phi + \frac{\partial\phi}{\partial X} = 0, \qquad X = 0, \qquad \tau > 0, \qquad (15.58)$$

$$\partial\phi/\partial X = 0, \qquad X = \tfrac{1}{2}, \qquad \tau > 0, \qquad (15.59)$$

$$\phi = 0, \qquad 0 < X < 1, \qquad \tau = 0. \qquad (15.60)$$

When expressed in terms of the new variables ϕ, X, and τ, the problem therefore contains only a single non-dimensional parameter kl/K, which we denote for convenience by h. A number of solutions of eqn (15.57) satisfying (15.58), (15.59), and (15.60) have been evaluated for different values of h,

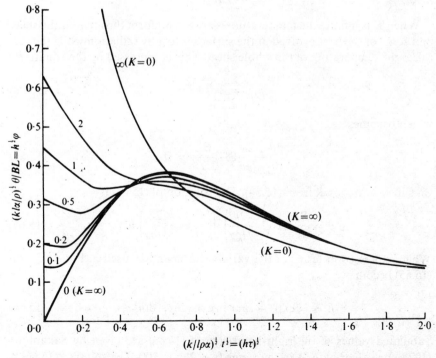

FIG. 15.5. Variation of surface temperature with (time)$^{\frac{1}{2}}$. Numbers on curves are values of $h = kl/K$.

using the numerical methods described in Chapter 8. The calculated surface temperatures are shown graphically in Fig. 15.5, and together with the two special cases for $K = \infty$ and $K = 0$ they constitute the solution of this problem for all values of the physical properties involved. For convenience of scale in Fig. 15.5, $(kl\alpha/\rho)^{\frac{1}{2}}(\theta/BL)$ is plotted against $(kt/l\rho\alpha)^{\frac{1}{2}}$. These variables are respectively $h^{\frac{1}{2}}\phi$ and $(h\tau)^{\frac{1}{2}}$ and hence are readily deduced from ϕ and τ for a given h.

15.7.2. *Special cases of zero and infinite conductivity*

While the case of zero heat conductivity may be of little practical significance it is nevertheless a useful limiting solution from the mathematical point of view. When $K = 0$, no heat penetrates the sheet and we are concerned only with eqn (15.51) which becomes

$$\tfrac{1}{2}\rho Ll\,dR/dt = k\theta. \tag{15.61}$$

For the particular case of R given by (15.54), integration of (15.61) leads immediately to

$$\theta = \frac{Bl\rho}{4kt^{\frac{1}{2}}}. \tag{15.62}$$

When K is infinite the temperature is always uniform throughout the sheet and some of the heat evolved at the surface is lost by radiation while the rest raises the temperature of the whole sheet. This is expressed by the equation

$$\frac{dR}{dt}\rho lL = 2k\theta + l\rho\alpha\frac{d\theta}{dt}, \tag{15.63}$$

which becomes

$$\frac{d\theta}{dt} + \frac{2k}{l\rho\alpha}\theta - \frac{BL}{2l\alpha t^{\frac{1}{2}}} = 0 \tag{15.64}$$

on substituting for R from (15.54). The solution of (15.64) is

$$\theta = \frac{BL}{l\alpha}\exp\left(-\frac{2kt}{l\rho\alpha}\right)\int_0^t \frac{1}{t_1^{\frac{1}{2}}}\exp\left(\frac{2k_1}{l\rho\alpha}\right)dt_1. \tag{15.65}$$

When expressed in terms of the variables $h^{\frac{1}{2}}\phi$ and $h\tau$ used in Fig. 15.5, eqn (15.65) becomes

$$h^{\frac{1}{2}}\phi = \exp(-2h\tau)\int_0^\tau \exp(2h\tau_1)\,d(h^{\frac{1}{2}}\tau_1^{\frac{1}{2}}). \tag{15.66}$$

Tabulated values of the right-hand side of (15.66) are given by Sakamoto (1928) for values of $(h\tau)^{\frac{1}{2}}$ in the range $0 < 2h\tau < 100$. The solution (15.66) is shown as the curve marked $K = \infty$ in Fig. 15.5.

15.7.3. *Calculated results. General solution in non-dimensional variables*

In Fig. 15.5, $h^{\frac{1}{2}}\phi$ is plotted against $(h\tau)^{\frac{1}{2}}$ for different values of $h = kl/K$. Since both variables and the parameter h are non-dimensional, the variation of surface temperature with time can be deduced from this figure for any system (subject to the basic assumptions) by inserting appropriate values of the physical constants involved.

The following general points are of interest.

(a) For any finite conductivity the surface temperature rises discontinuously at $t = 0$ to some finite value and immediately afterwards begins to fall.

(b) For the smaller values of h, i.e. larger conductivities, the surface temperature passes first through a minimum and later through a maximum before falling eventually to the temperature of the surroundings. The initial fall in temperature occurs while the sheet is behaving semi-infinitely with respect to heat flow. The temperature rises when an appreciable amount of the heat entering the sheet through its second surface reaches the surface under consideration. The later maximum in surface temperature occurs when heat is removed from the surface by conduction and radiation (and possibly convection) as quickly as it is given up by the vapour entering the sheet. It is to be remembered that the rate of uptake of vapour, and hence the rate of supply of heat to the surface, steadily decreases, being proportional to $1/\sqrt{t}$.

It is to be expected that the smaller the heat conductivity of the sheet the longer the time that elapses before the heat from one surface reaches the other and hence the later the minimum temperature occurs.

(c) For $h > 1$ no minimum or maximum temperatures occur, but the surface temperature falls continuously for $t > 0$, so that theoretically the highest temperature is achieved instantaneously at zero time.

(d) When the conductivity is infinite, the surface temperature rises continuously from that of the surroundings up to a maximum and then falls. For zero conductivity the surface temperature is theoretically infinite at $t = 0$ and then falls steadily.

(e) To within the accuracy of plotting, the maximum temperature can be deduced from the curve for $K = \infty$ even if the conductivity is finite, provided $kl/K < 0.2$ approximately.

(f) For a given sheet the temperature change in degrees is proportional to the heat of vaporization L and also to the constant B. We have seen in eqns (15.54) and (15.55) that B is the gradient of the linear plot of regain against $(t/l^2)^{\frac{1}{2}}$, and is also proportional to the product of the equilibrium regain and the square root of an average diffusion coefficient.

The following are two examples of the use of the general solution.

(i) *Effect of thickness.* As an example of the use of the curves of Fig. 15.5 we have calculated the way in which the temperature at the surface of a polymer sheet varies with time as it takes up vapour, and how this depends on thickness. The following values are taken as representative of a variety of polymers and vapours:

$$\rho = 1{\cdot}0, \qquad \alpha = 0{\cdot}35, \qquad L = 100, \qquad k = 1{\cdot}6 \times 10^{-4},$$
$$K = 4{\cdot}5 \times 10^{-4},$$

all in c.g.s. units. The value of k comes from Stefan's constant and relates to an ambient temperature of 35 °C. The value $B = 7{\cdot}10 \times 10^{-5}$ has been used. It corresponds to a system in which the equilibrium regain is 5 per cent and the average diffusion coefficient is

$$4 \times 10^{-7} \text{ cm}^2 \text{ s}^{-1}.$$

Temperature–time curves are plotted in Fig. 15.6 for a number of thicknesses. For a sheet 0·07 cm thick the maximum temperature change is approximately 1·4 °C at about 65 s. The maximum temperature is lower and occurs later, the thicker the sheet, but for a sheet as thick as 5·66 cm the highest temperature is 0·25 °C attained at $t = 0$. We note incidentally that in theory for all thicknesses the surface temperature changes instantaneously by 0·25 °C at the beginning of sorption. As we have already mentioned, corresponding curves for other equilibrium regains in the same system can be deduced from those of Fig. 15.6 since θ is directly proportional to B and hence to C_0.

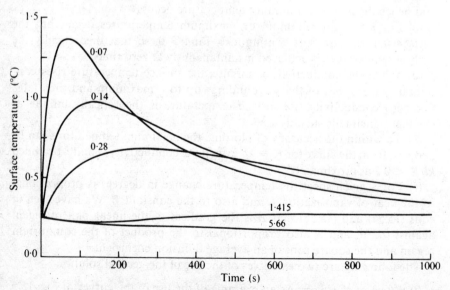

FIG. 15.6. Variation of surface temperature with time, shown as excess over 35 °C. Numbers on curves are sheet thicknesses in centimetres, $\alpha = 0{\cdot}35$, $L = 100$, $\rho = 1$, $B = 7{\cdot}1 \times 10^{-5}$, $k = 1{\cdot}6 \times 10^{-4}$, $K = 4{\cdot}5 \times 10^{-4}$. All units are c.g.s.

(ii) *Limiting rate of sorption for* 1 °*C rise in surface temperature.* G. S. Park suggested that, from an experimental point of view, it is useful to know what is the fastest sorption or alternatively the greatest regain for which the rise in surface temperature does not exceed 1 °C. Accordingly, in Fig. 15.7 are plotted curves which show how the limiting value of B for a 1 °C rise in temperature depends on the thickness of the sheet. The thicker the sheet the faster the sorption permissible.

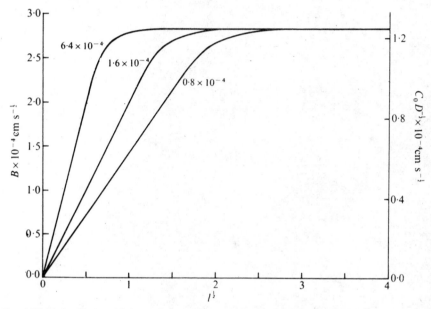

FIG. 15.7. Upper limits of B for surface temperature rise of 1 °C. Numbers on curves are values of k. l is the thickness of the sheet in centimetres. $\alpha = 0.35$, $L = 100$, $\rho = 1$, $K = 4.5 \times 10^{-4}$. All units are c.g.s.

The three curves of Fig. 15.7 are for different values of the emissivity constant k. The middle of the three corresponds to a surface loss purely by radiation when the temperature of the surroundings is 35 °C. Changes in k may be due to differences in ambient temperature θ_0 since $k = 4\sigma_0\theta_0^3$. Alternatively a higher value of k can correspond to some loss of heat by convection in the vapour, since both the convective term and the radiation term in the heat loss are directly proportional to temperature difference for small temperature excesses. Thus the curve on Fig. 15.7 for $k = 6.4 \times 10^{-4}$ could correspond to an experiment in which a stream of vapour is passed over the surfaces of the sheet.

15.8.

Ash and Barrer (1963) give expressions for the continued flow of matter and heat in the gas phase, surface phase, and intracrystalline phase simultaneously, using the equations of irreversible thermodynamics.

TABLES

TABLE 2.1
Table of the error function and associated functions

x	$e^{x^2}\,\text{erfc}\,x$	$4\pi^{-\frac{1}{2}}x\,e^{-x^2}$	$2\pi^{-\frac{1}{2}}e^{-x^2}$	$\text{erf}\,x$	$\text{erfc}\,x$	$2\,\text{ierfc}\,x$
0	1·0	0	1·1284	0	1·0	1·1284
0·05	0·9460	0·1126	1·1256	0·056372	0·943628	1·0312
0·1	0·8965	0·2234	1·1172	0·112463	0·887537	0·9396
0·15	0·8509	0·3310	1·1033	0·167996	0·832004	0·8537
0·2	0·8090	0·4336	1·0841	0·222703	0·777297	0·7732
0·25	0·7703	0·5300	1·0600	0·276326	0·723674	0·6982
0·3	0·7346	0·6188	1·0313	0·328627	0·671373	0·6284
0·35	0·7015	0·6988	0·9983	0·379382	0·620618	0·5639
0·4	0·6708	0·7692	0·9615	0·428392	0·571608	0·5043
0·45	0·6423	0·8294	0·9215	0·475482	0·524518	0·4495
0·5	0·6157	0·8788	0·8788	0·520500	0·479500	0·3993
0·55	0·5909	0·9172	0·8338	0·563323	0·436677	0·3535
0·6	0·5678	0·9447	0·7872	0·603856	0·396144	0·3119
0·65	0·5462	0·9614	0·7395	0·642029	0·357971	0·2742
0·7	0·5259	0·9678	0·6913	0·677801	0·322199	0·2402
0·75	0·5069	0·9644	0·6429	0·711156	0·288844	0·2097
0·8	0·4891	0·9520	0·5950	0·742101	0·257899	0·1823
0·85	0·4723	0·9314	0·5479	0·770668	0·229332	0·1580
0·9	0·4565	0·9035	0·5020	0·796908	0·203092	0·1364
0·95	0·4416	0·8695	0·4576	0·820891	0·179109	0·1173
1·0	0·4276	0·8302	0·4151	0·842701	0·157299	0·1005
1·1	0·4017	0·7403	0·3365	0·880205	0·119795	0·0729
1·2	0·3785	0·6416	0·2673	0·910314	0·089686	0·0521
1·3	0·3576	0·5413	0·2082	0·934008	0·065992	0·0366
1·4	0·3387	0·4450	0·1589	0·952285	0·047715	0·0253
1·5	0·3216	0·3568	0·1189	0·966105	0·033895	0·0172
1·6	0·3060	0·2791	0·0872	0·976348	0·023652	0·0115
1·7	0·2917	0·2132	0·0627	0·983790	0·016210	0·0076
1·8	0·2786	0·1591	0·0442	0·989091	0·010909	0·0049
1·9	0·2665	0·1160	0·0305	0·992790	0·007210	0·0031
2·0	0·2554	0·0827	0·0207	0·995322	0·004678	0·0020
2·1	0·2451	0·0576	0·0137	0·997021	0·002979	0·0012
2·2	0·2356	0·0393	0·0089	0·998137	0·001863	0·0007
2·3	0·2267	0·0262	0·0057	0·998857	0·001143	0·0004
2·4	0·2185	0·0171	0·0036	0·999311	0·000689	0·0002
2·5	0·2108	0·0109	0·0022	0·999593	0·000407	0·0001
2·6	0·2036	0·0068	0·0013	0·999764	0·000236	0·0001
2·7	0·1969	0·0042	0·0008	0·999866	0·000134	
2·8	0·1905	0·0025	0·0004	0·999925	0·000075	
2·9	0·1846	0·0015	0·0003	0·999959	0·000041	
3·0	0·1790	0·0008	0·0001	0·999978	0·000022	

TABLE 2.1 (*contd.*)

x	$4\,i^2\mathrm{erfc}\,x$	$6\,i^3\mathrm{erfc}\,x$	$8\,i^4\mathrm{erfc}\,x$	$10\,i^5\mathrm{erfc}\,x$	$12\,i^6\mathrm{erfc}\,x$
0	1·0	0·5642	0·25	0·0940	0·0313
0·05	0·8921	0·4933	0·2148	0·0795	0·0261
0·1	0·7936	0·4301	0·1841	0·0671	0·0217
0·15	0·7040	0·3740	0·1573	0·0564	0·0180
0·2	0·6227	0·3243	0·1341	0·0474	0·0149
0·25	0·5491	0·2805	0·1139	0·0396	0·0123
0·3	0·4828	0·2418	0·0965	0·0331	0·0101
0·35	0·4233	0·2079	0·0816	0·0275	0·0083
0·4	0·3699	0·1782	0·0687	0·0228	0·0068
0·45	0·3223	0·1522	0·0577	0·0189	0·0055
0·5	0·2799	0·1297	0·0484	0·0156	0·0045
0·55	0·2423	0·1101	0·0404	0·0128	0·0036
0·6	0·2090	0·0932	0·0336	0·0105	0·0029
0·65	0·1798	0·0787	0·0279	0·0086	0·0024
0·7	0·1541	0·0662	0·0231	0·0070	0·0019
0·75	0·1316	0·0555	0·0190	0·0057	0·0015
0·8	0·1120	0·0464	0·0156	0·0046	0·0012
0·85	0·0950	0·0386	0·0128	0·0037	0·0010
0·9	0·0803	0·0321	0·0104	0·0030	0·0008
0·95	0·0677	0·0265	0·0085	0·0024	0·0006
1·0	0·0568	0·0218	0·0069	0·0019	0·0005
1·1	0·0396	0·0147	0·0045	0·0012	0·0003
1·2	0·0272	0·0097	0·0029	0·0007	0·0002
1·3	0·0184	0·0063	0·0019	0·0004	0·0001
1·4	0·0122	0·0041	0·0011	0·0003	0·0001
1·5	0·0080	0·0026	0·0007	0·0002	
1·6	0·0052	0·0016	0·0004	0·0001	
1·7	0·0033	0·0010	0·0003		
1·8	0·0021	0·0006	0·0002		
1·9	0·0013	0·0003	0·0001		
2·0	0·0008	0·0002	0·0001		
2·1	0·0005	0·0001			
2·2	0·0003				
2·3	0·0002				
2·4	0·0001				
2·5					
2·6					
2·7					
2·8					
2·9					
3·0					

$$\mathrm{ierfc}\,x = \int_x^\infty \mathrm{erfc}\,\xi\,\mathrm{d}\xi = \frac{1}{\pi^{\frac{1}{2}}}\,e^{-x^2} - x\,\mathrm{erfc}\,x$$

$$i^2\mathrm{erfc}\,x = \int_x^\infty \mathrm{ierfc}\,\xi\,\mathrm{d}\xi = \tfrac{1}{4}\left\{(1+2x^2)\,\mathrm{erfc}\,x - \frac{2}{\pi^{\frac{1}{2}}}\,x\,e^{-x^2}\right\} = \tfrac{1}{4}(\mathrm{erfc}\,x - 2x\,\mathrm{ierfc}\,x)$$

$$2n\,i^n\mathrm{erfc}\,x = 2n\int_x^\infty i^{n-1}\mathrm{erfc}\,\xi\,\mathrm{d}\xi = i^{n-2}\mathrm{erfc}\,x - 2x\,i^{n-1}\mathrm{erfc}\,x.$$

TABLE 2.2
Table of Laplace transforms

$$\bar{v}(p) = \int_0^{\infty} e^{-pt} v(t)\, dt$$

We write $q = \sqrt{(p/D)}$. D, x and h are always positive. α is unrestricted.

	$\bar{v}(p)$	$v(t)$
1.	$\dfrac{1}{p}$	1
2.	$\dfrac{1}{p^{\nu+1}}, \quad \nu > -1$	$\dfrac{t^{\nu}}{\Gamma(\nu+1)}$
3.	$\dfrac{1}{p+\alpha}$	$e^{-\alpha t}$
4.	$\dfrac{\omega}{p^2+\omega^2}$	$\sin \omega t$
5.	$\dfrac{p}{p^2+\omega^2}$	$\cos \omega t$
6.	e^{-qx}	$\dfrac{x}{2\sqrt{(\pi D t^3)}}\, e^{-x^2/4Dt}$
7.	$\dfrac{e^{-qx}}{q}$	$\left(\dfrac{D}{\pi t}\right)^{\frac{1}{2}} e^{-x^2/4Dt}$
8.	$\dfrac{e^{-qx}}{p}$	$\operatorname{erfc}\dfrac{x}{2\sqrt{(Dt)}}$
9.	$\dfrac{e^{-qx}}{pq}$	$2\left(\dfrac{Dt}{\pi}\right)^{\frac{1}{2}} e^{-x^2/4Dt} - x\operatorname{erfc}\dfrac{x}{2\sqrt{(Dt)}}$
10.	$\dfrac{e^{-qx}}{p^2}$	$\left(t+\dfrac{x^2}{2D}\right)\operatorname{erfc}\dfrac{x}{2\sqrt{(Dt)}} - x\left(\dfrac{t}{\pi D}\right)^{\frac{1}{2}} e^{-x^2/4Dt}$
11.	$\dfrac{e^{-qx}}{p^{1+\frac{1}{2}n}}, \quad n = 0, 1, 2, \ldots$	$(4t)^{\frac{1}{2}n} i^n \operatorname{erfc}\dfrac{x}{2\sqrt{(Dt)}}$
12.	$\dfrac{e^{-qx}}{q+h}$	$\left(\dfrac{D}{\pi t}\right)^{\frac{1}{2}} e^{-x^2/4Dt} - hD\, e^{hx+Dth^2}$ $\times \operatorname{erfc}\left\{\dfrac{x}{2\sqrt{(Dt)}} + h\sqrt{(Dt)}\right\}$
13.	$\dfrac{e^{-qx}}{q(q+h)}$	$D\, e^{hx+Dth^2} \operatorname{erfc}\left\{\dfrac{x}{2\sqrt{(Dt)}} + h\sqrt{(Dt)}\right\}$

TABLE 2.2 (contd.)

	$\bar{v}(p)$	$v(t)$
14.	$\dfrac{e^{-qx}}{p(q+h)}$	$\dfrac{1}{h}\,\text{erfc}\,\dfrac{x}{2\sqrt{(Dt)}}-\dfrac{1}{h}e^{hx+Dth^2}$
		$\times\,\text{erfc}\left\{\dfrac{x}{2\sqrt{(Dt)}}+h\sqrt{(Dt)}\right\}$
15.	$\dfrac{e^{-qx}}{pq(q+h)}$	$\dfrac{2}{h}\left(\dfrac{Dt}{\pi}\right)^{\frac{1}{2}}e^{-x^2/4Dt}-\dfrac{(1+hx)}{h^2}\,\text{erfc}\,\dfrac{x}{2\sqrt{(Dt)}}$
		$+\dfrac{1}{h^2}\,e^{hx+Dth^2}\,\text{erfc}\left\{\dfrac{x}{2\sqrt{(Dt)}}+h\sqrt{(Dt)}\right\}$
16.	$\dfrac{e^{-qx}}{q^{n+1}(q+h)}$	$\dfrac{D}{(-h)^n}e^{hx+Dth^2}\,\text{erfc}\left\{\dfrac{x}{2\sqrt{(Dt)}}+h\sqrt{(Dt)}\right\}$
		$-\dfrac{D}{(-h)^n}\sum_{r=0}^{n-1}\{-2h\sqrt{(Dt)}\}^r i^r\,\text{erfc}\,\dfrac{x}{2\sqrt{(Dt)}}$
17.	$\dfrac{e^{-qx}}{(q+h)^2}$	$-2h\left(\dfrac{D^3t}{\pi}\right)^{\frac{1}{2}}e^{-x^2/4Dt}+D(1+hx+2h^2Dt)$
		$\times\,e^{hx+Dth^2}\,\text{erfc}\left\{\dfrac{x}{2\sqrt{(Dt)}}+h\sqrt{(Dt)}\right\}$
18.	$\dfrac{e^{-qx}}{p(q+h)^2}$	$\dfrac{1}{h^2}\,\text{erfc}\,\dfrac{x}{2\sqrt{(Dt)}}-\dfrac{2}{h}\left(\dfrac{Dt}{\pi}\right)^{\frac{1}{2}}e^{-x^2/4Dt}$
		$-\dfrac{1}{h^2}\{1-hx-2h^2Dt\}\,e^{hx+Dth^2}$
		$\times\,\text{erfc}\left\{\dfrac{x}{2\sqrt{(Dt)}}+h\sqrt{(Dt)}\right\}$
19.	$\dfrac{e^{-qx}}{p-\alpha}$	$\tfrac{1}{2}e^{\alpha t}\left[e^{-x\sqrt{(\alpha/D)}}\,\text{erfc}\left\{\dfrac{x}{2\sqrt{(Dt)}}-\sqrt{(\alpha t)}\right\}\right.$
		$\left.+\,e^{x\sqrt{(\alpha/D)}}\,\text{erfc}\left\{\dfrac{x}{2\sqrt{(Dt)}}+\sqrt{(\alpha t)}\right\}\right]$
20.	$\dfrac{1}{p^{\frac{1}{4}}}e^{-qx}$	$\dfrac{1}{\pi}\left(\dfrac{x}{2tD^{\frac{1}{2}}}\right)^{\frac{1}{2}}e^{-x^2/8Dt}K_{\frac{1}{4}}\left(\dfrac{x^2}{8Dt}\right)$
21.	$\dfrac{1}{p^{\frac{1}{2}}}K_{2v}(qx)$	$\dfrac{1}{2\sqrt{(\pi t)}}e^{-x^2/8Dt}K_v\left(\dfrac{x^2}{8Dt}\right)$
22.	$\left.\begin{array}{l}I_v(qx')K_v(qx),\quad x>x'\\ I_v(qx)K_v(qx'),\quad x<x'\end{array}\right\}$	$\dfrac{1}{2t}e^{-(x^2+x'^2)/4Dt}I_v\left(\dfrac{xx'}{2Dt}\right),\quad v\geqslant 0$
23.	$K_0(qx)$	$\dfrac{1}{2t}e^{-x^2/4Dt}$
24.	$\dfrac{1}{p}e^{x/p}$	$I_0[2\sqrt{(xt)}]$

TABLE 2.2 (contd.)

$\bar{v}(p)$	$v(t)$
25. $\dfrac{\exp[xp - x\{(p+a)(p+b)\}^{\frac{1}{2}}]}{\{(p+a)(p+b)\}^{\frac{1}{2}}}$	$e^{-\frac{1}{2}(a+b)(t+x)}I_0[\tfrac{1}{2}(a-b)\{t(t+2x)\}^{\frac{1}{2}}]$
26. $p^{\frac{1}{2}\nu-1}K_\nu(x\sqrt{p})$	$x^{-\nu}2^{\nu-1}\displaystyle\int_{x^2/4t}^{\infty}e^{-u}u^{\nu-1}\,du$
27. $\{p - \sqrt{(p^2-x^2)}\}^\nu, \quad \nu > 0$	$\nu x^\nu I_\nu(xt)/t$
28. $\dfrac{\exp[x\{(p+a)^{\frac{1}{2}}-(p+b)^{\frac{1}{2}}\}^2]}{(p+a)^{\frac{1}{2}}(p+b)^{\frac{1}{2}}[(p+a)^{\frac{1}{2}}+(p+b)^{\frac{1}{2}}]^{2\nu}},$ $\nu \geqslant 0$	$\dfrac{t^{\frac{1}{2}\nu}e^{-\frac{1}{2}(a+b)t}I_\nu\{\tfrac{1}{2}(a-b)t^{\frac{1}{2}}(t+4x)^{\frac{1}{2}}\}}{(a-b)^\nu(t+4x)^{\frac{1}{2}\nu}}$

TABLE 4.1
Roots of $\tan q_n = -\alpha q_n$

Fractional uptake	α	q_1	q_2	q_3	q_4	q_5	q_6
0	∞	1·5708	4·7124	7·8540	10·9956	14·1372	17·2788
0·1	9·0000	1·6385	4·7359	7·8681	11·0057	14·1451	17·2852
0·2	4·0000	1·7155	4·7648	7·8857	11·0183	14·1549	17·2933
0·3	2·3333	1·8040	4·8014	7·9081	11·0344	14·1674	17·3036
0·4	1·5000	1·9071	4·8490	7·9378	11·0558	14·1841	17·3173
0·5	1·0000	2·0288	4·9132	7·9787	11·0856	14·2075	17·3364
0·6	0·6667	2·1746	5·0037	8·0385	11·1296	14·2421	17·3649
0·7	0·4286	2·3521	5·1386	8·1334	11·2010	14·2990	17·4119
0·8	0·2500	2·5704	5·3540	8·3029	11·3349	14·4080	17·5034
0·9	0·1111	2·8363	5·7172	8·6587	11·6532	14·6870	17·7481
1·0	0	3·1416	6·2832	9·4248	12·5664	15·7080	18·8496

TABLE 4.2
Roots of $\beta \tan \beta = L$

L	β_1	β_2	β_3	β_4	β_5	β_6
0	0	3·1416	6·2832	9·4248	12·5664	15·7080
0·01	0·0998	3·1448	6·2848	9·4258	12·5672	15·7086
0·1	0·3111	3·1731	6·2991	9·4354	12·5743	15·7143
0·2	0·4328	3·2039	6·3148	9·4459	12·5823	15·7207
0·5	0·6533	3·2923	6·3616	9·4775	12·6060	15·7397
1·0	0·8603	3·4256	6·4373	9·5293	12·6453	15·7713
2·0	1·0769	3·6436	6·5783	9·6296	12·7223	15·8336
5·0	1·3138	4·0336	6·9096	9·8928	12·9352	16·0107
10·0	1·4289	4·3058	7·2281	10·2003	13·2142	16·2594
100·0	1·5552	4·6658	7·7764	10·8871	13·9981	17·1093
∞	1·5708	4·7124	7·8540	10·9956	14·1372	17·2788

TABLE 5.1
Roots of $\alpha q_n J_0(q_n) + 2J_1(q_n) = 0$

Fractional uptake	α	q_1	q_2	q_3	q_4	q_5	q_6
0	∞	2.4048	5.5201	8.6537	11.7915	14.9309	18.0711
0.1	9.0000	2.4922	5.5599	8.6793	11.8103	14.9458	18.0833
0.2	4.0000	2.5888	5.6083	8.7109	11.8337	14.9643	18.0986
0.3	2.3333	2.6962	5.6682	8.7508	11.8634	14.9879	18.1183
0.4	1.5000	2.8159	5.7438	8.8028	11.9026	15.0192	18.1443
0.5	1.0000	2.9496	5.8411	8.8727	11.9561	15.0623	18.1803
0.6	0.6667	3.0989	5.9692	8.9709	12.0334	15.1255	18.2334
0.7	0.4286	3.2645	6.1407	9.1156	12.1529	15.2255	18.3188
0.8	0.2500	3.4455	6.3710	9.3397	12.3543	15.4031	18.4754
0.9	0.1111	3.6374	6.6694	9.6907	12.7210	15.7646	18.8215
1.0	0	3.8317	7.0156	10.1735	13.3237	16.4706	19.6159

TABLE 5.2
Roots of $\beta J_1(\beta) - L J_0(\beta) = 0$

L	β_1	β_2	β_3	β_4	β_5	β_6
0	0	3.8137	7.0156	10.1735	13.3237	16.4706
0.01	0.1412	3.8343	7.0170	10.1745	13.3244	16.4712
0.1	0.4417	3.8577	7.0298	10.1833	13.3312	16.4767
0.2	0.6170	3.8835	7.0440	10.1931	13.3387	16.4828
0.5	0.9408	3.9594	7.0864	10.2225	13.3611	16.5010
1.0	1.2558	4.0795	7.1558	10.2710	13.3984	16.5312
2.0	1.5994	4.2910	7.2884	10.3658	13.4719	16.5910
5.0	1.9898	4.7131	7.6177	10.6223	13.6786	16.7630
10.0	2.1795	5.0332	7.9569	10.9363	13.9580	17.0099
100.0	2.3809	5.4652	8.5678	11.6747	14.7834	17.8931
∞	2.4048	5.5201	8.6537	11.7915	14.9309	18.0711

TABLE 5.3
Roots of $J_0(a\alpha_n)Y_0(b\alpha_n) - J_0(b\alpha_n)Y_0(a\alpha_n)$

b/a	$a\alpha_1$	$a\alpha_2$	$a\alpha_3$	$a\alpha_4$	$a\alpha_5$
1.2	15.7014	31.4126	47.1217	62.8302	78.5385
1.5	6.2702	12.5598	18.8451	25.1294	31.4133
2.0	3.1230	6.2734	9.4182	12.5614	15.7040
2.5	2.0732	4.1773	6.2754	8.3717	10.4672
3.0	1.5485	3.1291	4.7038	6.2767	7.8487
3.5	1.2339	2.5002	3.7608	5.0196	6.2776
4.0	1.0244	2.0809	3.1322	4.1816	5.2301

TABLE 6.1

$$Roots\ of\ \tan q_n = \frac{3q_n}{3 + \alpha q_n^2}$$

Fractional uptake	α	q_1	q_2	q_3	q_4	q_5	q_6
0	∞	3·1416	6·2832	9·4248	12·5664	15·7080	18·8496
0·1	9·0000	3·2410	6·3353	9·4599	12·5928	15·7292	18·8671
0·2	4·0000	3·3485	6·3979	9·5029	12·6254	15·7554	18·8891
0·3	2·3333	3·4650	6·4736	9·5567	12·6668	15·7888	18·9172
0·4	1·5000	3·5909	6·5665	9·6255	12·7205	15·8326	18·9541
0·5	1·0000	3·7264	6·6814	9·7156	12·7928	15·8924	19·0048
0·6	0·6667	3·8711	6·8246	9·8369	12·8940	15·9779	19·0784
0·7	0·4286	4·0236	7·0019	10·0039	13·0424	16·1082	19·1932
0·8	0·2500	4·1811	7·2169	10·2355	13·2689	16·3211	19·3898
0·9	0·1111	4·3395	7·4645	10·5437	13·6133	16·6831	19·7564
1·0	0	4·4934	7·7253	10·9041	14·0662	17·2208	20·3713

TABLE 6.2

$$Roots\ of\ \beta_n \cot \beta_n + L - 1 = 0$$

L	β_1	β_2	β_3	β_4	β_5	β_6
0	0	4·4934	7·7253	10·9041	14·0662	17·2208
0·01	0·1730	4·4956	7·7265	10·9050	14·0669	17·2213
0·1	0·5423	4·5157	7·7382	10·9133	14·0733	17·2266
0·2	0·7593	4·5379	7·7511	10·9225	14·0804	17·2324
0·5	1·1656	4·6042	7·7899	10·9499	14·1017	17·2498
1·0	1·5708	4·7124	7·8540	10·9956	14·1372	17·2788
2·0	2·0288	4·9132	7·9787	11·0856	14·2075	17·3364
5·0	2·5704	5·3540	8·3029	11·3349	14·4080	17·5034
10·0	2·8363	5·7172	8·6587	11·6532	14·6870	17·7481
100·0	3·1102	6·2204	9·3309	12·4414	15·5522	18·6633
∞	3·1416	6·2832	9·4248	12·5664	15·7080	18·8496

Table 7.1

Numerical solutions of $\dfrac{d^2 s}{d\eta^2} = -2\eta\, e^{s\ln(1+A)}\dfrac{ds}{d\eta}$ subject to $s.\ s = 1,\ \eta = 0$;

$$s = 0,\quad \eta = +\infty$$

A	0.500	1.000	2.000	4.000	6.000	10.00	25.00	100.0	200.0	500.0	1000.0	1500.0	3500.0
$\Delta\eta$	10^{-2}	10^{-2}	10^{-2}	10^{-2}	10^{-2}	10^{-2}	10^{-2}	10^{-2}	10^{-2}	$5\cdot10^{-3}$	$5\cdot10^{-3}$	$5\cdot10^{-3}$	10^{-3}
Iterations required	4	8	6	7	8	7	8	10	10	10	12	12	14
$-(ds/d\eta)_{\eta=0}$	1.212	1.286	1.399	1.565	1.699	1.895	2.389	3.576	4.464	6.091	7.787	9.027	12.38
η	s	s	s	s	s	s	s	s	s	s	s	s	s
0.0	1.000	1.000	1.000	1.000	1.000	1.000	1.000	1.000	1.000	1.000	1.000	1.000	1.000
0.1	0.879	0.872	0.861	0.846	0.833	0.815	0.775	0.695	0.649	0.589	0.545	0.521	0.474
0.2	0.762	0.749	0.730	0.703	0.682	0.655	0.596	0.502	0.458	0.406	0.371	0.353	0.319
0.3	0.651	0.634	0.609	0.576	0.552	0.521	0.460	0.375	0.339	0.297	0.271	0.257	0.232
0.4	0.548	0.528	0.501	0.467	0.442	0.413	0.357	0.285	0.256	0.223	0.203	0.192	0.173
0.5	0.455	0.435	0.408	0.375	0.352	0.326	0.277	0.218	0.195	0.169	0.155	0.146	0.132
0.6	0.373	0.353	0.328	0.299	0.277	0.256	0.214	0.167	0.150	0.129	0.119	0.112	0.101
0.7	0.301	0.283	0.260	0.236	0.217	0.200	0.165	0.127	0.115	0.098	0.091	0.085	0.077
0.8	0.240	0.223	0.204	0.184	0.168	0.155	0.126	0.097	0.088	0.075	0.070	0.065	0.059
0.9	0.188	0.174	0.158	0.143	0.128	0.119	0.096	0.073	0.067	0.056	0.054	0.050	0.045
1.0	0.146	0.134	0.121	0.110	0.097	0.091	0.072	0.055	0.051	0.042	0.041	0.037	0.034
1.1	0.112	0.102	0.092	0.083	0.073	0.069	0.053	0.041	0.038	0.031	0.031	0.028	0.025
1.2	0.084	0.076	0.069	0.063	0.054	0.052	0.039	0.030	0.029	0.028	0.024	0.021	0.019
1.3	0.063	0.056	0.051	0.048	0.040	0.040	0.028	0.021	0.021	0.017	0.018	0.016	0.014
1.4	0.047	0.041	0.037	0.036	0.028	0.030	0.020	0.015	0.016	0.012	0.015	0.012	0.011
1.5	0.035	0.030	0.027	0.027	0.020	0.023	0.014	0.010	0.012	0.008	0.012	0.009	0.008
1.6	0.025	0.022	0.020	0.021	0.014	0.018	0.010	0.007	0.009	0.006	0.009	0.007	0.006
1.7	0.019	0.016	0.014	0.016	0.010	0.014	0.007	0.005	0.007	0.004	0.008	0.005	0.005
1.8	0.014	0.011	0.011	0.013	0.007	0.011	0.004	0.004	0.006	0.002	0.007	0.004	
1.9	0.011	0.008	0.008	0.011	0.005	0.009	0.003	0.002					
2.0	0.009	0.006	0.006	0.011	0.003	0.008	0.002						
2.1	0.007	0.005	0.005	0.010	0.002								
2.2	0.006	0.004											

From Lee 1971.(a).

TABLE 7.2

Exponential diffusion in an infinite medium; numerical values of the function
$$1000\psi(y, g)$$

y \ g	0·0	0·2	0·4	0·6	0·8	1·0	1·2	1·4	1·6	1·8	2·0	2·2	2·4
−2·0	−995	−998	−1000										
−1·9	−993	−996	−999	−1000									
−1·8	−989	−994	−998	−999									
−1·7	−984	−990	−996	−998	−1000								
−1·6	−976	−984	−992	−996	−999	−1000							
−1·5	−966	−976	−985	−992	−997	−999	−1000						
−1·4	−952	−964	−976	−985	−993	−997	−999						
−1·3	−934	−948	−962	−975	−986	−994	−998	−1000					
−1·2	−910	−925	−942	−958	−974	−986	−995	−999					
−1·1	−880	−896	−914	−933	−953	−972	−987	−996	−1000				
−1·0	−843	−858	−876	−896	−920	−945	−969	−987	−997	−1000	−1000		
−0·9	−797	−810	−826	−845	−872	−901	−933	−963	−986	−998	−999		
−0·8	−742	−751	−765	−781	−804	−834	−870	−912	−952	−984	−997	−1000	
−0·7	−678	−682	−690	−700	−717	−741	−774	−817	−869	−928	−975	−996	−1000
−0·6	−604	−602	−602	−605	−612	−626	−647	−678	−721	−785	−866	−946	−993
−0·5	−520	−511	−504	−497	−494	−494	−498	−501	−528	−562	−615	−697	−831
−0·4	−429	−412	−397	−381	−366	−354	−343	−334	−329	−329	−336	−353	−393
−0·3	−329	−306	−282	−259	−237	−214	−191	−169	−148	−127	−107	−89	−76
−0·2	−223	−195	−117	−138	−109	−79	−50	−20	9	39	70	99	129
−0·1	−112	−81	−50	−18	14	46	78	111	143	176	209	240	272
0·0	0	32	64	97	129	161	193	225	256	288	320	349	380
+0·1	112	143	174	205	235	265	295	325	353	382	411	438	466
+0·2	223	250	278	305	332	359	385	411	436	462	487	511	535
+0·3	329	352	374	397	420	442	465	487	508	530	552	572	593
+0·4	429	445	462	481	499	517	535	553	570	589	607	624	642
+0·5	520	531	543	555	568	582	596	611	624	640	654	668	684
+0·6	604	608	614	622	631	640	650	661	672	684	696	709	720
+0·7	678	677	678	681	686	691	698	706	713	723	732	740	751
+0·8	742	737	733	733	734	736	740	745	750	757	764	771	779
+0·9	797	788	781	778	776	775	777	779	782	787	792	798	804
+1·0	843	831	823	816	812	810	809	809	810	813	817	820	825
+1·1	880	868	858	849	844	839	837	836	835	837	839	841	845
+1·2	910	899	886	878	871	865	861	859	857	858	858	859	862
+1·3	934	922	911	901	894	888	883	879	878	876	876	876	877
+1·4	952	941	931	921	913	906	901	897	894	892	891	890	891
+1·5	966	956	947	938	930	924	917	913	909	906	905	903	903
+1·6	976	968	959	951	943	937	931	926	922	919	917	915	915
+1·7	984	977	969	962	955	948	943	938	933	930	928	924	924
+1·8	989	984	977	971	964	958	953	948	943	940	937	934	933
+1·9	993	989	983	978	972	966	961	956	952	949	946	943	941
+2·0	995	992	988	983	978	973	968	964	959	956	952	950	948

TABLE 7.3

Exponential diffusion in a semi-infinite medium; numerical values of the function $1000c(y_s, g)$

y_s \ g	-2.4	-2.2	-2.0	-1.8	-1.6	-1.4	-1.2	-1.0	-0.8	-0.6	-0.4	-0.2	-0.00
0.0	0	0	0	0	0	0	0	0	0	0	0	0	0
0.1	78	81	84	87	90	93	96	99	102	105	107	110	112
0.2	182	185	189	193	196	200	204	207	210	214	217	220	223
0.3	330	325	323	322	321	322	322	323	324	325	325	328	329
0.4	600	520	497	479	466	457	449	443	439	435	433	430	429
0.5	877	775	709	660	624	600	580	564	551	541	534	527	520
0.6	995	960	898	833	778	737	704	678	657	639	626	614	604
0.7	1000	997	981	944	896	851	811	777	750	727	708	692	678
0.8		1000	998	987	962	928	891	857	827	800	779	759	742
0.9			1000	998	989	970	944	915	886	860	836	816	797
1.0				1000	998	989	974	953	929	905	883	862	843
1.1					1000	996	989	976	958	939	919	899	880
1.2						1000	996	988	977	961	945	928	910
1.3						1000	998	995	987	977	964	949	934
1.4							999	998	994	986	977	965	952
1.5							1000	999	997	992	986	977	966
1.6								1000	999	996	992	985	976
1.7									1000	998	995	990	984
1.8										999	998	994	989
1.9										1000	999	997	993
2.0											1000	998	995

y_s \ g	$+0.2$	$+0.4$	$+0.6$	$+0.8$	$+1.0$	$+1.2$	$+1.4$	$+1.6$	$+1.8$	$+2.0$	$+2.2$	$+2.4$
0.0	0	0	0	0	0	0	0	0	0	0	0	0
0.1	115	117	120	122	124	126	128	130	132	134	136	138
0.2	226	228	231	233	236	238	240	242	244	246	248	250
0.3	330	331	333	334	335	336	338	339	340	341	342	344
0.4	427	426	425	424	424	423	423	422	422	422	422	422
0.5	516	512	508	505	502	499	497	495	494	492	491	489
0.6	595	588	582	576	571	567	563	559	556	553	550	548
0.7	666	656	647	639	632	626	620	615	610	606	602	599
0.8	728	715	704	694	685	677	670	664	658	653	648	644
0.9	781	766	754	742	732	723	715	707	700	694	688	683
1.0	826	811	797	784	773	763	754	745	738	731	724	718
1.1	863	848	833	820	809	798	788	779	771	763	756	749
1.2	896	879	865	851	839	828	818	808	800	792	784	777
1.3	919	905	891	878	866	855	844	835	825	817	809	802
1.4	939	926	913	900	889	878	867	857	848	840	832	824
1.5	955	943	931	919	910	897	887	877	868	860	852	844
1.6	967	956	946	935	925	914	904	895	886	878	870	862
1.7	976	967	958	948	938	929	919	910	902	893	885	878
1.8	983	976	969	959	950	941	932	924	915	903	900	892
1.9	988	982	975	968	960	952	944	935	928	920	912	905
2.0	992	987	981	975	968	961	953	945	938	929	923	916

TABLE 7.4

Linear diffusion in an infinite medium; concentration gradients
$$-(dC/dy_a)/(C_1 - C_2) \text{ against } y_a = x/\{2\sqrt{(D_a t)}\}$$

b = ratio of the diffusion coefficient at concentration C_1 to that at C_2

y_a	$b = 1$†	$b = 0.8806$	$b = 0.7228$	$b = 0.5506$	$b = 0.3270$	$b = 0.1407$
2·8	..	..	..	0·001	0·002	0·003
2·6	0·001	0·001	0·001	0·002	0·004	0·006
2·4	0·002	0·002	0·003	0·005	0·007	0·011
2·2	0·005	0·005	0·007	0·009	0·013	0·017
2·0	0·010	0·012	0·015	0·018	0·023	0·028
1·8	0·022	0·024	0·028	0·032	0·039	0·043
1·6	0·044	0·047	0·050	0·055	0·062	0·066
1·4	0·079	0·083	0·086	0·089	0·092	0·095
1·2	0·134	0·136	0·135	0·134	0·134	0·132
1·0	0·208	0·207	0·200	0·195	0·188	0·179
0·8	0·298	0·292	0·280	0·268	0 250	0·233
0·6	0·394	0·384	0·367	0·348	0·321	0·296
0·4	0·481	0·468	0·452	0·430	0·398	0·365
0·2	0·542	0·533	0·522	0·504	0·474	0·439
0·0	0·564	0·561	0·560	0·555	0·539	0·513
−0·2	0·542	0·547	0·558	0·569	0·579	0·578
−0·4	0·481	0·491	0·511	0·535	0·577	0·621
−0·6	0·394	0·406	0·423	0·455	0·514	0·607
−0·8	0·298	0·308	0·319	0·339	0·38	0·479
−1·0	0·208	0·212	0·215	0·221	0·232	0·226
−1·2	0·134	0·135	0·130	0·125	0·108	0·049
−1·4	0·079	0·077	0·071	0·061	0·039	0·005
−1·6	0·044	0·040	0·035	0·027	0·011	...
−1·8	0·022	0·019	0·015	0·010	0·003	..
−2·0	0·010	0·008	0·006	0·003	..	..
−2·2	0·005	0·003	0·002	0·001	..	..
−2·4	0·002	..	..	..	..	..
−2·6	0·001	..	..	..	..	..
−2·8	..	..	..	..	..	..

† The data for $b = 1$ represent the Gaussian curve corresponding to ideal diffusion.

TABLE 7.5
Linear diffusion in an infinite medium; relative concentration
$$(C - C_2)/(C_1 - C_2) \text{ against } y_a = x/\{2\sqrt{(D_a t)}\}$$

b = ratio of diffusion coefficient at concentration C_1 to that at C_2

y_a	$b = 1$	$b = 0.8806$	$b = 0.7228$	$b = 0.5506$	$b = 0.3270$	$b = 0.1407$
2·6	..	..	..	..	0·001	0·001
2·4	..	..	..	0·001	0·002	0·003
2·2	..	0·001	0·001	0·002	0·004	0·006
2·0	0·002	0·003	0·004	0·005	0·007	0·010
1·8	0·005	0·006	0·008	0·010	0·013	0·017
1·6	0·012	0·013	0·015	0·019	0·024	0·028
1·4	0·024	0·026	0·029	0·033	0·039	0·044
1·2	0·045	0·048	0·051	0·055	0·061	0·068
1·0	0·079	0·082	0·084	0·088	0·093	0·099
0·8	0·129	0·131	0·132	0·134	0·137	0·139
0·6	0·198	0·199	0·197	0·195	0·194	0·192
0·4	0·286	0·284	0·279	0·273	0·266	0·258
0·2	0·389	0·385	0·376	0·367	0·353	0·339
0·0	0·500	0·495	0·485	0·473	0·455	0·434
−0·2	0·611	0·606	0·597	0·586	0·567	0·543
−0·4	0·714	0·710	0·705	0·697	0·683	0·664
−0·6	0·802	0·800	0·799	0·797	0·793	0·785
−0·8	0·871	0·872	0·873	0·876	0·884	0·898
−1·0	0·921	0·923	0·927	0·933	0·946	0·970
−1·2	0·955	0·958	0·961	0·967	0·980	0·995
−1·4	0·976	0·979	0·981	0·986	0·994	0·999
−1·6	0·988	0·990	0·992	0·995	0·998	1·000
−1·8	0·995	0·996	0·997	0·998	1·000	1·000
−2·0	0·998	0·999	0·999	1·000	1·000	1·000
−2·2	1·000	1·000	1·000	1·000	1·000	1·000

TABLE 7.6

Amount of material M that has diffused through unit area of plane x = 0 up to time t, expressed as dimensionless ratio $M/C_0(D_a t)^{\frac{1}{2}}$ (Wilkins 1963)

Infinite medium

$\varepsilon \to 0$: $\quad \pi^{-\frac{1}{2}}\left\{1 - \dfrac{\pi + 2 - 3^{\frac{1}{2}}}{16\pi}\varepsilon^2(1 - \varepsilon + \ldots)\right\}$

$\qquad = 0.56419(1 + 0.0010854\varepsilon^2 - 0.0010854\varepsilon^3 + \ldots)$

$\varepsilon = \infty$: $\quad -2y_2^3 u(0)u'(0) = 0.56486$

$\varepsilon = -1$: $\quad -2y_2^3 u(0)u'(0) = 0.56486$

Sorption

$\varepsilon \to 0$: $\quad 2\pi^{-\frac{1}{2}}\left\{1 + \dfrac{4 - \pi}{4\pi}\varepsilon + \dfrac{7\pi^2 - 8(11 - 3^{\frac{1}{2}})\pi + 64}{32\pi^2}\varepsilon^2 + \ldots\right\}$

$\qquad = 1.1284(1 + 0.06832\varepsilon - 0.04048\varepsilon^2 + \ldots)$

$\varepsilon = \infty$: $\quad -2y_1 u'(0) = 1.2551$

$\varepsilon = -1$: $\quad 2^{-\frac{1}{2}}f''(0) = 0.93922$

Desorption

$\varepsilon \to 0$: $\quad 2\pi^{-\frac{1}{2}}\left\{1 - \dfrac{4 - \pi}{4\pi}\varepsilon + \dfrac{64 - 8(7 - 3^{\frac{1}{2}})\pi - \pi^2}{32\pi^2}\varepsilon^2 + \ldots\right\}$

$\qquad = 1.1284(1 - 0.06832\varepsilon + 0.02784\varepsilon^2 + \ldots)$

$\varepsilon = \infty$: $\quad 2^{-\frac{1}{2}}f''(0) = 0.93922$

$\varepsilon = -1$: $\quad -2y_1 u'(0) = 1.2551$

TABLE 7.7

Some exact solutions with D concentration dependent

No.	D	ϕ	Remarks
1	$D_0 + \frac{1}{2}n\theta^n\left(1 - \frac{\theta^n}{1+n}\right)$ $\pm \frac{1}{2}D_0^{\frac{1}{2}}\left\{\frac{2\theta}{B}\left(1 - \frac{\theta^n}{1+n}\right) + Bn\theta^{n-1}\right\}$	$2D_0^{\frac{1}{2}}$ inverfc $\theta \pm (1 - \theta^n)$	$n > 0$. For minus sign $D_0 \geqslant \pi n^2/16$
2	$\frac{n}{2(n+1)}\{(1-\theta)^{n-1} - ((1-\theta)^{2n}\}$ $\pm \frac{1}{2}D_0^{\frac{1}{2}}\left[\frac{2}{(n+1)B}\{1 - (1-\theta)^{n+1}\}\right.$ $\left. + Bn(1-\theta)^{n-1}\right]$	$2D_0^{\frac{1}{2}}$ inverfc $\theta \pm (1-\theta)^n$	$n > 0$. For minus sign D_0 has lower limit
3	$D_0 + \frac{1}{2}\sin^2 \frac{1}{2}\pi\theta \pm \frac{1}{4}D_0^{\frac{1}{2}}\sin \frac{1}{2}\pi\theta \cdot \left(\frac{8}{\pi B} + \pi B\right)$	$2D_0^{\frac{1}{2}}$ inverfc $\theta \pm \cos \frac{1}{2}\pi\theta$	For minus sign $D_0 \geqslant \pi/16$
4	$D_0 + \frac{1}{2}e^{n\theta}(1 - e^{n\theta} + n\theta e^n)$ $\pm \frac{1}{2}D_0^{\frac{1}{2}}\left\{\frac{2}{nB}(1 - e^{n\theta} + n\theta e^n) + Bn e^{n\theta}\right\}$	$2D_0^{\frac{1}{2}}$ inverfc $\theta \pm (e^n - e^{n\theta})$	$n > 0$. For minus sign $D_0 \geqslant \pi n^2 e^{2n}/16$

From Philip 1960.

TABLE 8.1

		$i = 0$ $X = 0$	$i = 1$ 0·1	$i = 2$ 0·2	$i = 3$ 0·3	$i = 4$ 0·4	$i = 5$ 0·5	$i = 6$ 0·6
$(j = 0)T =$	0·000	0	0·2000	0·4000	0·6000	0·8000	1·0000	0·8000
$(j = 1)$	0·001	0	0·2000	0·4000	0·6000	0·8000	0·9600	0·8000
$(j = 2)$	0·002	0	0·2000	0·4000	0·6000	0·7960	0·9280	0·7960
$(j = 3)$	0·003	0	0·2000	0·4000	0·5996	0·7896	0·9016	0·7896
$(j = 4)$	0·004	0	0·2000	0·4000	0·5986	0·7818	0·8792	0·7818
$(j = 5)$	0·005	0	0·2000	0·3999	0·5971	0·7732	0·8597	0·7732
	⋮							
$(j = 10)$	0·01	0	0·1996	0·3968	0·5822	0·7281	0·7867	0·7281
	⋮							
$(j = 20)$	0·02	0	0·1938	0·3781	0·5373	0·6486	0·6891	0·6486

From Smith 1965.

TABLE 8.2
F.D.S. = Finite-difference solution: A.S. = Analytical solution

T	$X = 0.3$			$X = 0.5$		
	F.D.S.	A.S.	Percentage difference	F.D.S.	A.S.	Percentage difference
0·005	0·5971	0·5966	0·08	0·8597	0·8404	2·3
0·01	0·5822	0·5799	0·4	0·7867	0·7743	1·6
0·02	0·5373	0·5334	0·7	0·6891	0·6809	1·2
0·10	0·2472	0·2444	1·1	0·3056	0·3021	1·2

TABLE 8.3

		$X = 0$	0·1	0·2	0·3	0·4	0·5
	$T = 0.00$	0	0·2	0·4	0·6	0·8	1·0
	$T = 0.01$	0	0·1989	0·3956	0·5834	0·7381	0·7691
	$T = 0.02$	0	0·1936	0·3789	0·5400	0·6461	0·6921
	⋮	0 ⋮					
	$T = 0.10$	0	0·0948	0·1803	0·2482	0·2918	0·3069
Analytical solution	$T = 0.10$	0	0·0934	0·1776	0·2444	0·2873	0·3021

From Smith 1965.

TABLE 9.1†
Intersections of $c(y)$ for given a with $c(y)$ when $a = 0$.
$$c = C/C_0, \qquad D = D_0(1 + aC/C_0), \qquad y = x/(4D_a t)^{\frac{1}{2}}$$

	Infinite medium		Sorption	Desorption
	a small, tending to zero			
y	0·6586	−0·6586	1·0188	1·0188
c	0·1758	0·8242	0·1496	0·8504
	$a = \infty$			
y	0·6653	−0·6766	1·0012	1·0474
c	0·1733	0·8307	0·1568	0·8615
	$a = -1$			
y	0·6766	−0·6653	1·0474	1·0012
c	0·1693	0·8267	0·1385	0·8432

† Taken from Wilkins (1963).

TABLE 10.1

Diffusion coefficients for chloroform in polystyrene at 25°C. All diffusion coefficients are in $cm^2\ s^{-1} \times 10^{-10}$

% Regain at equilibrium = C_0	$(t/l^2)_{\frac{1}{2}}$ $scm^{-2} \times 10^{10}$	$\bar{D}$ experimental	1st Approximation			2nd Approximation		
			$\frac{1}{C_0}\int_0^{C_0} D\,dC$ $(cm^2 s^{-1}$ $10^{-10})$	D	$\bar{D}$ calculated	$\frac{1}{C_0}\int_0^{C_0} D\,dC$ $(cm^2 s^{-1}$ $10^{-10})$	D	$\bar{D}$ calculated
5·0	..	0·024 (extrap.)	0·024	0·024 (extrap.)	0·024	0·024	0·024	0·024
7·5	1·130	0·0437	0·0437	0·116	0·0504	0·039	0·12	0·044
9·9	0·620	0·0797	0·0797	0·288	0·103	0·062	0·29	0·080
12·9	0·288	0·171	0·171	0·780	0·238	0·125	0·40	0·17
13·2	0·248	0·199	0·199	0·970	0·276	0·144	0·54	0·20
15·1	0·151	0·326	0·326	3·36	0·585	0·216	1·6	0·33
16·3	0·0583	0·846	0·846	10·5	1·20	0·583	6·6	0·85
16·8	0·0481	0·972	0·972	16·0 approx.	..	..	..	..

TABLE 10.2

exp K		Desorption			Absorption	
	$T_{\frac{1}{2}}$	$T_{\frac{1}{4}}$	$(F_d)_{\frac{1}{2}}$	$(F_d)_{\frac{1}{4}}$	$T_{\frac{1}{2}}$	$(F_a)_{\frac{1}{2}}$
10^0	1.94×10^{-1}	4.90×10^{-2}	1.00	1.00	1.94×10^{-1}	1.00
2	1.51×10^{-1}	3.80×10^{-2}	1.56	1.55	1.26×10^{-1}	1.3
5	1.05×10^{-1}	2.56×10^{-2}	2.70	2.61	6.60×10^{-2}	1.7
10^1	7.80×10^{-2}	1.88×10^{-2}	4.0	3.84	3.90×10^{-2}	2.01
10^2	2.60×10^{-2}	5.00×10^{-3}	13.4	10.2	6.40×10^{-3}	3.30
10^3	8.40×10^{-3}	1.13×10^{-3}	43.3	23.1	9.40×10^{-4}	4.85
10^4	2.69×10^{-3}	2.32×10^{-4}	138.7	47.4	1.19×10^{-4}	6.14
10^5	8.60×10^{-4}	4.36×10^{-5}	443	89.0	1.48×10^{-5}	7.63
10^6	2.66×10^{-4}	7.87×10^{-6}	370	160.5	1.74×10^{-6}	8.97
10^7	8.36×10^{-5}	1.42×10^{-6}	4300	290	2.05×10^{-7}	10.60
10^8	2.65×10^{-5}	2.48×10^{-7}	13670	506	2.35×10^{-8}	12.10

From Hansen 1967.

TABLE 13.1

Values of $(Dt/a^2)^{\frac{1}{2}}$ are tabulated; M_t is the total uptake at time t; M_∞ the corresponding amount after infinite time

	M_t/M_∞	Plane sheet			Cylinder			Sphere		
s/c_B		0.25	0.50	0.75	0.25	0.50	0.75	0.25	0.50	0.75
Unlimited	1	0.275	0.549	0.825	0.146	0.306	0.508	0.102	0.216	0.367
amount	2	0.325	0.650	0.976	0.174	0.363	0.598	0.119	0.253	0.428
of solute	5	0.446	0.891	1.337	0.237	0.495	0.814	0.160	0.348	0.584
	10	0.596	1.191	1.787	0.316	0.660	1.080	0.213	0.460	0.772
50% equili-	1	0.225	0.480	0.778	0.119	0.261	0.468	0.077	0.183	0.333
brium	2	0.292	0.619	0.994	0.156	0.341	0.606	0.102	0.236	0.427
uptake of	5	0.436	0.915	1.462	0.220	0.504	0.784	0.153	0.349	0.636
of solute	10	0.601	1.265	2.006	0.316	0.702	1.218	0.209	0.483	0.867
90% equili-	1	0.176	0.403	0.750	0.094	0.219	0.443	0.057	0.147	0.309
brium	2	0.255	0.587	1.058	0.137	0.320	0.634	0.092	0.220	0.449
uptake of	5	0.423	0.942	1.675	0.219	0.522	1.010	0.149	0.356	0.720
solute	10	0.603	1.340	2.361	0.314	0.743	1.435	0.213	0.514	1.027

TABLE 13.2
Values of $(Dt/a^2)^{\frac{1}{2}}$ are tabulated

				Plane sheet		
	X/a	0·75	0·50	0·25		0
s/c_B						
Unlimited amount	1	0·202	0·403	0·605		0·806
of solute	2	0·269	0·538	0·808		1·076
	5	0·408	0·816	1·223		1·631
	10	0·568	1·136	1·705		2·273
50% equilibrium	1	0·208	0·430	0·668		0·923
uptake of solute	2	0·280	0·582	0·908		1·270
	5	0·425	0·889	1·397		1·970
	10	0·596	1·248	1·972		2·786
90% equilibrium	1	0·219	0·482	0·818		1·328
uptake of solute	2	0·293	0·644	1·105		1·850
	5	0·444	0·980	1·700		2·928
	10	0·618	1·375	2·380		4·125

TABLE 13.3
Values of $(Dt/a^2)^{\frac{1}{2}}$ are tabulated

		Cylinder			Sphere		
	r_0/a	0·75	0·50	0·25	0·75	0·50	0·25
s/c_B							
Unlimited amount	1	0·195	0·371	0·523	0·187	0·344	0·468
of solute	2	0·259	0·491	0·689	0·246	0·453	0·608
	5	0·390	0·740	1·028	0·375	0·678	0·910
	10	0·543	1·024	1·427	0·522	0·942	1·242
50% equilibrium	1	0·206	0·410	0·601	0·202	0·390	0·548
uptake of solute	2	0·274	0·553	0·811	0·268	0·527	0·736
	5	0·419	0·847	1·247	0·412	0·808	1·123
	10	0·587	1·190	1·750	0·580	1·132	1·569
90% equilibrium	1	0·223	0·509	0·853	0·230	0·533	0·850
uptake of solute	2	0·300	0·683	1·167	0·309	0·718	1·171
	5	0·454	1·046	1·796	0·467	1·098	1·806
	10	0·638	1·459	2·549	0·655	1·538	2·535

TABLE 14.1
Values of M_t/M_∞ for $\mu a^2/D = 0.01$

$Dt/(R+1)a^2$	Plane sheet		Cylinder		Sphere	
R	10	100	10	100	10	100
0·005	0·024	0·009	0·043	0·012	0·057	0·013
0·01	0·034	0·013	0·058	0·017	0·073	0·018
0·02	0·048	0·021	0·074	0·026	0·086	0·027
0·04	0·067	0·036	0·088	0·043	0·093	0·046
0·06	0·079	0·051	0·093	0·060	0·096	0·064
0·08	0·086	0·065	0·096	0·077	0·098	0·081
0·10	0·091	0·079	0·099	0·094	0·100	0·099
0·15	0·099	0·114	0·104	0·133	0·105	0·140
0·2	0·104	0·147	0·108	0·172	0·109	0·180
0·3	0·114	0·210	0·118	0·243	0·119	0·254
0·4	0·123	0·268	0·128	0·308	0·129	0·322
0·5	0·133	0·322	0·137	0·368	0·138	0·383
1·0	0·178	0·537	0·183	0·597	0·184	0·616
1·5	0·220	0·683	0·226	0·743	0·227	0·761
2·0	0·261	0·782	0·267	0·836	0·269	0·851
2·5	0·299	0·850	0·306	0·895	0·307	0·907
5·0	0·463	0·977	0·471	0·989	0·473	0·991
7·5	0·588	0·996	0·597	0·999	0·599	0·999
10	0·685	0·999	0·693	1·000	0·695	1·000
15	0·815	1·000	0·822	..	0·823	..
20	0·891	..	0·896	..	0·898	..
40	0·987	..	0·988	..	0·989	..
60	0·998	..	0·999	..	0·999	..
80	1·000	..	1·000	..	1·000	.

TABLE 14.2

Values of M_t/M_∞ for $\mu a^2/D = 0.1$

$Dt/(R+1)a^2$	Plane sheet			Cylinder			Sphere		
R	1	10	100	1	10	100	1	10	100
0.005	0.057	0.025	0.017	0.108	0.044	0.029	0.155	0.059	0.037
0.01	0.080	0.035	0.032	0.149	0.060	0.054	0.210	0.076	0.068
0.02	0.113	0.052	0.062	0.205	0.081	0.101	0.279	0.095	0.127
0.04	0.160	0.076	0.116	0.277	0.106	0.188	0.361	0.117	0.232
0.06	0.197	0.095	0.166	0.327	0.125	0.265	0.409	0.136	0.324
0.08	0.227	0.112	0.211	0.365	0.143	0.333	0.441	0.153	0.404
0.10	0.254	0.128	0.253	0.395	0.159	0.394	0.463	0.171	0.474
0.15	0.310	0.164	0.343	0.445	0.200	0.519	0.494	0.212	0.613
0.2	0.354	0.198	0.418	0.476	0.238	0.615	0.508	0.252	0.712
0.3	0.418	0.261	0.536	0.507	0.309	0.748	0.522	0.325	0.838
0.4	0.459	0.320	0.626	0.524	0.373	0.832	0.532	0.391	0.907
0.5	0.488	0.374	0.696	0.535	0.432	0.887	0.542	0.451	0.946
1.0	0.559	0.585	0.889	0.579	0.652	0.983	0.585	0.672	0.996
1.5	0.601	0.724	0.959	0.619	0.786	0.997	0.624	0.804	1.000
2.0	0.638	0.816	0.985	0.655	0.868	1.000	0.660	0.883	..
2.5	0.671	0.877	0.994	0.687	0.919	..	0.692	0.930	..
5.0	0.797	0.983	1.000	0.809	0.993	..	0.813	0.995	..
7.5	0.875	0.998	..	0.884	0.999	..	0.886	1.000	..
10	0.923	1.000	..	0.929	1.000	..	0.931	..	..
15	0.971	..	..	0.974	..	..	0.974	..	..
20	0.989	..	..	0.990	..	..	0.990	..	..
40	1.000	..	..	1.000	..	..	1.000	..	..

TABLE 14.3
Values of M_t/M_∞ for $\mu a^2/D = 1.0$

$Dt/(R+1)a^2$	Plane sheet			Cylinder			Sphere		
R	1	10	100	1	10	100	1	10	100
0.005	0.057	0.028	0.045	0.109	0.051	0.085	0.155	0.070	0.120
0.01	0.081	0.045	0.081	0.151	0.079	0.151	0.211	0.105	0.212
0.02	0.115	0.075	0.136	0.208	0.129	0.250	0.283	0.166	0.346
0.04	0.164	0.131	0.210	0.285	0.217	0.378	0.371	0.273	0.510
0.06	0.203	0.181	0.264	0.341	0.295	0.465	0.429	0.366	0.613
0.08	0.237	0.226	0.309	0.385	0.363	0.531	0.471	0.445	0.687
0.10	0.267	0.268	0.348	0.423	0.423	0.587	0.504	0.514	0.743
0.15	0.334	0.358	0.430	0.495	0.547	0.692	0.564	0.648	0.840
0.2	0.390	0.449	0.497	0.549	0.638	0.767	0.608	0.743	0.899
0.3	0.481	0.549	0.607	0.630	0.768	0.866	0.676	0.860	0.959
0.4	0.553	0.637	0.691	0.691	0.848	0.923	0.732	0.922	0.983
0.5	0.613	0.707	0.758	0.741	0.898	0.955	0.777	0.956	0.993
1.0	0.804	0.895	0.927	0.891	0.986	0.997	0.912	0.997	1.000
1.5	0.900	0.962	0.978	0.954	0.998	1.000	0.965	1.000	
2.0	0.948	0.986	0.994	0.980	1.000		0.986		
2.5	0.973	0.993	0.998	0.992			0.995		
5.0	0.999	1.000	1.000	1.000			1.000		
7.5	1.000								

TABLE 14.4
Values of M_t/M_∞ for $\mu a^2/D = 10$

$Dt/(R+1)a^2$ R	Plane sheet 1	Cylinder 1	Sphere 1
0·005	0·059	0·112	0·160
0·01	0·085	0·159	0·224
0·02	0·126	0·230	0·316
0·04	0·190	0·336	0·448
0·06	0·242	0·420	0·546
0·08	0·288	0·489	0·623
0·10	0·329	0·547	0·686
0·15	0·414	0·660	0·796
0·2	0·484	0·740	0·865
0·3	0·595	0·845	0·939
0·4	0·680	0·906	0·972
0·5	0·746	0·943	0·987
1·0	0·920	0·995	1·000
1·5	0·975	1·000	..
2·0	0·992	..	..
2·5	0·998	..	..
5·0	1·000	..	..

TABLE 15.1

Diffusion coefficients and amplitudes

Top half (diffusion coefficients). Arrangement in each cell:

$$\begin{array}{cc} D & D_1 \\ \mathscr{D} & D_2 \end{array}$$

	ρ = 0.2 g cm⁻³			ρ = 0.5 g cm⁻³		
	20% R.H.	65% R.H.	90% R.H.	20% R.H.	65% R.H.	90% R.H.
20°C	4.5 34 / 4.1 130	3.1 12 / 2.5 140	0.93 3.2 / 0.72 160	1.4 16 / 1.3 63	0.92 5.7 / 0.80 66	0.28 1.5 / 0.24 62
50°C	28 41 / 18 180	21 16 / 9.3 230	5.0 4.0 / 2.3 250	8.4 20 / 6.4 78	5.7 7.7 / 3.4 99	1.5 2.0 / 0.86 110
80°C	140 49 / 39 420	90 20 / 17 690	D D_1 / $\mathscr{D}$ D_2	41 24 / 17 140	27 9.8 / 74 210	

Bottom half (amplitudes). Arrangement in each cell: p (top), n (bottom).

	ρ = 0.2 g cm⁻³			ρ = 0.5 g cm⁻³		
	20% R.H.	65% R.H.	90% R.H.	20% R.H.	65% R.H.	90% R.H.
20°C	0.094 / 0.0037	0.192 / 0.0040	0.22 / 0.0013	0.050 / 0.0016	0.13 / 0.0019	0.15 / 0.0006
50°C	0.34 / 0.050	0.39 / 0.041	0.42 / 0.0010	0.25 / 0.027	0.38 / 0.023	0.39 / 0.0062
80°C	0.21 / 0.20	0.17 / 0.11	p / n	0.30 / 0.147	0.24 / 0.087	

The figures in each cell of the top half are for diffusion coefficients, the arrangement being as shown in the cell for 80°C and 90% R.H. The unit is 10^{-5} cm² s⁻¹. The significance of the figures in the bottom half is shown in the corresponding cell.

REFERENCES

ABRAMOWITZ, M. and STEGUN, I. A. (1964). Handbook of mathematical functions with formulas, graphs and mathematical tables, p. 319. National Bureau of Standards, Washington.

ALEXANDER, A. E. and JOHNSON, P. (1949). *Colloid science* Vol. 1, Chap. 10. Oxford University Press.

ALFREY, A., GURNEE, E. F., and LLOYD, W. G. (1966). *J. Polym. Sci.* **C12**, 249.

ALLEN, D. N. de G. and SEVERN, R. T. (1962). *Q. Jl. Mech. appl. Math.* **15**, 53.

ANDERSON, J. S. and SADDINGTON, K. (1949). *J. chem. Soc.*, Suppl. No. 2, S381.

ARIS, R. (1956). *Proc. R. Soc.* **A235**, 67.

ARMSTRONG, A. A. and STANNETT, V. (1966a). *Makromolek. Chem.* **90**, 145.

———, WELLONS, J. D., and STANNETT, V. (1966b). *Makromolek. Chemie* **95**, 78.

ASH, R., BAKER, R. W., and BARRER, R. M. (1968). *Proc. R. Soc.* **A 304**, 407.

——— and BARRER, R. M. (1963). *Trans. Faraday Soc.* **59**, 2260.

——— and ——— (1971). *J. Phys.* D **4**, 888.

———, ——— and NICHOLSON, D. (1963). *Z. phys. Chem.* **37**, 257.

———, ——— and PALMER, D. G. (1965). *Br. J. appl. Phys.* **16**, 873.

———, ——— and PETROPOULOS, J. H. (1963). *Br. J. appl. Phys.* **14**, 854.

BAGLEY, E. and LONG, F. A. (1955). *J. Am. chem. Soc.* **77**, 2172.

BANKOFF, S. G. (1964). *Advances in chemical Engineering* (Eds. T. B. Drew, J. W. Hoopes, and T. Vermeulen.) Vol. 5 p. 75. Academic Press, New York.

BARNES, C. (1934). *Physics*, **5**, 4.

BARRER, R. M. (1944). *Phil. Mag.* **35**, 802.

——— (1946). *Proc. phys. Soc.* **58**, 321.

——— (1951). *Diffusion in and through solids.* Cambridge University Press.

——— (1968). *Diffusion in polymers* (Eds. J. Crank and G. S. Park). Chap. 6. Academic Press, New York.

———- (1969). *Surface area determination*, Proceedings of the International Symposium, Bristol, p. 227. Butterworths, London.

———, BARRIE, J. A., and ROGERS, M. G. (1962). *Trans. Faraday. Soc.* **58**, 2473.

——— and BROOK, D. W. (1953). *Trans. Faraday Soc.* **49**, 1049.

——— and FERGUSON, R. R. (1958). *Trans. Faraday Soc.* **54**, 989.

BARRIE, J. A., LEVINE, J. D., MICHAELS, A. S., and WONG, P. (1963). *Trans. Faraday Soc.*, **59**, 869.

——— and MACHIN, D. (1967). *J. Polym. Sci.* A2, **5**, 1300.

——— and ———. (1971). *Trans. Faraday Soc.* **67**, 2970.

BAXLEY, A. L., NICHOLAS, C. H., and COUPER, J. R. (1968). *7th Conference on Thermal Conductivity*, p. 685. Natn. Bur. Stand. U.S.A., Spec. Publ. No. 302.

BEARMAN, R. J. (1961). *J. phys. Chem.* **65**, 1961.

BELL, G. E. (1973). British Gas Corporation, Midlands Research Station. Internal Rep. No. I, 1041.

——— and CRANK, J. (1973). *J. Inst. Math. and its Appl.* **12**, 37.

——— and ——— (1974). *J. Chem. Soc. Faraday Trans. II*, **70**, 1259.

BELL, R. P. (1945). *Proc. phys. Soc.* **57**, 45.

BERTHIER, G. (1952). *J. Chim. phys.* **49**, 527.

BIOT, M. A. (1970). *Variational principles in heat transfer.* Oxford University Press.

——— and DAUGHADAY, H. (1962). *J. Aero/Space Sci.* **29**, 227.

BOLEY, B. A. (1961). *J. math. Phys.* **40**, 300.

BOLTZMANN, L. (1894). *Annln Phys.* **53**, 959.

BOOTH, F. (1948). *Trans. Faraday Soc.* **44**, 796.

BROMWICH, T. J. I'A. (1921). *Proc. Camb. phil. Soc. math. phys. Sci.*, **20**, 411.

BRUCK, J. C. and ZYVOLOSKI, G. (1973). *The mathematics of finite elements and applications* (Ed. J. Whiteman) p. 201. Academic Press, New York.

BUDHIA, H. and KREITH, F. (1973) *Int. J. Heat Mass Transfer* **16**, 195.

BURGER, H. C. (1919). *Phys. Z.* **20**, 73.

CARMAN, P. C. and HAUL, R. A. W. (1954) *Proc. R. Soc.* A **222**, 109.

—— and STEIN, L. H. (1956). *Trans. Faraday Soc.* **52**, 619.

CARRIER, G. F. (1956). *Q. appl. Math.* **14**, 108.

CARSLAW, H. S. (1921). *Proc. Camb. phil. Soc. math. phys. Sci.* **20**, 399.

—— (1909). *Plane trigonometry* p. 275. Macmillan, London.

—— and JAEGER, J. C. (1939). *Proc. Camb. phil. Soc. math. phys. Sci.* **35**, 394.

—— and —— (1941). *Operational methods in applied mathematics.* Clarendon Press, Oxford.

CARSLAW, H. S. and JAEGER, J. C. (1959). *Conduction of heat in solids.* Clarendon Press, Oxford.

CARSTEN, H. R. F. and McKERROW, N. W. (1944). *Phil. Mag.* (7) **35**, 812.

CASSIE, A. B. D. (1940a). *J. Text. Inst.* **31**, T17.

—— (1940b). *Trans. Faraday Soc.* **36**, 453.

—— and BAXTER, S. (1940). *Trans. Faraday Soc.* **36**, 458.

CHENG, S. C. and VACHON, R. I. (1969). *Int. J. Heat Mass Transfer* **12**, 149.

CHERNOUS'KO, F. L. (1969). *Zh. prikl. Mekh. tekhn. Fiz.* No. 2, 6.

——. (1970). *Int. chem. Engng Process Inds* **10**, No. 1, 42.

CHRISTIAN, W. J. (1961). *J. Aero/Space Sci.* **28**, 911.

CHURCHILL, R. V. (1944). *Modern operational mathematics in engineering.* McGraw-Hill, New York.

CITRON, S. J. (1960). *J. Aero/Space Sci.* **27**, 219, 317, 470.

CLACK, B. W. (1916). *Proc. phys. Soc.* **29**, 49.

—— (1921). *ibid* **33**, 259.

COHEN, A. M. (1973). *Numerical analysis.* McGraw-Hill, New York.

CRANK, J. (1948a). *J. Soc. Dyers Colour.* **64**, 386.

—— (1948b). *Phil. Mag.* **39**, 140.

—— (1948c). *Phil. Mag.* **39**, 362.

—— (1950). *Proc. phys. Soc.* **63**, 484.

—— (1953). *J. Polym. Sci.* **11**, 151.

—— (1957a). *Q. Jl. Mech. appl. Math.* **10**, 220.

—— (1957b). *Trans. Faraday Soc.* **53**, 1083.

—— (1974). *Moving Boundary Problems in Heat Flow and Diffusion* (Eds. J. R. Ockendon and R. Hodgkins). Clarendon Press, Oxford.

—— and GODSON, S. M. (1947). *Phil. Mag.* (7), **38**, 794.

—— and GUPTA, R. S. (1972a). *J. Inst. Math. and its Appl.* **10**, 19.

—— and —— (1972b). *J. Inst. Math. and its Appl.* **10**, 296.

—— and HENRY, M. E. (1949a). *Trans. Faraday Soc.* **45**, 636.

—— and —— (1949b). *Trans. Faraday Soc.* **45**, 1119.

—— and —— (1949c). *Proc. phys. Soc.* **42B**, 257.

—— and —— (1951). *Trans. Faraday Soc.* **47**, 450.

—— and NICOLSON, P. (1947). *Proc. Camb. phil. Soc. math. phys. Sci.* **43**, 50.

—— and PARK, G. S. (1949). *Trans. Faraday Soc.* **45**, 240.

—— and —— (1951). *Trans. Faraday Soc.* **47**, 1072.

—— and —— (1968). *Diffusion in polymers* (Eds. J. Crank and G. S. Park). Chap. 1. Academic Press, New York.

—— and PARKER, I. B. (1966). *Q. Jl. Mech. appl. Math.* **19**, 167.

—— and PHAHLE, R. D. (1973). *Bull. Inst. Math. Appl.* **9**, 12.

DANCKWERTS, P. V. (1950*a*). *Trans. Faraday Soc.* **46**, 300.

—— (1950*b*). *Trans. Faraday Soc.* **46**, 701.

—— (1951). *Trans. Faraday Soc.* **47**, 1014.

DANIELS, H. E. (1941). *Trans. Faraday Soc.* **37**, 506.

DARKEN, L. S. (1948). *Trans. Am. Inst. Minmetall. Engrs* **175**, 184.

DAYNES, H. (1920). *Proc. R. Soc.* A **97**, 286.

DIX, R. C. and CIZEK, J. (1970). *Heat Transfer 1970* (Eds. U. Grigull and E. Hahne). Vol. 1, p. Cu 1.1. Elsevier, Amsterdam.

DODSON, M. H. (1973). *Contributions to Mineralogy and Petrology* **40**, 259.

DOUGLAS, J. and GALLIE, T. M. (1955). *Duke math. J.* **22**, 557.

DRESCHEL, P., HOARD, J. L., and LONG, F. A. (1953). *J. Polym. Sci.* **10**, 241.

DUDA, J. L. and VRENTAS, J. S. (1971). *A.I.Ch.E.Jl* **17**, 464.

EHRLICH, L. W. (1958). *J. Ass. comput. Mach.* **5**, 161.

EUCKEN, A. (1932). Forsch. Geb. Ingenieurw, *3B*, Forschungsheft No. 353.

EVERSOLE, W. G., PETERSON, J. D., and KINDSWATER, H. M. (1941). *J. phys. Chem.* **45**, 1398.

EVNOCHIDES, S. K. and HENLEY, E. J. (1970). *J. Polym. Sci.* A2, **8**, 1987.

EYRES, N. R., HARTREE, D. R., INGHAM, J., JACKSON, R., SARJANT, R. J., and WAGSTAFF, S. M. (1946). *Phil. Trans. R. Soc.* A **240**, 1.

FELS, M. and HUANG, R. Y. M. (1970). *J. appl. Polym. Sci.* **14**, 523.

FICK, A. (1855). *Annln Phys.* **170**, 59.

FIDELLE, P. T. and KIRK, R. S. (1971). *A.I.Ch.E.Jl* **17**, 1427.

FLORY, P. J. (1953). *Principles of polymer chemistry* p. 576. Cornell University Press, Ithaca, New York.

FOURIER, J. B. (1822). *Théorie analytique de la chaleur.* English translation by A. Freeman, Dover Publ., New York, 1955.

FOX, E. N. (1934). *Phil. Mag.* **18**, 210.

FOX, L. (1962). *Numerical solution of ordinary and partial differential equations.* Pergamon Press, Oxford.

—— (1974). *Moving Boundary Problems in Heat Flow and Diffusion* (Eds. J. R. Ockendon and R. Hodgkins). Clarendon Press, Oxford.

—— and MAYERS, D. F. (1968). *Computing methods for scientists and engineers.* Clarendon Press, Oxford.

—— and PARKER, I. B. (1968). *Chebyshev polynomials in numerical analysis.* Clarendon Press, Oxford.

—— and SANKAR, R. (1969). *J. Inst. Math. and its Appl.* **5**, 340.

FRANK, F. C. (1950). *Proc. R. Soc.* A **201**, 586.

FRENSDORFF, H. K. (1964). *J. Polym. Sci.* A **2**, 341.

FRICKE, H. (1924). *Phys. Rev.* **24**, 575.

FRISCH, H. L. (1957). *J. phys. Chem.* **62**, 93.

—— (1958). *ibid* **62**, 401.

—— (1959). *ibid* **63**, 1249.

—— (1962*a*). *J. chem. Phys.* **37**, 2408.

—— (1962*b*). *ibid* **36**, 510.

—— (1964). *ibid* **41**, 3679.

—— (1966). *Non-equilibrium thermo-dynamics, variational techniques and stability* (Eds. R. J. Donnelly, R. Herman, and I. Prigogine) p. 277. University of Chicago Press.

——; WANG, T. T., and KWEI, T. K. (1969). *J. Polym. Sci.* A2, **7**, 879.

FUJITA, H. (1951). *Mem. Coll. Agric. Kyoto Univ.* **59**, 31.

—— (1952). *Text. Res. J.* **22**, 757, 823

—— (1954). *Text. Res. J.* **24**, 234.

—— (1961). *Fortschr. Hochpolym.-Forsch.* **3**, 1.

—— and GOSTING, L. J. (1956). *J. Am. chem. Soc.* **78**, 1101.

—— and KISHIMOTO, A. (1952). *Text. Res. J.* **22**, 94.

GARG, D. R. and RUTHVEN, D. M. (1972). *Chem. Engng Sci.* **27**, 417.

GLASSTONE, S., LAIDLER, K. J., and EYRING, H. (1941). *The theory of rate processes*, Chap. 9. McGraw-Hill, New York.

GOLDENBERG, H. (1951). *Br. J. appl. Phys.* **2**, 233.

—— (1952). *ibid* **3**, 296.

—— (1963). *Q. Jl Mech. appl. Math.* **16**, 483.

GOODMAN, T. R. (1958). *Trans. Am. Soc. mech. Engrs* **80**, 335.

—— (1964). *Advances in heat transfer* (Eds. T. F. Irvin and J. P. Hartnett) Vol. 1 p. 51. Academic Press, New York.

GORDON, A. R. (1937). *J. Chem. Phys.* **5**, 522.

GOURLAY, A. R. (1970). *J. Inst. Math. and its Appl.* **6**, 375.

—— (1971). *J. Inst. Math. and its Appl.* **7**, 216.

GRIGULL, I. U. (1961). *Die Grundgesetze der Warmeuberleitung* s. 111, Springer Verlag, Berlin.

DE GROOT, S. R. (1961). *Thermodynamics of irreversible processes*, Chap. 7. North Holland Publishing Co., Amsterdam.

—— and MAZUR, P. (1962). *Non equilibrium thermodynamics*. North Holland Publishing Co., Amsterdam.

GURNEY, H. P. and LURIE, J. (1923). *Ind. Engng Chem.* **15**, 1170.

HAASE, R. (1969). *Thermodynamics of irreversible processes*. Addison Wesley, New York.

HALL, L. D. (1953). *J. chem. Phys.* **21**, 87.

HAMILTON, R. L. and CROSSER, O. K. (1962). *Ind. Engng Chem., Fundam.* **3**, 187.

HANSEN, C. M. (1967). *Ind. Engng Chem., Fundam.* **6**, 609.

—— (1968). *J. Oil. Col. Chem. Assoc.* **51**, 27.

HARTLEY, G. S. (1946). *Trans. Faraday Soc.* **42B**, 6.

—— (1948). *Discuss. Faraday Soc.* No. 3, 223.

—— and CRANK, J. (1949). *Trans. Faraday Soc.* **45**, 801.

HEASLET, M. A. and ALKSNE, A. (1961). *J. SIAM* **9**, 584.

HELFFERICH, F. (1963). *J. Polym. Sci., Pt B* **1**, 87.

HENRY, P. S. H. (1939). *Proc. R. Soc.* A **171**, 215.

—— (1948). *Discuss. Faraday Soc.* No. 3, 243.

HERMANS, J. J. (1947). *J. Colloid Sci.*, **2**, 387.

HERMANS, P. H. (1948). *A contribution to the physics of cellulose fibres* p. 23. Elsevier, Amsterdam.

HIGUCHI, W. I. and HIGUCHI, T. (1960). *J. Am. pharm. Ass.* (Scientific Edition) **49**, 598.

HILL, A. V. (1928). *Proc. R. Soc.* B **104**, 39.

HOLLIDAY, L. (1963). *Chemy. Ind.* 18 May. 794.

HOPFENBERG, H. B. and STANNETT, V. (1973). *The physics of glassy polymers* (Ed. R. N. Haward) Applied Science Publishers, Banking.

HORVAY, G. and CAHN, J. W. (1961). *Acta Metall.* **9**, 695.

HOWARTH, L. (1938). *Proc. R. Soc.* **164** A, 547.

HUBER, A. (1939). *Z. angew. Math. Mech.* **19**, 1.

IVANTSOV, G. P. (1947). *Dokl. Akad. Nauk.* SSSR **58**, 567. Mathematical Physics. Translated by G. Horvay, Report No. 60-RL-(2511M), G. E. Research Lab. Schenectady, New York, 1960.

JACKSON, R. A., OLDLAND, S. R. D., and PAJACZKOWSKI, A. (1968). *J. appl. Polym. Sci.* **12**, 1297.

JACOBS, M. H. (1967). *Diffusion processes.* Springer-Verlag, Berlin.

JAEGER, J. C. (1940). *Proc. R. Soc.* **74**, 342.

―――― (1944). *Proc. phys. Soc. Lond.* **56**, 197.

―――― (1949). *An introduction to the Laplace transformation.* Methuen, London.

―――― (1950a). *Proc. Camb. phil. Soc. math. phys. Sci.* **46**, 634.

―――― (1950b) *Q. appl. Math.* **8**, 187.

―――― and BECK, A. (1955). *Br. J. appl. Phys.* **6**, 15.

―――― and CLARKE, M. (1942). *Proc. R. Soc. Edinb.* **61A**, 229.

―――― and ―――― (1947). *Phil. Mag.* **38**, 504.

JASON, A. C. and PEiERS, G. R. (1973). *J. Phys. D* **6**, 512.

JEANS, J. H. (1940). *Kinetic theory of gases* Cambridge University Press.

JEFFERSON, T. B., WITZELL, O. W., and SIBBETT, W. L. (1958). *Ind. Engng Chem.* **50**, 1589.

JENKINS, R. C. Ll., NELSON, P. M., and SPIRER, L. (1970). *Trans. Faraday Soc.* **66**, 1391.

JOHNSON, W. A. (1942). *Trans. Am. Inst. Min. Metall. Engrs* **47**, 331.

JOST, W. (1952). *Diffusion in solids, liquids, gases* p. 63. Academic Press, New York.

KATZ, S. M., KUBU, E. T., and WAKELIN, J. H. (1950). *Text. Res. J.* **20**, 754.

KAWALKI, W. (1894). *Weidemanns Ann.* **52**, 166.

KEDEM, O. and KATCHALSKY, A. (1963). *Trans. Faraday Soc.* **59** 1918, 1931, 1941.

KELLER, J. B. (1963). *J. appl. Phys.* **34**, 991.

―――― and SACHS, D. (1964). *J. appl. Phys.* **35**, 537.

KIDDER, R. E. (1957). *J. appl. Mech.*, **24**, 329.

KINC, G. (1945). *Trans. Faraday Soc.* **41**, 325.

―――― and CASSIE, A. B. D. (1940). *Trans. Faraday Soc.* **36**, 445.

KIRCHHOFF, G. (1894). Vorlesungen über de theorie der Wärme. Barth, Leipzig.

KIRKENDALL, E. O. (1942). *Trans. Am. Inst. Min. Metall. Engrs* **147**, 104.

KIRKWOOD, J. G., BALDWIN, R. L., DUNLOP, P. J., GOSTING, L. J., and KEGELES, G. (1960), *J. chem. Phys* **33**, 1505.

KISHIMOTO, A. and ENDA, Y. (1963). *J. Polym. Sci.* A. **1**, 1799.

KLUG, A., KREUZER, F., and ROUGHTON, F. J. W. (1956). *Proc. R. Soc.* B **145**, 452.

KNIGHT, J. H. (1973a). *Bull. Aust. Math. Soc.* **9**, 477.

―――― (1973b). Thesis. Australian National University.

―――― and PHILIP, J. R. (1973). *Soil Science* **116**, 407.

KOKES, R. J., LONG, F. A., and HOARD, J. L. (1952). *J. chem. Phys* **20**, 1711.

KREITH, F. and ROMIE, F. E. (1955). *Proc. phys. Soc.* **68**, 277.

KUBIN, M. and ŠPAČEK, P. (1967). Collect. Czech. Comm. **32**, 2733.

KUUSINEN, J. (1935). *Annln. Phys.* **5**, 447.

KWEI, T. K. and ZUPKO, H. M. (1969). *J. Polym. Sci.* A2, **7**, 867.

――――, WANG, T. T., and ZUPKO, H. M. (1972). *Macromolecules* **5**, (5) 645.

LAMM, O. (1943). *Ark. Kemi Miner. Geol.* **17A**, 1.

LANDAU, H. G. (1950). *Q. Appl. Math.* **8**, 81.

LANGFORD, D. (1967). *Q. Appl. Math.* **24**, 315.

LAZARIDIS, A. (1970). *Int. J. Heat Mass Transfer*, **13**, 1459.

LEBEDEV, YA. S. (1966). *Kinet. Katal.* **6**, 522.

LEE, C. F. (1969). Thesis. University of Queensland.

―――― (1971a). *J. Inst. Math. and its Appl.* **8**, 251.

―――― (1971b). *Atti Accad. naz. Lincei Rc.* **50**, 60.

――――― (1972). *J. Inst. Math. and its Appl.* **10**, 129.

LIN HWANG, J. (1952). *J. Chem. Phys.* **20**, 1320.

LONG, F. A. and RICHMAN, D. (1960). *J. Am. chem. Soc.* **82**, 513.

MACEY, H. H. (1940). *Proc. phys. Soc.* **52**, 625.
—— (1942). *ibid* **54**, 128
MANDELKERN, L. and LONG, F. A. (1951). *J. Polym. Sci.* **6**, 457.
MARCH, H. and WEAVER, W. (1928). *Phys. Rev.* **31**, 1081.
MATANO, C. (1932–3) *Jap. J. Phys* **8**, 109.
MAXWELL, C. (1873). *Treatise on electricity and magnetism* Vol. I, p. 365. Oxford University Press.
MCKAY, A. T. (1930). *Proc. phys. Soc.* **42**, 547.
MEADLEY, C. K. (1971). *Q. Jl Mech. appl. Math.* **24**, 43.
MEARES, P. (1958). *J. Polym. Sci.* **27**, 405.
—— (1965). *J. appl. Polym. Sci.* **9**, 917.
MEYER, O. E. (1899). *Kinetic theory of gases.* Longmans Green, London.
MILLS, R. (1963). *J. phys. Chem.* **67**, 600.
MILNE-THOMSON, L. M., and COMRIE, L. J. (1944). Standard four-figure mathematical tables. Macmillan, London.
MITCHELL, A. R. (1969). *Computational methods in partial differential equations.* Wiley, New York.
MOORE, R. S. and FERRY, J. D. (1962). *J. phys. Chem.* **66**, 2699.
—— and —— (1968). *Macromolecules* **1**, 270.
MOTZ, H. (1946). *Quartz. J. appl. Maths.* **4**, 371.
MUEHLBAUER, J. C. and SUNDERLAND, J. E. (1965). *Appl. Mech. Rev.* **8**, 951.
MURRAY, W. D. and LANDIS, F. (1959). *J. Heat Transfer* **81**, 106.
MYERS, G. E. (1971). *Analytical methods in conduction heat transfer.* McGraw-Hill, New York.
NEALE, S. M. and STRINGFELLOW, W. A. (1933). *Trans. Faraday Soc.* **29**, 1167.
NEWMAN, A. B. (1931). *Trans. Am. Inst. chem. Engrs* **27**, 203.
NEWNS, A. C. (1950). *J. Text. Inst.* **41**, T269.
—— (1956). *Trans. Faraday Soc.* **52**, 1533.
NICOLSON, P. and ROUGHTON, F. J. W. (1951). *Proc. R. Soc.* B **138**, 241.
NOONEY, G. C. (1973). *J. chem. Soc. Faraday Trans II*, **3**, 330.
ODIAN, G. and KRUSE, R. L. (1969). *J. Polym. Sci.* C**22**, 691.
ÖLÇER, N. Y. (1968). *Q. appl. Math.* **26**, 355.
OLSON, F. C. and SCHULZ. O. T. (1942). *Ind. Engng. Chem.* **34**, 874.
OTT, R. J. and RYS, P. (1973). *J. phys. Chem. Faraday Trans. I*, **9**, 1694, 1705.
PARK, G. S. (1950). *Trans. Faraday Soc.* **46**, 684.
—— (1951). *Trans. Faraday Soc.* **48**, 11.
—— (1953). *J. Polym. Sci.* **11**, 97.
—— (1954). *Radioisotope conference* (Eds. J. E. Johnston, R. A. Faires, and R. J. Millet). Vol. 2, p. 11. Butterworth, London.
—— (1961). *Trans. Faraday Soc.* **57**, 2314.
—— (1968). *Diffusion in polymers* (Eds. J. Crank and G. S. Park.) Chap. 5. Academic Press, New York.
PARLANGE, J.-Y. (1971a). *Soil Sci.* **111**, 134, 170.
—— (1917b) *ibid.* **112**, 313.
PASTERNAK, R. A., SCHIMSCHEIMER, J. F., and HELLER, J. (1970). *J. Polym. Sci.* A2, **8**, 167.
PATTLE, R. E. (1959). *Q. Jl Mech. appl. Math.* **12**, 407.
——, SMITH, P. J. A., and HILL, R. W. (1967). *Trans. Faraday Soc.* **63**, 2389.
PAUL, D. R. (1969). *J. Polym. Sci.* A2, **7**, 1811.
PAUL, D. R. and DIBENEDETTO, A. T. (1965). *J. Polym. Sci.* C**10**, 17.
PEACEMAN, D. W. and RACHFORD, H. H. (1955). *J. Soc. ind. appl. Math.* **3**, 28.
PEKERIS, C. L. and SLICHTER, L. B. (1939). *J. appl. Phys.* **10**, 135.

PETERLIN, A. and WILLIAMS, J. L. (1971). *J. appl. Polym. Sci.* **15**, 1493.

PETROPOULOS, J. H. (1970). *J. polym. Sci.* A2, **8**, 1797.

—— and ROUSSIS, P. P. (1967). *J. chem. Phys* **47**, 1491; **47**, 1496; **48**, 4619.

—— and —— (1969a). *J. Chem. phys.* **50**, 3951.

—— and —— (1969b). *Organic solid state chemistry* (Ed. G. Adler) Chap. 19. Gordon and Breach, London.

PHILIP, J. R. (1955). *Trans. Faraday Soc.* **51**, 885.

—— (1957). *Aust. J. Phys* **10**, 29.

—— (1960). *Aust. J. Phys* **13**, 1, 13.

—— (1963). *Aust. J. Phys* **16**, 287.

—— (1966). *Proc. UNESCO Netherlands Symp. Water Unsaturated Zone, Wageningen.*

—— (1968). *Aust. J. Soil Res.* **6**, 249.

—— (1969a). *Adv. Hydrosci.* **5**, 215.

—— (1969b). *Circulatory and respiratory mass transport* (Eds. G. E. W. Wolstenholme and J. Knight) Churchill, London.

—— (1973a). *Aust. J. Phys.* **26**, 513.

—— (1973b). *Soil Sci.* **116**, 328.

PICARD, E. (1893). *J. Math* **9**, 217.

PLESHANOV, A. S. (1962). *Soviet Phys. tech. Phys* (English Translation **7**, 70).

POLLACK, H. O. and FRISCH, H. L. (1959). *J. phys. Chem.* **63**, 1022.

POOTS, G. (1962a). *Int. J. Heat Mass Transfer* **5**, 339.

—— (1962b). *Int. J. Heat Mass Transfer* **5**, 525.

PORTER, A. W. and MARTIN, T. R. (1910). *Phil. Mag.* **20**, 511.

PRAGER, S. (1951). *J. chem. Phys* **19**, 537.

—— (1953). *J. chem. Phys* **21**, 1344.

—— and LONG, F. A. (1951). *J. Am. chem. Soc.* **73**, 4072.

PRIGOGINE, I. (1967). *Introduction to the thermodynamics of irreversible processes.* Interscience, New York.

RAYLEIGH, LORD (1892). *Phil. Mag.* **34**, 481.

REESE, C. E. and EYRING, H. (1950). *Text. Res. J.* **20**, 743.

REISS, A. and LA MER, V. K. (1950). *J. chem. Phys.* **18**, 1.

RIDEAL, E. K. and TADAYON, J. (1954). *Proc. R. Soc.* A **225**, 357.

ROBINSON, C. (1946). *Trans. Faraday Soc.* B **42**, 12.

—— (1950). *Proc. R. Soc.* A **204**, 339

ROGERS, C. E. (1965). *Physics and chemistry of the organic solid state* (Eds. D. Fox, M. M. Labes, and A. Weissberger) Vol. II Chap. 6. Interscience, New York.

ROGERS, W. A., BURITZ, R. S. and ALPERT, D. (1954). *J. appl. Phys.* **25**, 868.

ROGERS, C. E. and STERNBERG, S. (1971). *J. Macromol. Sci.*, Pt. B **5(1)**, 189

ROUGHTON, F. J. W. (1959). *Prog. Biophys and Biochem.* **9**, 55.

RUBENSTEIN, L. I. (1971). *The Stefan problem.* Trans. of Math. Monographs. Vol. 27. American Mathematical Society.

RUNGE, I. (1925). *Z. tech. Phys.* **6**, 61.

SAITO, Y. (1968). *Rev. Polarogr., Kyoto* **15**, 177.

SAKAMOTO, S. (1928). Leip. Ber. **80**, 217.

SANDERS, R. W. (1960). *J. Ann. Rocket Soc.* **30**, 1030.

SEITZ, F. (1948). *Phys. Rev.* **74**, 1513.

SHAMPINE, L. F. (1973a). *Q. appl. Math.* **30**, 441.

—— (1973b). *Q. appl. Math.* **31**, 287.

DA SILVA, L. C. C. and MEHL, R. F. (1951). *J. Metals, N.Y.* **3**, 155.

SMITH, G. D. (1965). *Numerical solution of partial differential equations.* Oxford University Press.

ŠPAČEK, P. and KUBIN, M. (1967). *J. Polym. Sci.* **16**, 705.

SPALDING, D. B. and GIBSON, M. (1971). *Diffusion processes* (Eds. J. N. Sherwood, A. V. Chadwick, W. M. Muir, and F. L. Swinton.) Vol. 2, p. 561. Gordon and Breach, London.

STANDING, H. A., WARWICKER, J. O., and WILLIS, H. F. (1948). *J. Text. Inst.* **38**, T335.

STANNET, V., HOPFENBERG, H. B. and PETROPOULOS, J. H. (1972). *Macromolecular science* (Ed. C. E. H. Bawn) Physical Chemistry Series One of the MTP International Review of Science. Butterworths, London.

STEFAN, J. (1879). *Sber. Akad. Wiss. Wien.* **79**, 161.

―――― (1890). *Sber. Akad. Wiss. Wien* **98**, 965.

―――― (1891). *Ann. Phys. u. Chem.* (Wiedemann) (N.F.) **42**, 269.

STERNBERG, S. and ROGERS, C. E. (1968). *J. appl. Polym. Sci.* **12**, 1017.

STEVENSON, J. F. (1974). *A.I.Ch.E. Journal* **20**, 461.

STOKES, R. H. (1952). *Trans. Faraday Soc.* **48**, 887.

STORM, M. L. (1951). *J. appl. Phys.* **22**, 940.

SYMM, G. T. (1966). *Num. Math.* **9**, 250.

―――― (1973). National Physical Laboratory, Report NAC 31.

TALBOT, A. and KITCHENER, J. A. (1956). *Br. J. appl. Phys.* **7**, 96.

TAYLOR, G. I. (1953). *Proc. R. Soc.* **A219**, 186.

―――― (1954). *Proc. R. Soc.* **A225**, 473.

TRANTER, C. J. (1951). *Integral transforms in mathematical physics.* Methuen, London.

TSANG, T. (1960). *Ind. Engng Chem.* **52**, 707.

―――― (1961). *J. Appl. Phys.* **32**, 1518.

TSAO, G. T. (1961). *Ind. Engng Chem.* **53**, 395.

TYRRELL, H. J. V. (1963). *J. chem. Soc. II* 1599.

―――― (1971). *Diffusion processes* (Eds. J. N. Sherwood, A. V. Chadwick, W. M. Muir, and F. L. Swinton) Vol. 1, Section 2. Gordon and Breach, London.

VIETH, W. R. and SLADEK, K. J. (1965). *J. Colloid Sci.* **20**, 1014.

DE VRIES, D. A. (1952a). *The thermal conductivity of granular materials.* Bull. Inst. Intern. du Froid, Paris.

―――― (1952b). *Meded. LandbHoogesch. Wageningen* **52**, (1), 1.

WAGNER, C. (1950). *J. chem. Phys* **18**, 1227.

―――― (1952). *J. Metals* **4**, 91.

―――― (1968). *Physics Chem. Solids* **29**, 1925.

WANG, T. T., KWEI, T. K., and FRISCH, H. L. (1969). *J. Polym. Sci.* A2, **7**, 2019.

WEISZ, P. B. (1967). *Trans. Faraday Soc.* **63**, 1801.

WEISZ, P. B. and HICKS, J. S. (1967). *Trans. Faraday Soc.* **63**, 1807.

WHIPPLE, R. T. P. (1954). *Phil. Mag.* (7) **45**, 1225.

WHITEMAN, J. R. (1970). *Q. Jl Mech. appl. Math.* **21**, 41; **23**, 449.

―――― (1973). *The mathematics of finite elements and applications,* Academic Press, New York.

―――― and PAPAMICHAEL, N. (1972). *Z. Angew. Math. Phys.* **23**, 655.

WILKINS, J. E. (1963). *J. Soc. ind. appl. Maths.* **11**, 632.

WILLIAMSON, E. D. and ADAMS, L. H. (1919). *Phys. Rev.* (2) **14**, 99.

WILSON, A. H. (1948). *Phil. Mag.* **39**, 48.

WOODS, L. C. (1953). *Quart. J. Mech. appl. Math.* **6**, 163.

WORKS PROJECT ASSOCIATION (1941). *Tables of the probability function.* New York.

WRIGHT, P. G. (1972). *J. chem. Soc. Faraday Trans.* II, **68**, 1951, 1955, 1959.

YAMADA, H. (1947). *Reports Research Inst. Fluid Eng.* Kyushu University, **3** (3) 29.

ZENER, C. (1949). *J. appl. Phys* **20**, 950.

ZIENKIEWICZ, O. C. (1967 and 1971). *The finite element method in engineering science.* McGraw-Hill, New York.

AUTHOR INDEX

SUBJECT INDEX